T0375785

Simulation and Analysis of Circuits for Power Electronics

Other related titles:

You may also like

- PBPO067 | Uriarte | Multicore Simulation of Power System Transients | 2013
- PBPO123 | Watson, Arrillaga | Power Systems Electromagnetic Transients Simulation, 2nd Edition | 2018
- PBCE118 | Rahman, Dwivedi | Modeling, Simulation and Control of Electrical Drives | 2019

We also publish a wide range of books on the following topics:
Computing and Networks
Control, Robotics and Sensors
Electrical Regulation
Electromagnetics and Radar
Energy Engineering
Healthcare Technologies
History and Management of Technology
IET Codes and Guidance
Materials, Circuits and Devices
Model Forms
Nanomaterials and Nanotechnologies
Optics, Photonics and Lasers
Production, Design and Manufacturing
Security
Telecommunications
Transportation

All books are available in print via https://shop.theiet.org or as eBooks via our Digital Library https://digital-library.theiet.org.

IET ENERGY ENGINEERING 220

Simulation and Analysis of Circuits for Power Electronics

Muhammad H. Rashid

The Institution of Engineering and Technology

About the IET

This book is published by the Institution of Engineering and Technology (The IET).

We inspire, inform and influence the global engineering community to engineer a better world. As a diverse home across engineering and technology, we share knowledge that helps make better sense of the world, to accelerate innovation and solve the global challenges that matter.

The IET is a not-for-profit organisation. The surplus we make from our books is used to support activities and products for the engineering community and promote the positive role of science, engineering and technology in the world. This includes education resources and outreach, scholarships and awards, events and courses, publications, professional development and mentoring, and advocacy to governments.

To discover more about the IET please visit https://www.theiet.org/.

About IET books

The IET publishes books across many engineering and technology disciplines. Our authors and editors offer fresh perspectives from universities and industry. Within our subject areas, we have several book series steered by editorial boards made up of leading subject experts.

We peer review each book at the proposal stage to ensure the quality and relevance of our publications.

Get involved

If you are interested in becoming an author, editor, series advisor, or peer reviewer please visit https://www.theiet.org/publishing/publishing-with-iet-books/ or contact author_support@theiet.org.

Discovering our electronic content

All of our books are available online via the IET's Digital Library. Our Digital Library is the home of technical documents, eBooks, conference publications, real-life case studies and journal articles. To find out more, please visit https://digital-library.theiet.org.

In collaboration with the United Nations and the International Publishers Association, the IET is a Signatory member of the SDG Publishers Compact. The Compact aims to accelerate progress to achieve the Sustainable Development Goals (SDGs) by 2030. Signatories aspire to develop sustainable practices and act as champions of the SDGs during the Decade of Action (2020–30), publishing books and journals that will help inform, develop, and inspire action in that direction.

In line with our sustainable goals, our UK printing partner has FSC accreditation, which is reducing our environmental impact to the planet. We use a print-on-demand model to further reduce our carbon footprint.

Published by The Institution of Engineering and Technology, London, United Kingdom

The Institution of Engineering and Technology (the "**Publisher**") is registered as a Charity in England & Wales (no. 211014) and Scotland (no. SC038698).

First published 2025

The Institution of Engineering and Technology
Futures Place,
Kings Way,
Stevenage,
Herts, SG1 2UA, United Kingdom

www.theiet.org

British Library Cataloguing in Publication Data
A catalogue record for this product is available from the British Library

ISBN 978-1-83953-607-6 (hardback)
ISBN 978-1-83953-608-3 (PDF)

Typeset in India by MPS Limited

Cover image credit: dowell/Moment by Getty Images

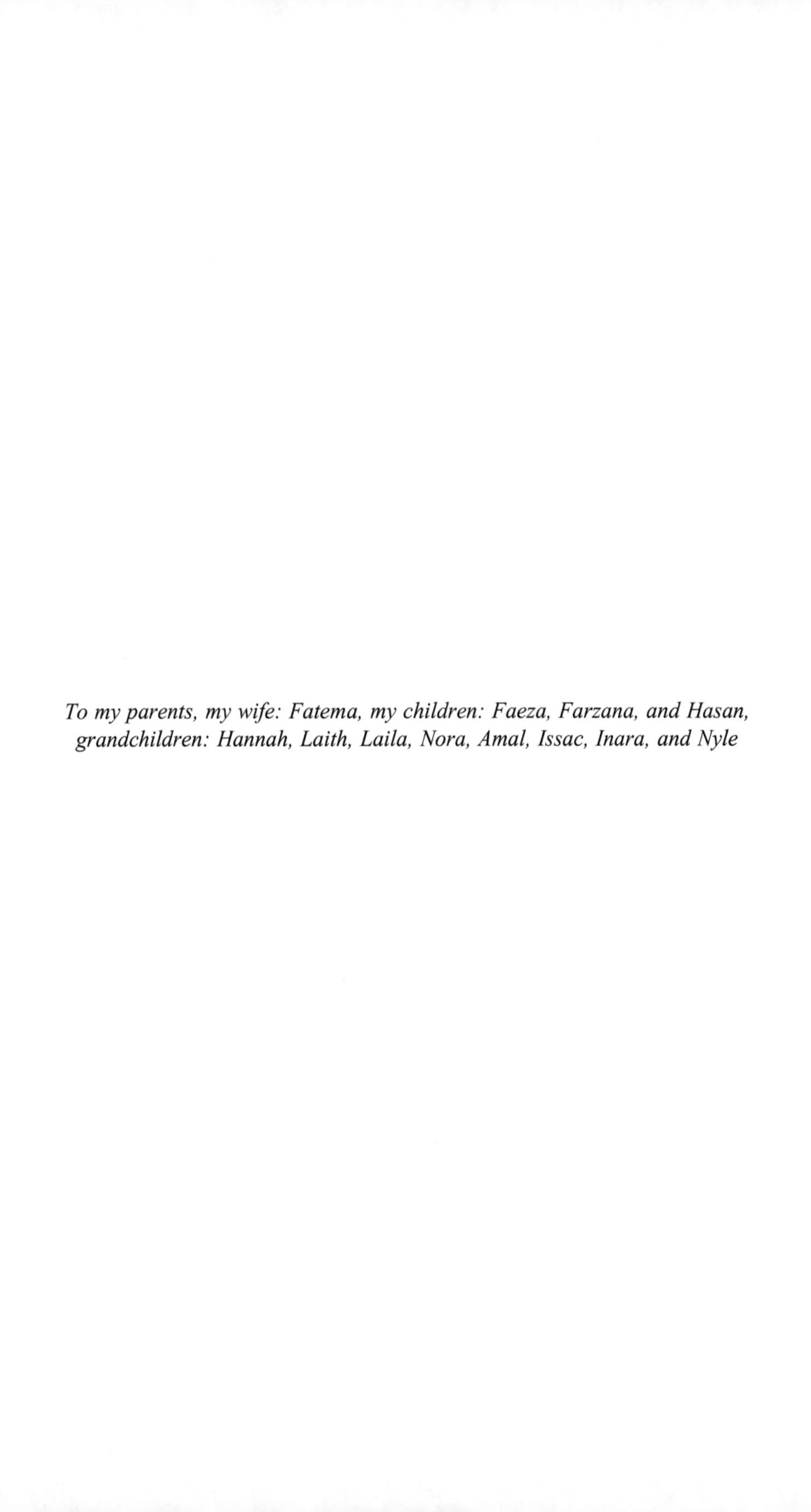

To my parents, my wife: Fatema, my children: Faeza, Farzana, and Hasan, grandchildren: Hannah, Laith, Laila, Nora, Amal, Issac, Inara, and Nyle

Contents

Preface

Power electronics is an application-oriented emerging area and finding applications in every day of our life in providing electric energy to sustain the usages of our modern electronic devices and equipment. The electric motors and auxiliary power of electric vehicles, solar energy, and other renewable energy sources require power electronics. With the increasing emphasis on renewable energy and energy savings in home, transportation, and industrial applications, power electronics education is becoming essential for supporting manpower needs for sustaining technological growth and development. Although power electronics is essential for powering modern devices, machines, and industrial equipment,

Power electronics is normally offered in universities and colleges as a technical elective. It is an application-oriented and interdisciplinary course that requires a background in mathematics, electrical circuits, control systems, analog and digital electronics, microprocessors, electric power, and electrical machines. Power electronics deals with power processing from AC to DC, DC to AC, DC to DC, and AC to AC. Therefore, it requires a clear understanding of the performance parameters of the AC sources and loads, DC sources and loads, and the parameters of power electronic switching devices.

Power electronics use power semiconductor devices as switching elements for transferring energy from the source to the load and vice versa. Due to the switching actions of the power electronic devices, the voltages and currents of the power electronic circuits are pulsing DC or DC waveforms.

The understanding of the operation of a power electronics circuit requires a clear knowledge of the transient behavior of current and voltage waveforms for each circuit element at every instant of time. These features make power electronics a difficult course for students to understand and for professors to teach. However, Power electronics is playing a key role in power processing and control.

This book is based on the author's experience in teaching power electronics for many years and integrating design content and LTspice on power electronics. The book is intended for undergraduate students of engineering and engineering technology programs. It could be used by working engineers with some background in electric circuits and differential equations. The book considers power electronics consisting of piece-wise circuits. The book focuses on the LTspice circuit simulation and the performance analysis of the power electron circuits.

The LTspice software, which is available free to students and professionals, is ideal for classroom use and for assignments requiring computer-aided simulation and analysis. Without any additional resources and lecture time, LTspice can also be integrated into power electronics. The objective of this book is to integrate the LTspice simulator with a power electronics course at the junior level or senior level with a minimum amount of time and effort. This book assumes no prior knowledge about the LTspice simulator.

The simulation requires (a) a clear knowledge and understanding of the characteristics of the switching services, (b) the gating signal requirements of the devices to turn on and off of the devices to ensure the devices are fully turned on or off, (c) the generation of the gating signals to produce the desired waveforms of output voltages and currents, and (d) the insolation of the low-level gating signals from the high voltage and current circuits.

The performance analysis requires (a) the determination of the modes of operation during a complete switching cycle or period, (b) the circuit models during the modes of operation, (c) developing mathematical models describing the voltage and current relations, (d) solving explicit equations of the circuit parameters, the boundary conditions and the switching time of the devices, the switching frequency, and the output frequency, and (d) determining the performance parameters of the converter, the input side parameters, and the output side parameters.

This book can be divided into eight parts: (1) DC and AC Fundamentals – Chapter 1; (2) Switched RLC Circuits – Chapter 2; (3) Diode Rectifiers – Chapters 3; (4) DC–DC converters – Chapter 4; (5) AC–DC converters – Chapters 5; (6) Resonant Pulse Inverters – Chapters 6; (7) Controlled Rectifiers – Chapter 7; and (8) AC voltage controllers – Chapter 8.

The book is designed for intended audiences of professionals such as design and applications engineers and seniors including postgraduate students not for lower-level students. The supplemental material in the manuscript includes LTspice simulation circuit files and Mathcad files to solve problems as illustrated in the example problems in the book.

The key features of his book are as follows:

- The book considers power electronics consisting of piece-wise circuits.
- Integrates (a) the free-version LTspice for drawing schematics and simulating to generate waveforms of the output voltage and currents, (b) mathematical derivations of the circuits, and (c) Mathcad software tool for computation of the mathematical derivations to determine the performance parameters.
- Develops a clear understanding of the converter switching operation and the voltage and current waveforms to determine the converter performance parameters.
- Identifies and evaluates the performance parameters of the converters in three parts: (a) Output side parameters, (b) converter parameters, and (c) input side parameters.
- Circuit model of the converters.
- Design guidelines for L-filter and LC-filter.
- Illustrates the effects of filters on the output waveforms and performances.

- Illustrates the performance parameter and LTspice simulation results through numerical examples.
- LTspice can be downloaded from

https://www.analog.com/en/resources/design-tools-and-calculators/ltspice-simu-lator.html

https://ltspice-iv.software.informer.com/

Any comments and suggestions regarding this book are welcome and should be sent to the author at the following address:

Dr. Muhammad H. Rashid
Professor of Electrical Engineering
Florida Polytechnic University
4700 Research Way, Lakeland, FL 33805
E-mail: mrashid@floridapoly.edu

About the author

Muhammad H. Rashid is a professor in the Department of Electrical, Computer, and Cybersecurity Engineering at Florida Polytechnic University, USA. Prior, he held professorships at the University of West Florida, Pensacola and Purdue University Fort Wayne, and was an associate professor of electrical engineering at Concordia University, Canada. He also worked as a design and development engineer with Brush Electrical Machines Ltd, Lucas Group Research Centre, UK. Rashid has published 29 books and more than 160 technical papers and serves as an editorial advisor for electric power and energy to international publishers.

Chapter 1

Fundamentals of DC and AC sources

1.1 Introduction

Power electronics uses one or more semiconductor switches to connect the input source to a load and controls the power flow from the source to the load and vice versa. The input to power electronic circuits can be either a direct current (DC) source or an alternative current (AC) source. The switches should be capable of allowing current flow in either direction when these are tuned on and should be able to withstand voltages when they are turned off. Power electronics uses energy storage elements such as inductors to limit the range to the change of current flow, capacitors to the range to the change of voltages, and transformers to adjust the voltage levels to desired levels. As a result of the switching action of the power semiconductor devices, the power electronics circuits become nonlinear, and the simulating of power electronics requires simulating switched circuits consisting of resistances (Rs), capacitors (Cs), and inductors (Ls).

1.2 DC source

A DC source voltage source is, by definition, a direct current source, and its magnitude is expected to remain a fixed or a constant value. That is, the current flows in one direction from the source to the load. For example, V_B is a battery voltage, and the instantaneous voltage $v_s(t)$ can be expressed as

$$v_s(t) = V_B \text{ for } t \geq 0 \tag{1.1}$$

A DC voltage source has a constant contact value over the time interval. Its peak, average, and root-mean-square (rms) values remain constant at a peak value of V_m, as shown in Figure 1.1

$$v_s(t) == V_m \tag{1.2}$$

We can find the rms value of a DC waveform as given by

$$V_s = \frac{1}{T}\sqrt{\int_0^T v_s^2(t)dt} = \frac{1}{T}\sqrt{\int_0^T V_m^2 dt} = \frac{V_m}{T}\sqrt{\int_0^T dt} = V_m \tag{1.3}$$

which is the same as the peak or the DC value.

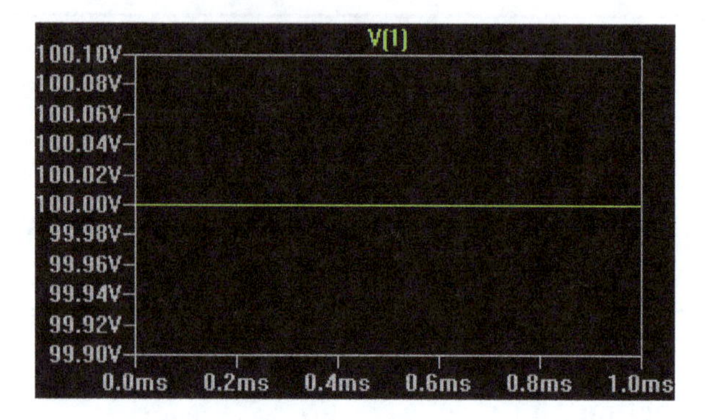

Figure 1.1 Typical plot of a DC voltage

1.3 AC source

In an AC, as the name stands, the current flows from the source to the load and vice versa. That is, the polarity of the current changes from positive to negative and vice versa. The voltage of the AC voltage source varies in general sinusoidally. If the voltage has the peak value of V_m and varies at a frequency f Hertz per second, the instantaneous voltage $v_s(t)$ can be expressed as

$$v_s(t) = V_m \sin(2\pi ft) \tag{1.4}$$

where f is the supply frequency, Hz.

The frequency can also be expressed as an angular or rotational frequency of ω as given by

$$\omega = 2\pi f(\text{rad/s}) \tag{1.5}$$

As "AC" stands for alternative current, the current in an AC circuit flows from a positive to a negative direction and vice versa, as shown in Figure 1.2 with a certain period, T. The frequency corresponding to a period T is given by

$$f = \frac{1}{T}, \text{Hz} \tag{1.6}$$

An AC current is caused due to an AC voltage, as shown in Figure 1.3. The AC voltage source that causes a current flow through the load, and its value depends on the type of load. The utility supply voltage that is normally sinusoidal voltage of a peak value at a frequency can be expressed as

$$v_s(t) = V_m \sin(\omega t) = V_m \sin(2\pi ft) \tag{1.7}$$

where V_m is the peak value and $\omega = 2\pi f$ is the angular frequency in rad/s.

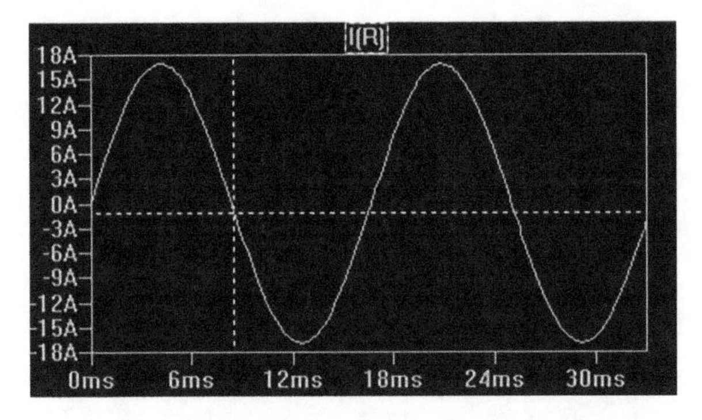

Figure 1.2 Typical alternative current (AC)

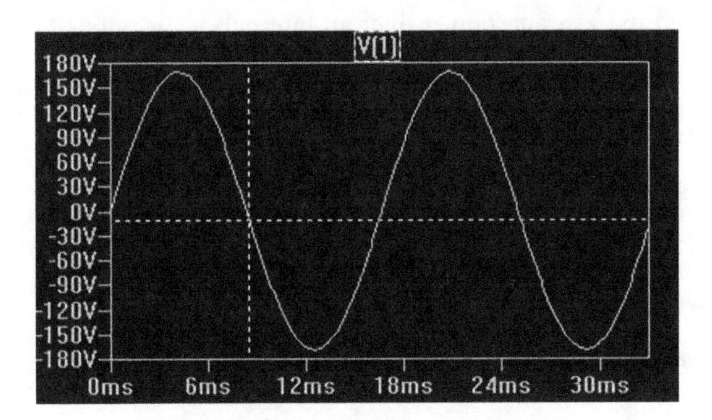

Figure 1.3 Typical AC voltage

1.3.1 AC source with resistive load

For a source voltage of $v_s(t) = V_m \sin(\omega t)$ to a resistive load of *R*, as shown in Figure 1.4, the load current $i_s(t)$ is given by

$$i_s(t) = \frac{v_s(t)}{R} = \frac{V_m}{R} \sin(\omega t) \tag{1.8}$$

which can be written in a generalized current as

$$i_s(t) = \frac{V_m}{R} \sin(\omega t) = I_m \sin(\omega t) \tag{1.9}$$

where $I_m = V_m/R$ is the peak value of the sine-wave current.

1.3.2 AC load power with resistive load

The instantaneous power *P(t)* dissipated in the load resistance is

$$p_o(t) = i_s{}^2(t)R = I_m^2 \sin^2(\omega t)R \tag{1.10}$$

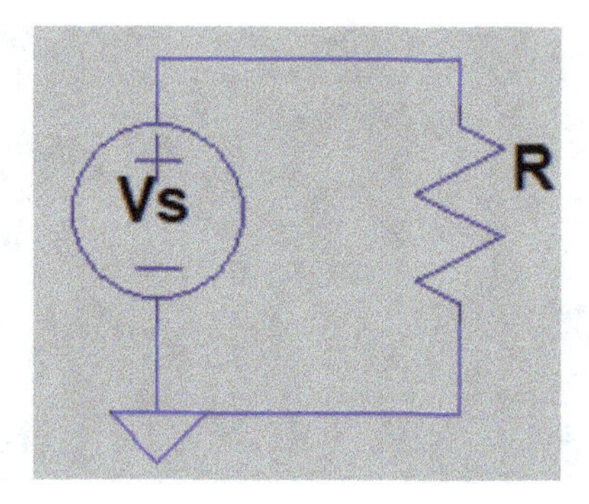

Figure 1.4 An AC source supplying a resistive load, R

We can find the average power over an internal of 2π rad as given by

$$P_o = \frac{1}{2\pi}\int_0^{2\pi} {i_s}^2(t)R = \frac{RI_m^2}{2\pi}\int_0^{2\pi} \sin^2(\omega t)d(\omega t) = \frac{RI_m^2}{2\pi}\left|\frac{\omega t}{2} - \frac{\sin(2\omega t)}{4}\right|_0^{2\pi} \tag{1.11}$$

which can be simplified to

$$P_o = \frac{RI_m^2}{2\pi}\left|\frac{\omega t}{2} - \frac{\sin(2\omega t)}{4}\right|_0^{2\pi} = \frac{RI_m^2}{2\pi}\left|\frac{2\pi}{2} - 0\right| = \frac{RI_m^2}{2} \tag{1.12}$$

which can be rewritten as a more generalized form of

$$P_o = I_o^2 R \tag{1.13}$$

where I_o the rms of the current is defined by

$$I_o = \sqrt{\frac{1}{2\pi}\int_0^{2\pi} I_m^2 \sin^2(\omega t)d(\omega t)} = \frac{I_m}{\sqrt{2}} \tag{1.14}$$

This is an important relationship to remember that the rms value of a sinusoidal voltage or a current is that the peak voltage V_m or the peak current I_m divided by a factor of $\sqrt{2}$ or multiplying a factor of $1/\sqrt{2} = 0.707$

Similarly, we can find the instantaneous power $p(t)$ supplied by the source as

$$P_s(t) = v_s(t)\ \ i_s(t) = V_m \sin(\omega t) I_m \sin(\omega t) = V_m I_m \sin^2(\omega t) \tag{1.15}$$

which gives the peak power as

$$P_{\max} = V_m I_m \tag{1.16}$$

The average power delivered by the source over an interval of 2π rad is given by

$$\begin{aligned} P_s &= \frac{1}{2\pi}\int_0^{2\pi} V_m \sin(\omega t) I_m \sin(\omega t) d(\omega t) \\ &= \frac{V_m I_m}{2\pi}\int_0^{2\pi} \sin^2(\omega t)d(\omega t) = \frac{V_m I_m}{2\pi}\left|\frac{\omega t}{2} - \frac{\sin(2\omega t)}{4}\right|_0^{2\pi} = \frac{V_m I_m}{2\pi} \times \pi = \frac{V_m I_m}{2} \end{aligned} \tag{1.17}$$

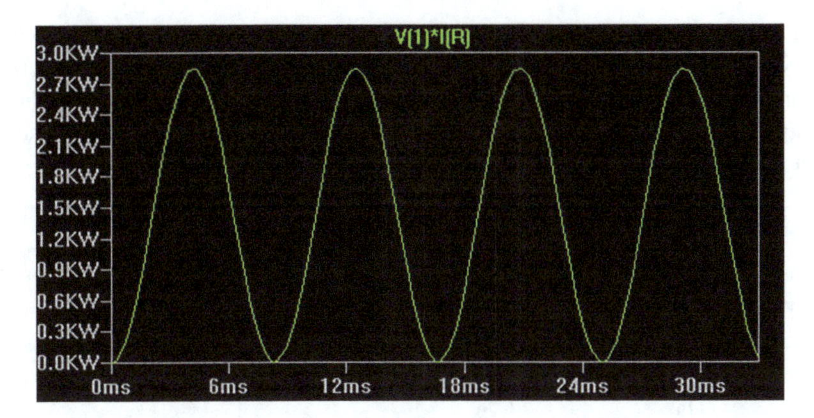

Figure 1.5 Typical instantaneous source power

which can be rewritten as a more generalized form of

$$P_s = \frac{V_m}{\sqrt{2}} \times \frac{I_m}{\sqrt{2}} = V_s I_s \tag{1.18}$$

where V_s and I_s are the rms of the source voltage and current, respectively.

A typical plot of the instantaneous source power for the instantaneous voltage in Figure 1.2 and the current in Figure 1.3 is shown in Figure 1.5. Note that the power frequency is double the supply frequency.

The input power factor, PF, is defined as the ratio of power dissipated into the load to the power supplied by the input source as given by

$$PF = \frac{P_o}{P_s} \tag{1.19}$$

1.4 Input power

The calculation of the input power will depend on the type of source. If $v_s(t)$ is the input source voltage and $i_s(t)$ is the corresponding input current drawn from the source, the average input power can be found from

$$P_{in} = \frac{1}{T}\int_0^T v_s(t) i_s(t) dt \tag{1.20}$$

For a DC source, $v_s(t)$ has a contact value, $v_s(t) = V_S$. Thus, (1.20) becomes

$$P_{in} = V_S \frac{1}{T}\int_0^T i_s(t) dt = V_S I_S \tag{1.21}$$

where $I_S = \frac{1}{T}\int_0^T i_s(t) dt$ is the average source or supply current.

For an AC source, let us assume that the input is a sinusoidal voltage with a peak value V_m at a frequency of ω rad/s and the corresponding input current, $i_s(t)$ is also a sinusoidal voltage with a peak value of I_m with a phase delay of ϕ due to energy storage elements. Thus, we can write the instantaneous input voltage and the input current as

$$v_s(t) = V_m \sin(\omega t) \tag{1.22}$$

$$i_s(t) = I_m \sin(\omega t - \phi) \tag{1.23}$$

Therefore, the average power is given by

$$P_{in} = \frac{1}{T}\int_0^T v_s(t) i_s(t) dt = \frac{1}{T}\int_0^T V_m I_m \sin(\omega t) \sin(\omega t - \phi) dt \tag{1.24}$$

Using trigonometric relation

$$\sin(A)\sin(B) = \frac{1}{2}[\cos(A - B) - \cos(A + B)] \tag{1.25}$$

$$\begin{aligned} P_{in} &= \frac{1}{T}\int_0^T v_s(t) i_s(t) dt = \frac{1}{T}\int_0^T V_m I_m \frac{\cos(\omega t - \omega t + \phi) - \cos(\omega t + \omega t - \phi)}{2} dt \\ &= \frac{V_m I_m}{2} \times \frac{1}{T}\int_0^T [\cos(\phi) - \cos(2\omega t - \phi)] dt \\ &= \frac{V_m I_m}{2}\left[\cos(\phi) - \frac{1}{T}\int_0^T [\cos(2\omega t - \phi)] dt\right] \\ &= \frac{V_m I_m}{2}\cos(\phi) = V_{rms} I_{rms} \cos(\phi) \end{aligned} \tag{1.26}$$

where $V_{rms} = \frac{V_m}{\sqrt{2}}$, $I_{rms} = \frac{I_m}{\sqrt{2}}$, and $\cos(\phi)$ is called the *power factor*, PF.

1.5 Output power

There is no power dissipation in capacitors and inductors. Power is dissipated only in resistors. For an instantaneous output current of $i_o(t)$, the average power dissipated in a resistor R is given by

$$\begin{aligned} P_o &= \frac{1}{T}\int_0^T i_o^2(t) R dt = R\frac{1}{T}\int_0^T i_o^2(t) dt \\ &= I_o^2 R \end{aligned} \tag{1.27}$$

where $I_0 = \sqrt{\frac{1}{T}\int_0^T i_o^2(t) dt}$ is the rms value of the load or output current.

(1.28)

1.6 Complex domain analysis of RLC load

With a sinusoidal AC source, the circuit analysis in the time domain involved the operation of sinusoidal functions. Analysis in the time domain is not the most convenient method. The analysis is normally done in the complex domain. A sinusoidal voltage v of rms value V_s can be represented in the time domain as $v(t) = V_m \sin(\omega t) = \sqrt{2}V_s \sin(\omega t)$, and the corresponding representation in the complex domain is $V = V_s\angle 0^o$. The impedance for resistance R is

$$Z_R = R \tag{1.29}$$

The impedance for inductance L is

$$Z_L = j\omega L \tag{1.30}$$

The impedance for capacitance C is

$$Z_C = \frac{1}{j\omega C} = -\frac{j}{\omega C} \tag{1.31}$$

For an RLC series circuit, the total load impedance Z_T is

$$Z_T = R + j\omega L - \frac{j}{\omega C} = Z\angle\theta \tag{1.32}$$

where Z is the magnitude of the load impedance and θ is the impedance angle as given by

$$\theta = \tan^{-1}\frac{R}{\omega L - 1/\omega C} \tag{1.33}$$

The current through the circuit can be found from

$$I_o = \frac{V_s}{Z_T} = \frac{V_s}{Z}\angle -\theta \tag{1.34}$$

Power dissipated in the resistance R is

$$P_o = I_o^2 R = \left(\frac{V_s}{Z}\right)^2 R \tag{1.35}$$

The input power factor is

$$PF_i = \cos(-\theta) \text{ lagging} \tag{1.36}$$

The input power is drawn from the source

$$P_i = P_o = V_s I_o PF_i \tag{1.37}$$

1.7 Electrical and electronics symbols

The voltage or the current in electrical and electronic circuits is normally represented by a symbol with a subscript. The symbol and the subscript can be either uppercase or lowercase, according to the conventions shown in Table 1.1. For example, consider the circuit in Figure 1.6(a), whose input consists of a DC voltage $V_B = 80$ V (note both the symbol and the subscript are upper case) and an AC voltage $v_s = 70 \times \sin(2\pi \times 60t)$ (note both the symbol and the subscript are lower case). The instantaneous voltage $v_S = V_B + v_s = 80 + 70 \times \sin(2\pi \times 60t)$ (note the symbol is upper case and the subscript is lower case) as shown in Figure 1.6(b) has a DC component with an AC voltage superimposed on it. The resultant rms value of a DC source superimposed with an AC source can be found from

$$V_{rms} = \sqrt{V_B^2 + V_s^2}$$

$$= \sqrt{80^2 + \left(\frac{70}{\sqrt{2}}\right)^2} = 94.074\ V$$

Note: The factor $\sqrt{2}$ converts the peak value to a rms value. We cannot add rms values but can add the square of the rms values.

Table 1.1 Definition of symbols and subscripts

Definition	Quantity	Subscript	Example
DC value of the signal	Uppercase	Uppercase	V_D
AC value of the signal	Lowercase	Lowercase	v_d
Total instantaneous value of the signal	Lowercase	Uppercase	v_D
Complex variable, phasor, or rms value of the signal	Uppercase	Lowercase	V_d

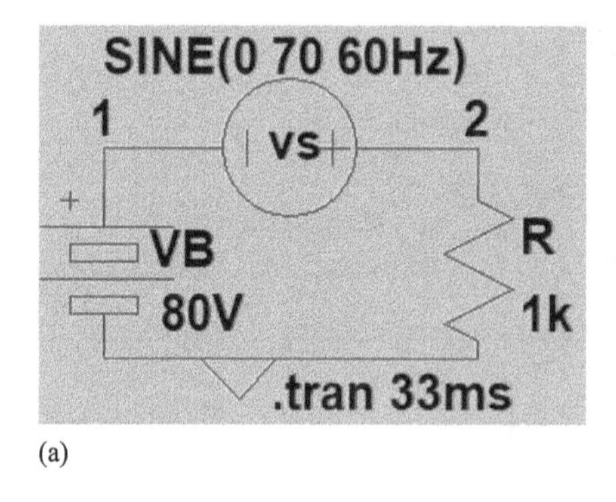

(a)

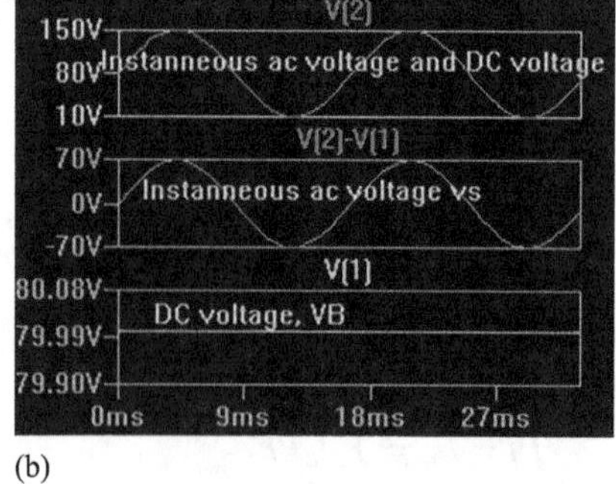

(b)

Figure 1.6 Circuit with DC and AC voltages: _(a) circuit schematic and (b) waveforms of DC, AC, and AC–DC voltages

1.8 Summary

There are two types of input sources: DC and AC. A DC source voltage source is, by definition, a direct current source, and its magnitude is expected to remain a fixed or a constant value. An alternating current (AC), as the name stands, the current flows from the source to the load and vice versa. That is, the polarity of the current changes from positive to negative and vice versa. The AC voltage source varies in general sinusoidally. There is no power dissipation in capacitors and inductors. Power is dissipated only in resistors. The power dissipated in a resistance may vary with time, but the power is measured as the average power over a period. The power dissipation is calculated from the rms current or the rms voltage of the resistor for both DC and AC sources. For a DC source, the rms quantity is the same value as the average or the DC quality.

Problems

1. The DC voltage of a battery is $V_B = 110$ V. Determine its V_{rms}.
2. A voltage source is sinusoidal, $v_s = 170\sin(\omega t)$. Determine its average value (a) V_{avg} and (b) its rms value, V_{rms}.
3. A voltage source has a DC component and an AC component, $v_S = 170 + 15\sin(\omega t)$. Determine (a) its average value V_{avg} and (b) its rms value, $V_{s(rms)}$
4. Determine the average value V_{avg} of the half-sinewave as shown in Figure P1.4 for $V_m = 100$ V.

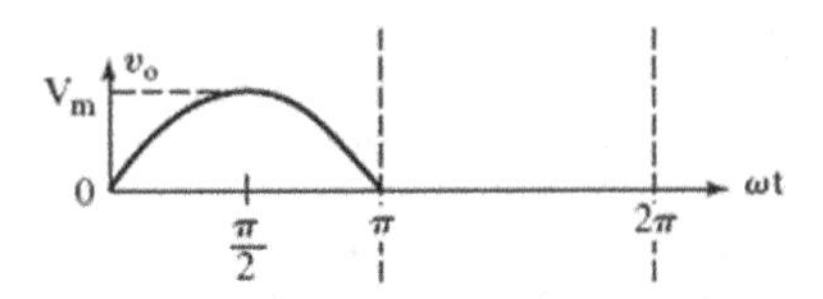

Figure P1.4 Half-sinewave

5. Determine the rms value V_{rms} of the half-sinewave as shown in Figure P1.4 for $V_m = 100$ V.
6. Determine the magnitude and phase angle after converting from rectangular to polar coordinates
   - Impedance, $Z = 3 + j4$ Admittance, $Y = \frac{1}{Z} =$
   - Impedance, $Z = 3 - j4$ Admittance, $Y = \frac{1}{Z} =$
   - Impedance, $Z = -3 - j4$ Admittance, $Y = \frac{1}{Z} =$

7. For a supply frequency of $f = 60$ Hz, determine the inductive and capacitive reactance
   - $L = 10.61$ mH $X_L ==$
   - $C = 331.6$ μF $X_C ==$

8. Determine the magnitude and phase angle of an impedance with a resistance $R = 3\Omega$ in series with an inductance $L = 10.61$ mH and a capacitance $C = 331.6$ μF. Assume a supply frequency of $f = 60$ Hz

   $Z = R + X_L + X_C ==$ Admittance, $Y = \frac{1}{Z} =$

9. Determine the magnitude and phase angle of an impedance with a resistance $R = 3\Omega$ in series with an inductance $L = 10.61$ mH and a capacitance $C = 331.6$ μF. Assume a supply frequency of $f = 0$ Hz

   $Z = R + X_L + X_C ==$ Admittance, $Y = \frac{1}{Z} =$

10. The instantaneous source voltage and the current are given by

$$v_s(\omega t) = \sqrt{2} \times 120 \cos(\omega t + \phi_v)$$

$$i_s(\omega t) = \sqrt{2} \times 10 \cos(\omega t + \phi_i)$$

   - Find the expression of the instantaneous power *p(t)* and simplify.

$$p(\omega t) = v_s(\omega t) i_s(\omega t) =$$

   - Find the average power P_{avg}

$$P_{avg} = \frac{1}{2\pi} \int_0^{2\pi} v_s(\omega t) i_s(\omega t) d(\omega t) =$$

   - For $\omega = 377\ rad/s$, $\phi_v = 30^0$ and $\phi_i = 60^0$, find the average power P_{avg}

11. The line-to-line voltages of a three-phase balanced supply is $V_L = 208$ V. Express the line-to-line voltages as phasor quantities for a positive sequence of *abc*.

   - $V_{ab} = 208\angle 0^o$ reference voltage $V_{bc} =$ $V_{ca} =$
     $V_a =$ $V_b =$ $V_c =$
   - $V_{ab} =$ $V_{bc} = 208\angle 0^o$ $V_{ca} =$
     $V_a =$ $V_b =$ $V_c =$
   - $V_{ab} =$ $V_{bc} =$ $V_{ca} = 208\angle 0^o$
     $V_a =$ $V_b =$ $V_c =$

12. The line-to-line voltages of a three-phase balanced supply is $V_L = 208$ V. Express the line-line voltages as phasor quantities for a negative sequence of *acb*.

- $V_{ab} = 208\angle 0^o$ reference voltage $V_{bc} =$ $V_{ca} =$
  $V_a =$ $V_b =$ $V_c =$
- $V_{ab} =$ $V_{bc} = 208\angle 0^o$ $V_{ca} =$
  $V_a =$ $V_b =$ $V_c =$
- $V_{ab} =$ $V_{bc} =$ $V_{ca} = 208\angle 0^o$
  $V_a =$ $V_b =$ $V_c =$

Chapter 2
Switched RLC circuits

2.1 Introduction

Power electronics uses energy storage elements such as inductors to limit the range to the change of current flow, capacitors to the range to the change of voltages, and transformers to adjust the voltage levels to desired levels. As a result of the switching action of the power semiconductor devices, the power electronics circuits become nonlinear, and the simulating power electronics requires simulating switched circuits consisting of resistances (R's), capacitors (C's), and inductors (L's).

2.2 Electrical circuit elements

In addition to power semiconductor devices, power electronic circuits consist of three types of circuit elements – resistor R, capacitor C, and inductor L. The electrical quantities are voltage v (in volts, V), current i (in amperes, A), and charge q (in coulombs, C). These elements are related to the voltage across the element $v(t)$ and current through the element $i(t)$ as a function of time t as follows [1].

Voltage across a resistor R

$$v(t) = Ri(t), \quad i(t) = \frac{v(t)}{R} \tag{2.1}$$

Voltage across a capacitor C

$$v(t) = \frac{1}{C}\int_0^t i(t)\,dt, \quad i(t) = C\frac{dv(t)}{dt} \tag{2.2}$$

Voltage across an inductor L

$$v(t) = L\frac{di(t)}{dt}, \quad i(t) = \frac{1}{L}\int_0^t v(t)dt \tag{2.3}$$

The current i is caused due to the flow of electric charge q, and it is the charge velocity that causes the current due to charge q as given by

$$i(t) = \frac{dq(t)}{dt} \quad q(t) = \int_0^t i(t)dt \tag{2.4}$$

If the flux φ vary with the time t, the voltage $v(t)$ is related to the flux as given by

$$v(t) = \frac{d\phi(t)}{dt}, \phi(t) = \int_0^t v(t)dt \tag{2.5}$$

Dividing the time-dependent voltage $v(t)$ in (2.5) by the time-dependent current in (2.4) gives the time-dependent resistance known as the memristor M as given by [2]

$$M(q) = \frac{v(t)}{i(t)} = \frac{d\phi(t)}{dq(t)} \tag{2.6}$$

Memresistance M is the rate of change of flux with the charge. The memristor is a circuit element in which the magnetic flux φ is a function of the accumulated charge. M is a nonlinear resistance. If the relation between the flux φ and the charge q is linear, then memristor M becomes the same as resistance R.

An *inductor* is an energy-storage element that tries to maintain the current flow constant and opposes any change by inducing a voltage of appropriate polarity, whereas a *capacitor* is also an energy-storage element that tries to maintain the voltage level constant for any changes in current flow. A *resistor* limits the current flow and consumes or dissipates power. Power electronic circuit in general consists of a semiconductor switching element connecting the source to the load. The load falls in general in one of the following types:

- RC circuit: Resistance R is connected in series with a capacitor, C, to limit the charging or discharging of a capacitor.
- RL circuit: Resistance R is connected in series with an inductor L to maintain the current continuously.
- LC circuit: Inductor L is connected in series with a capacitor, C, to create a resonant circuit.
- RLC circuit: Resistance R is connected in series with an inductor L and a capacitor, C, to create an underdamped circuit to obtain an output voltage or current of desired magnitude and shape.

2.3 Electrical circuit elements

2.3.1 RC circuit

Resistance R is connected in series with a capacitor C to limit the charging or discharging of a capacitor, as shown in Figure 2.1(a). The charging current $i(t)$ can be described as [3]

$$V_S = Ri + \frac{1}{C}\int idt + V_c(t=0) \tag{2.7}$$

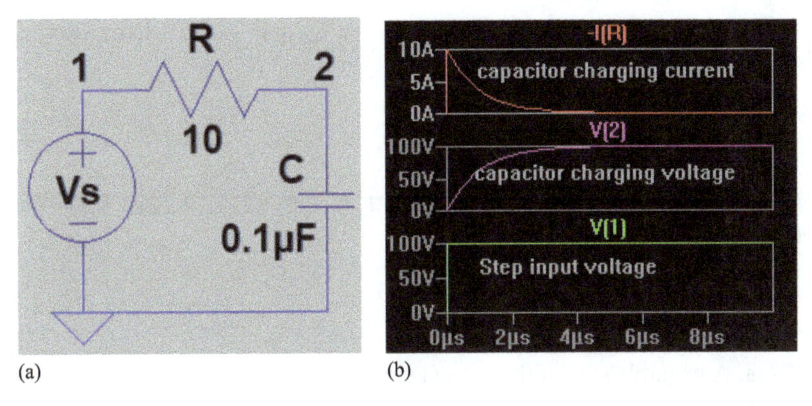

(a) (b)

Figure 2.1 RC-circuit charging: (a) RC circuit and (b) switch current and capacitor voltage

Assuming an initial capacitor voltage, $V_c(t = 0) = 0$, we can solve for the charging current as given by

$$i(t) = \frac{V_S}{R} e^{-t/(RC)} \tag{2.8}$$

where $\tau = RC$ is the time constant of the circuit.

The voltage across the capacitor is given by

$$v_C(t) = \frac{1}{C}\int i(t)dt = \frac{1}{C}\int \frac{V_S}{R} e^{-tRC} dt = V_S(1 - e^{-t/RC}) \tag{2.9}$$

As $t \to \infty$, $i(t) \to 0$ and $v_C(t) \to V_S$.

The rate of rise of the voltage at $t = 0$ is given by

$$\left.\frac{dv_C}{dt}\right|_{t=0} = \frac{V_S}{RC} \tag{2.10}$$

The energy stored in the capacitor is given by

$$W_C = \frac{1}{2} C V_s^2 \tag{2.11}$$

The same amount of energy dissipated in the resistance R is given by

$$W_R = W_C = \frac{1}{2} C V_s^2 \tag{2.12}$$

The plots of the step input voltage, the capacitor voltage $v_c(t)$, and the charging current $i(t)$ are shown in Figure 2.1(b).

Once the capacitor is charged to the supply voltage V_S, ideally, it maintains its charge and the voltage. For any circuit action, therefore, we can consider that the capacitor has an initial value of $V_c(t=0) = V_S$ and can discharge through a resistance R. Applying (2.7) gives

$$0 = Ri + \frac{1}{C}\int i dt + V_c(t=0) \tag{2.13}$$

Assuming an initial capacitor voltage, $V_c(t=0) = V_S$, we can solve for the discharging current as given by

$$i(t) = -\frac{V_S}{R}e^{-t/(RC)} \tag{2.14}$$

Note: A negative sign of the current signifies that the current is flowing off the capacitor.

The voltage across the capacitor is the same as the voltage drops across R. That is,

$$v_C(t) = R\frac{V_S}{R}e^{-t/(RC)} = V_S e^{-t/(RC)} \tag{2.15}$$

As $t \to \infty$, $i(t) \to 0$ and $v_C(t) \to 0$.

Note: $V_c(t=0) = V_S$ is the initial value of the capacitor.

Figure 2.2(a) shows the discharging of a capacitor with an initial voltage of 100 V. The DC source is set to 0 V. The plots of the input voltage, the capacitor voltage $v_c(t)$, and the discharging current $i(t)$ are shown in Figure 2.2(b). As expected, the capacitor voltage, and the current fall exponentially with a time constant of $\tau = RC$.

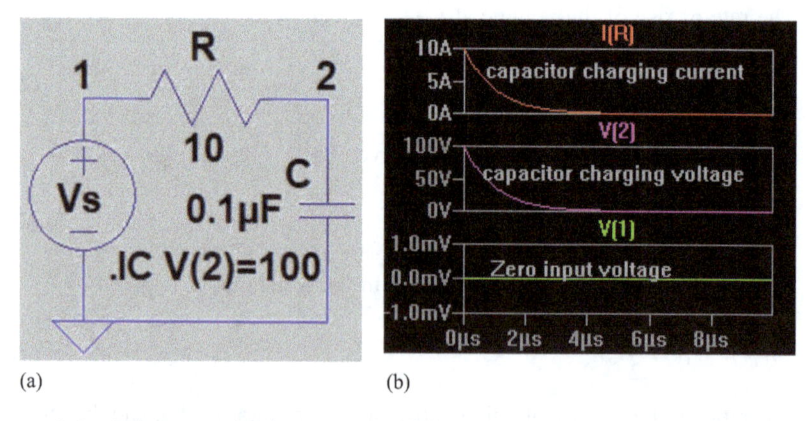

(a) (b)

Figure 2.2 RC-circuit discharging: (a) RC circuit and (b) discharging capacitor current and voltage

2.3.2 RL circuit

Inductor L is connected in series with a resistance R to limit the rate-of-rise of the current, as shown in Figure 2.3(a). The charging current $i(t)$ can be described as [3]

$$V_S = Ri + L\frac{di}{dt} \tag{2.16}$$

which we can solve for the current $i(t)$ as given by

$$i(t) = \frac{V_S}{R}\left(1 - e^{-(R/L)t}\right) \tag{2.17}$$

where the time constant is $\tau = L/R$. The rate of rise of the voltage at $t = 0$ is given by

$$\left.\frac{di}{dt}\right|_{t=0} = \frac{V_S}{R} \times \frac{R}{L} = \frac{V_S}{L} \tag{2.18}$$

As $t \to \infty$, $i(t) \to \frac{V_S}{R}$ and the energy stored in the inductor is given by

$$W_L = \frac{1}{2}L\left(\frac{V_S}{R}\right)^2 \tag{2.19}$$

This energy would remain stored forever, ideally. The same amount of energy stored in the inductor L is dissipated in the resistance R as given by

$$W_R = W_L = \frac{1}{2}L\left(\frac{V_S}{R}\right)^2 \tag{2.20}$$

The voltage across the inductor $v_L(t)$ is given by

$$v_L(t) = L\frac{di}{dt} = V_S e^{-(R/L)t} \tag{2.21}$$

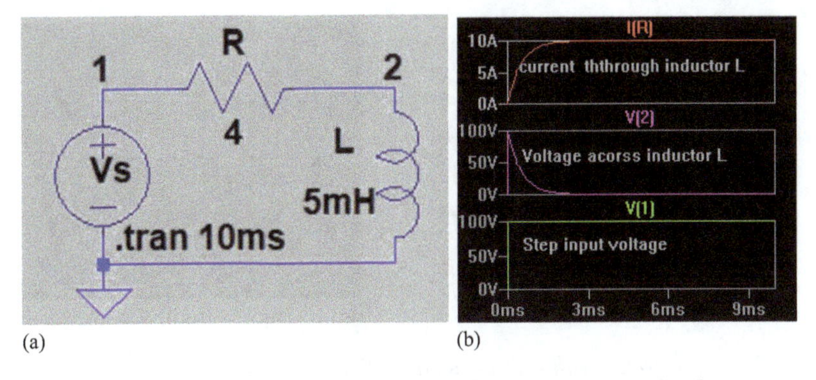

Figure 2.3 (a) RL circuit and (b) inductor current and voltage

The plots of the inductor voltage $v_L(t)$ and the current $i(t)$ are shown in Figure 2.3(b).

Once the inductor current reaches the steady-state value $I_L = \frac{V_S}{R}$, ideally there is no change in current, and there is no voltage across the inductor. The supply voltage appears across the resistance R. Energy is stored in the inductor L. If we want to break the circuit, the inductor will oppose any changes and induce an opposite voltage. An anti-parallel diode is generally connected across an RL load, as shown in Figure 2.4(a). If the supply source is disconnected, the inductor induces an opposing voltage causing the diode to turn on, and the inductor current continues to conduct. This diode is commonly known as the *freewheeling* diode. We can consider that the inductor has an initial value of $I_L = \frac{V_S}{R}$. Applying (2.16) gives

$$0 = Ri + L\frac{di}{dt} \tag{2.22}$$

which we can solve the current $i(t)$ for an initial current of I_L as given by

$$i(t) = I_L e^{-(R/L)t} \tag{2.23}$$

The voltage across the inductor as

$$v(t) = Ri(t) = RI_L e^{-(R/L)t} \tag{2.24}$$

As $t \to \infty$, $i(t) \to 0$, $v(t) \to 0$ and the energy stored in the inductor is dissipated across R if the inductor current is allowed to fall close to zero. Figure 2.4(a) shows the freewheeling of the inductor current with an initial current of 10 A. The DC source is a pulse of 100 to 0 V. The plots of the input voltage, the inductor voltage $v(t)$, and the inductor current $i(t)$ are shown in Figure 2.4(b). As expected, the inductor voltage and the current falls exponentially with a time constant of $\tau = L/R$.

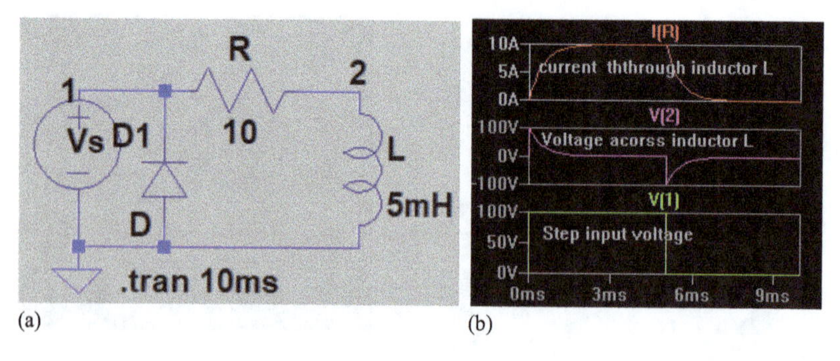

Figure 2.4 RL-circuit with a freewheeling diode: (a) RL-circuit with a diode and (b) inductor voltage and current

2.3.3 LC circuit

An inductor L is connected in series with a capacitor C to limit the charging or discharging of a capacitor, as shown in Figure 2.5(a). The charging current $i(t)$ can be described as given by [3]

$$V_S = L\frac{di}{dt} + \frac{1}{C}\int idt + V_c(t = 0) \tag{2.25}$$

Assuming an initial capacitor voltage, $V_c(t = 0) = 0$, we can solve for the charging current as given by

$$i(t) = V_S\sqrt{\frac{C}{L}}\sin(\omega_o t) \tag{2.26}$$

where $\omega_o = \sqrt{LC}$ is the resonant frequency in rad/s.

As $\omega_o t \to \frac{\pi}{2}$, $i(t)$ reaches its peak current as given by

$$I_p = V_S\sqrt{\frac{C}{L}} \tag{2.27}$$

The voltage across the capacitor is given by

$$v_C(t) = \frac{1}{C}\int i(t)dt = \frac{1}{C}\int V_S\sqrt{\frac{C}{L}}\sin(\omega_o t)dt = V_S(1 - \cos(\omega_o t)) \tag{2.28}$$

As $\omega_o t \to \pi$, $v_C(t)$ reaches its the peak voltage as given by

$$i(t) \to 0 \text{ and } V_{C(peak)} = V_S(1 - \cos(\pi)) = 2V_S \tag{2.29}$$

The plots of the capacitor voltage $v_c(t)$ and the charging current $i(t)$ are shown in Figure 2.5(b). If we add a diode to the LC circuit in the forward direction of the current, as shown in Figure 2.6(a), the diode will stop any current flow in the reverse direction, and the current flow will stop at $\omega_o t = \pi$. The capacitor will be charged $2V_S$, and the peak current through the circuit is $I_p = V_S\sqrt{C/L}$, as shown in Figure 2.6(b).

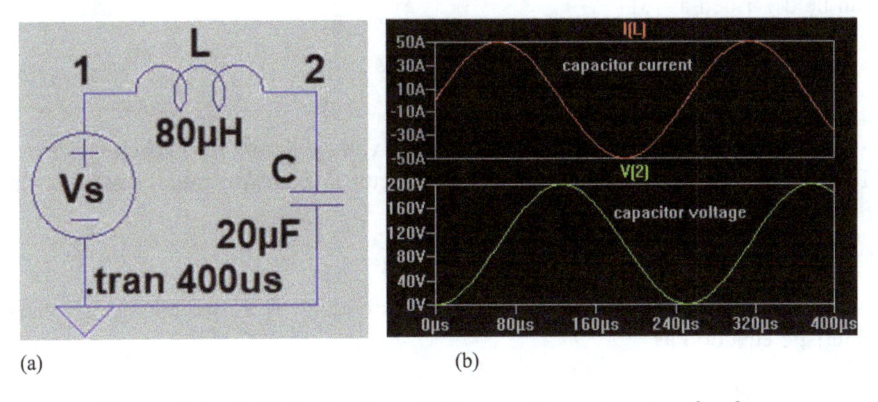

(a) (b)

Figure 2.5 (a) LC circuit and (b) capacitor current and voltage

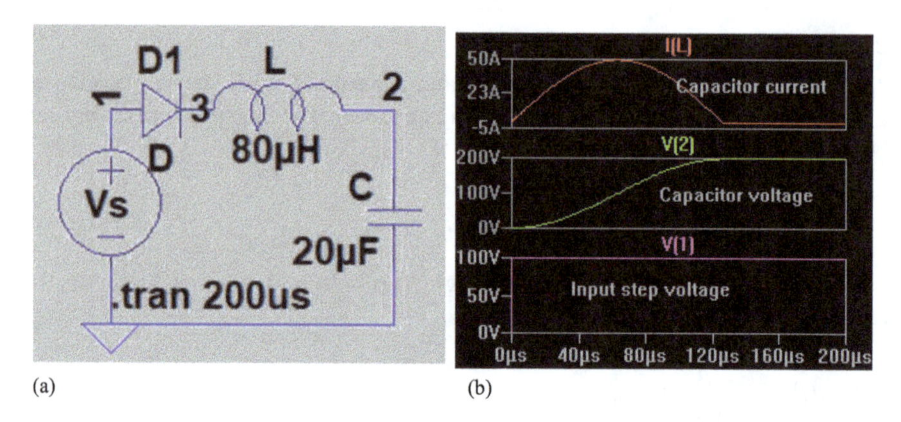

(a) (b)

Figure 2.6 LC-circuit with a series diode: (a) LC circuit with a diode and (b) capacitor current and voltage [4,5]

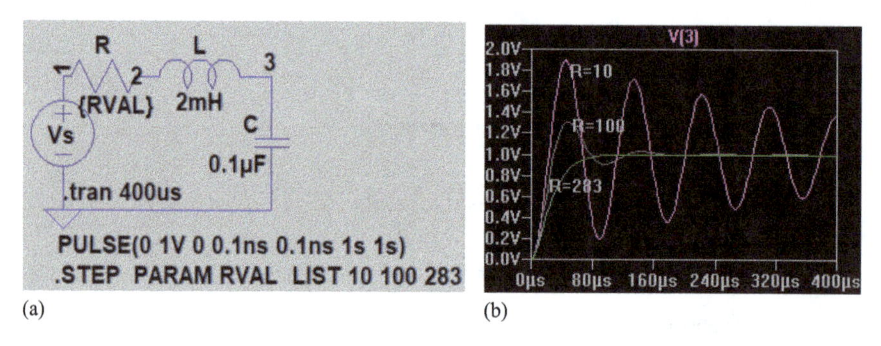

(a) (b)

Figure 2.7 RLC circuit: (a) schematic and (b) output voltage for R = 10, 100, 283 [6,7]

2.3.4 RLC circuit

Inductor L is connected in series with a resistance R and a capacitor C to limit the rate of the rise of the current, as shown in Figure 2.7(a). The charging current $i(t)$ can be described as [3]

$$L\frac{di}{dt} + Ri + \frac{1}{C}\int_{t_0}^{t} idt + v_c(t=0) = V_S \tag{2.30}$$

At $t = 0+$, the current $i(t = 0) = 0$, and the voltage across the capacitor is zero. The source voltage V_s appears across the inductor L. Thus, the initial conditions at $t = 0+$, $L\frac{di}{dt}(t=0) = V_s$

$\frac{di}{dt}(t=0) = \frac{V_s}{L}$, $i(t=0) = 0$, and $v_c(t=0) = 0$

Differentiating both sides of (2.30) and dividing by L, we obtain the characteristic equation as

$$\frac{d^2i}{dt^2} + \frac{R}{L}\frac{di}{dt} + \frac{i}{LC} = 0 \tag{2.31}$$

The characteristic equation in Laplace's domain of s

$$s^2 + \frac{R}{L}s + \frac{i}{LC} = 0 \tag{2.32}$$

We can find the roots of the characteristic equation as

$$\begin{aligned} s_{1,2} &= -\frac{R}{2L} \mp \sqrt{\left(\frac{R}{2L}\right)^2 - \frac{1}{LC}} \\ &= -\alpha \pm \sqrt{\alpha^2 - \omega_o^2} \end{aligned} \tag{2.33}$$

where the damping factor is

$$\alpha = \frac{R}{2L} \tag{2.34}$$

and the resonant frequency is

$$\omega_o = \frac{1}{\sqrt{LC}} \tag{2.35}$$

Depending on the value of α and ω_o, there are three possible cases for the solutions of (2.33).

Case 1 $\alpha = \omega_o$ and $s_1 = s_2$. The circuit is critically damped. The solution has the form

$$i(t) = (A_1 + A_2 t)e^{s_1 t} \tag{2.36}$$

For $V_S = 100$ V, $L = 2$ mH $C = 0.1$ μF, $R = 283.843\Omega$

$$\alpha = R/L = 7.071 \times 10^4 \text{ rad/s}, \quad \omega_o = 1/\sqrt{LC} = 7.071 \times 10^4 \text{ rad/s}$$

For the initial condition at $t = 0$, $i(t) = 0$, and $A_1 = 0$.
Using the derivative of the current in (2.36), we obtain

$$\frac{di(t)}{dt} = A_2 t s_1 e^{s_1 t} + e^{s_1 t} A_2 = A_2 \text{ at } t = 0 \tag{2.37}$$

$$\text{At } t = 0 \text{ for an RLC, we get } \frac{di}{dt} = \frac{V_S}{L} = A_2 \tag{2.38}$$

Substituting A_1 and A_2 in (2.36), we obtain

$$\begin{aligned} i(t) &= A_2 t e^{s_1 t} = \frac{V_s}{L} t e^{-\alpha t} \\ &= \frac{100}{2 \times 10^{-3}} t e^{-7.071 \times 10^4 \; t} \end{aligned} \tag{2.39}$$

Case 2 $\alpha > \omega_o$. The roots are real, and the circuit is over-damped. The solution has the form

$$i(t) = A_1 e^{s_1 t} + A_2 e^{s_2 t} \tag{2.40}$$

For the initial condition at $t = 0$, $i(t) = 0$, $A_1 + A_2 = 0$, or $A_2 = -A_1$. Using the derivative of the current in (2.40), we obtain

$$\frac{di(t)}{dt} = \frac{d}{dt}(A_1 e^{s_1 t} + A_2 e^{s_2 t}) = A_1 s_1 e^{s_1 t} + A_2 s_2 e^{s_2 t} = A_1 s_1 + A_2 s_2$$

$$= A_1 (s_1 - s_2) \tag{2.41}$$

At t = 0 for an RLC,

$$\frac{di}{dt} = \frac{V_S}{L} = A_1 (s_1 - s_2) \text{ or } A_1 = \frac{V_S}{L(s_1 - s_2)} \tag{2.42}$$

Substituting A_1 and A_2 in (2.40), we obtain

$$i(t) = \frac{V_S}{L(s_1 - s_2)}(e^{s_1 t} - e^{s_2 t}) = \frac{V_S}{L(s_1 - s_2)}(e^{s_1 t} - e^{s_2 t}) \tag{2.43}$$

Case 3 $\alpha < \omega_o$. The roots are complex, and the circuit is under-damped. The roots are

$$s_{1,2} = -\alpha \pm j\omega_r \tag{2.44}$$

The solution has the form of damped or decaying sinusoidal as given by

$$i(t) = e^{-\alpha t}(A_1 \cos \omega_r t + A_2 \sin \omega_r t) \tag{2.45}$$

where ω_r is the ringing frequency or damped resonant frequency given by

$$\omega_r = \sqrt{\omega_o^2 - \alpha^2} \tag{2.46}$$

For $V_S = 220$ V, $L = 2$ mH, $C = 0.05$ μF, $R = 160$ Ω.

$$\alpha = R/L = 40{,}000 \text{ rad/s}, \quad \omega_o = 1/\sqrt{LC} = 10^5 \text{ rad/s}$$

Since $\alpha < \omega_o$, the circuit is under-damped. At $t = 0$ and $A_1 = 0$, the current $i(t)$ in (2.45) becomes

$$i(t) = e^{-\alpha t} A_2 \sin \omega_r t \tag{2.47}$$

Using the derivative of the current in (2.47), we obtain

$$\frac{di}{dt} = \omega_r \cos \omega_r t \; A_2 e^{-\alpha t} - \alpha \sin \omega_r t \; A_2 e^{-\alpha t} \tag{2.48}$$

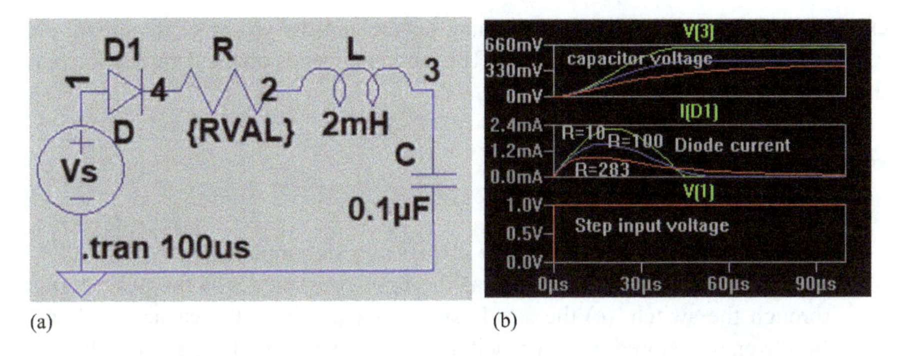

(a) (b)

Figure 2.8 RLC circuit with a series diode: (a) schematic and (b) output voltage for R = 10, 100, and 283

which for $t = 0$ becomes

$$\left.\frac{di}{dt}\right|_{t=0} = \omega_r A_2 = \frac{V_S}{L} \tag{2.49}$$

We obtain A_2 as

$$A_2 = \frac{V_S}{\omega_r\ L} = 1.2A$$

Using (2.47), we obtain the current $i(t)$

$$i(t) = 1.2\sin(91,652t)e^{-40,000t}$$

Figure 2.8(b) shows the effects of the resistance on the transient response of an RLC circuit. An underdamped circuit exhibits an overshot. Depending on the damping, the capacitor voltage overshoots the input voltage, and the current flows from the capacitor to the input source, causing the oscillation to continue. Adding a diode to the circuit shown in Figure 2.7(a) is shown in Figure 2.8(a) and stops the reverse current flow. The diode stops conducting when the current falls to zero, and the capacitor charges and retains its charge. As expected, a lower value of resistance reduces the damping, the diode conduction is reduced, and the capacitor reaches its final value in a shorter time. A series diode is often connected to an RLC series to charge the capacitor and transfer energy from the source to a capacitor.

2.4 Summary

Inductors and capacitors are energy storage elements. The energy stored in an inductor depends on the current flowing through it, and the energy stored in a capacitor depends on the voltage across it. An RL, RC, or RLC circuit can be switched through a power semiconductor switching device to a DC source to transfer energy from a DC source and store it in an inductor or a capacitor. The amount of stored energy can be controlled by the switching device and the values of the circuit elements. The current flowing through an RLC circuit depends on the damping factor of the circuit.

Problems

1. An RC circuit as shown in Figure 2.1(a) has a step input of $V_S = 100$ V, $R = 5$ Ω, and C = 0.1 μF. Determine (a) the time constant τ, (b) the peak current through the switch, (c) the steady-state voltage across the capacitor V_c, and (d) the energy stored in the capacitor. Assume zero initial capacitor voltage of the capacitor.
2. An RC circuit as shown in Figure 2.1(a) has a step input of $V_S = 98$ V, $R = 4$ Ω, and C = 20 μF. Determine (a) the time constant τ, (b) the peak current through the switch, (c) the steady-state voltage across the capacitor V_c, and (d) the energy stored in the capacitor. Assume zero initial capacitor voltage of the capacitor.
3. An RC circuit as shown in Figure 2.2(a) has an initial capacitor voltage of $V_S = 100$ V, $R = 5$ Ω, and $C = 0.1$ μF. Determine (a) the time constant τ, (b) the peak current through the switch, (c) the steady-state voltage across the capacitor V_c, and (d) the energy dissipated in the resistor R.
4. An RC circuit as shown in Figure 2.2(a) has an initial capacitor voltage of $V_S = 98$ V, $R = 2$ Ω, and $C = 20$ μF. Determine (a) the time constant τ, (b) the peak current through the switch, (c) the steady-state voltage across the capacitor V_c, and (d) the energy dissipated in the resistor R.
5. An RL circuit as shown in Figure 2.3(a) has a step input of $V_S = 98$ V, $R = 4$ Ω, and $L = 4$ mH. Determine (a) the time constant τ, (b) the steady-state current through the circuit, (c) the steady-state voltage across the inductor V_L, and (d) the energy stored in the inductor. Assume zero initial inductor current.
6. An RL circuit as shown in Figure 2.3(a) has a step input of $V_S = 100$ V, $R = 5$ Ω, and $L = 4$ mH. Determine (a) the time constant τ, (b) the steady-state current through the circuit, (c) the steady-state voltage across the inductor V_L, and (d) the energy stored in the inductor. Assume zero initial inductor current.
7. The inductor of the RL circuit as shown in Figure 2.4(a) has an initial current of 10 A. The effective step input of $V_S = 0$ V due to the freewheeling action, $R = 4$ Ω, and $L = 4$ mH. Determine (a) the time constant τ, (b) the steady-state current through the circuit, (c) the steady-state voltage across the inductor V_L, and (d) the energy stored in the inductor.
8. The inductor of the RL circuit as shown in Figure 2.4(a) has an initial current of 10 A. The effective step-input of $V_S = 0$ V due to the freewheeling action, $R = 5$ Ω, and $L = 5$ mH. Determine (a) the time constant τ, (b) the steady-state current through the circuit, (c) the steady-state voltage across the inductor V_L, and (d) the energy stored in the inductor.
9. The parameters of the LC circuit as shown in Figure 2.6(a) are: step input of $V_S = 98$ V, $L = 80$ μH, and $C = 20$ μF. Determine (a) the peak current through

the diode, I_p, (b) the conduction of the diode, t_c, and (c) the energy stored in the capacitor *C*.

10. The parameters of the LC circuit as shown in Figure 2.6(a) are: step input of $V_S = 110$ V, $L = 80$ μH, and $C = 20$ μF. Determine (a) the peak current through the diode, I_p, (b) the conduction of the diode, t_c, and (c) the energy stored in the capacitor *C*.
11. The parameters of the LC circuit as shown in Figure 2.6(a) are: step input of $V_S = 98$ V, $L = 40$ μH, and $C = 10$ μF. Determine (a) the peak current through the diode, I_p, (b) the conduction of the diode, t_c, and (c) the energy stored in the capacitor *C*.
12. The parameters of the LC circuit as shown in Figure 2.6(a) are: step input of $V_S = 110$ V, $L = 40$ μH, and $C = 10$ μF. Determine (a) the peak current through the diode, I_p, (b) the conduction of the diode, t_c, and (c) the energy stored in the capacitor *C*.
13. The parameters of the diode LC circuit as shown in Figure 2.6(a) are step input of $V_S = 100$ V, and $C = 10$ μF. The peak current through the inductor is $I_p = 100A$. Assuming an ideal switch and an ideal diode, determine, once the switch is closed, (a) the value of the inductor *L* in μH, (b) the energy stored in the capacitor *C*, W_c, and (c) the conduction of the diode, t_c (μs).
14. The parameters of the switched RLC diode circuit as shown in Figure 2.8(a) are: DC step voltage $Vs = 98$ V, inductance $L = 4$ mH, capacitance $C = 0.1$ μF, and resistance $R = 8$ Ω. Assume that the capacitor and the inductor have no initial values. Determine (a) the expression for the diode current, (b) the peak diode current I_p, and (c) the conduction time of diode t_c.
15. The parameters of the switched RLC diode circuit as shown in Figure 2.8(a) are: DC step-voltage $Vs = 100$ V, inductance $L = 4$ mH, capacitance $C = 0.1$ μF, and resistance $R = 4$ Ω. Assume that the capacitor and the inductor have no initial values. Determine (a) the expression for the diode current, (b) the peak diode current I_p, and (c) the conduction time of diode t_c.
16. The parameters of the switched RLC diode circuit as shown in Figure 2.8(a) are: DC step voltage $Vs = 100$ V, inductance $L = 4$ mH, capacitance $C = 0.1$ μF, and resistance $R = 8$ Ω. Assume that the capacitor and the inductor have no initial values. Determine (a) the expression for the diode current, (b) the peak diode current I_p, and (c) the conduction time of diode t_c.
17. The parameters of the switched RLC diode circuit as shown in Figure 2.8(a) are: DC step voltage $Vs = 100$ V, inductance $L = 4$ mH, capacitance $C = 0.1$ μF, and resistance $R = 100$ Ω. Assume that the capacitor and the inductor have no initial values. Determine (a) the expression for the diode current, (b) the peak diode current I_p, and (c) the conduction time of diode t_c.
18. The parameters of the switched RLC diode circuit as shown in Figure 2.8(a) are: DC step voltage $Vs = 100$ V, inductance $L = 4$ mH, capacitance $C = 0.1$ μF, and resistance $R = 50$ Ω. Assume that the capacitor and the inductor have no initial values. Determine (a) the expression for the diode current, (b) the peak diode current I_p, and (c) the conduction time of diode t_c.

References

[1] M. H. Rashid, *Control Systems, Analysis and Design*. Noida: Cengage Learning India, 2022, Chapter 4.
[2] M. H. Rashid. *Microelectronic Circuits: Analysis and Design*, Noida: Cengage Publishing, 2017, Chapter 1.
[3] M. H. Rashid, *Power Electronics: Circuits, Devices, and Applications*, 4th ed., Englewood Cliffs, NJ: Prentice-Hall, 2014, Chapter 3.
[4] M. H. Rashid, *Introduction to PSpice Using OrCAD for Circuits and Electronics*, 3rd ed., Englewood, NJ: Prentice-Hall, 2003.
[5] M. H. Rashid, *Electronics Analysis and Design Using Electronics Workbench*. Boston, MA: PWS Publishing, 1997.
[6] M. H. Rashid, *SPICE and LTspice for Power Electronics and Electric Power*, Boca Raton, FL: CRC Press, 4/e, 2024.
[7] *LTspice IV Manual*, Milpitas, CA: Linear Technology Corporation, 2013.

Chapter 3
Diode rectifiers

3.1 Introduction

A rectifier converts an alternating current (AC) voltage to a direct current (DC) voltage. It uses one or more diodes to make a unidirectional current flow during the positive half-cycle of the input voltage and also during the negative half-cycle of the input voltage. Thus, the voltage and the current on the input side are of AC types, and the voltage and the current on the output side are of DC types.

3.2 Diodes

A diode is a semiconductor device and has two terminals: anode and cathode. If the anode voltage is held positive with respect to the cathode terminal, the diode conducts and offers a small forward resistance. The diode is then said to be *forward biased*, and it behaves as a short circuit. If the anode voltage is kept negative with respect to the cathode terminal, the diode offers a high resistance. The diode is then said to be *reverse biased*, and it behaves as an open circuit. The characteristic of a practical diode that distinguishes it from an ideal one is that the practical diode experiences a finite voltage drop when it conducts. This drop is typically in the range of 0.5–0.7 V. If the input voltage to a diode circuit is high enough, this small drop can be ignored for most applications.

The voltage drop across the diode is a function of the current flowing through it. The diode current can be related to the diode voltage by the Shockley equation as given by

$$i_D = I_S(e^{-v_D/V_T} - 1) \tag{3.1}$$

where

$$V_T = \frac{kT_K}{q} \tag{3.2}$$

V_T = thermal voltage;

I_S = leakage (or reverse saturation) current, typically in the range of 10^{-6} A to 10^{-15} A;

i_D = diode current, A;

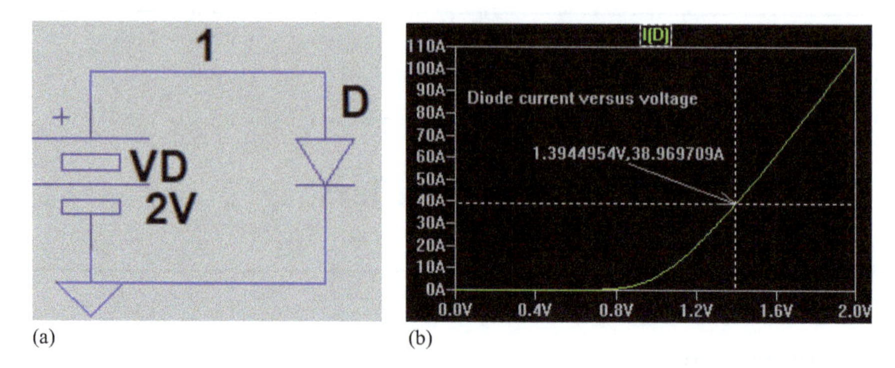

(a) (b)

Figure 3.1 Diode V–I characteristics: (a) diode test circuit and (b) V–I characteristics [2,3]

v_D = diode voltage with the anode positive with respect to the cathode, V;

η = empirical constant is known as an *emission constant* or the *ideality factor* whose value varies from 1 to 2;

q = electron charge = = 1.6022×10^{-19} coulomb (C);

T_k = absolute temperature in kelvins = $273 + T_{Celsius}$; and

k = Boltzmann's constant = = 1.3806×10^{-23} J per kelvin.

At a junction temperature of 25 °C, (3.2) gives the value of V_T as

$$V_T = \frac{kT_K}{q} = \frac{(1.3806 \times 10^{-23})(273 + 25)}{1.6022 \times 10^{-19}} = \frac{T_K}{11605.1} \approx 25.6\text{mV}$$

An LTspice schematic of a diode test circuit is shown in Figure 3.1(a), and the voltage versus current characteristic for a diode of $I_S = 2.54 \times 10^{-9}$ A and $\eta = 1.48$ is shown in Figure 3.1(b).

Note: The input voltage to diode rectifiers in power electronic circuits is significantly higher than the input voltage of low-signal diode rectifiers in electronic circuits. The voltage ratings of the diodes are also selected accordingly. For the analysis of diode rectifiers for high-power applications, we will assume ideal diodes, that is, voltage drop across the diode is negligible, $v_D = 0$.

3.3 Single-phase half-wave rectifier

A half-wave rectifier uses only one diode as shown in the LTspice schematic (Figure 3.2(a)) for a resistive load with an AC input voltage is given by

$$v_s(t) = V_m \sin(\omega t) = V_m \sin(2\pi f t) \tag{3.3}$$

where V_m = peak value of the AC input voltage, V;

f = supply frequency, Hz; and

$\omega = 2\pi f$ = supply frequency, rad/s.

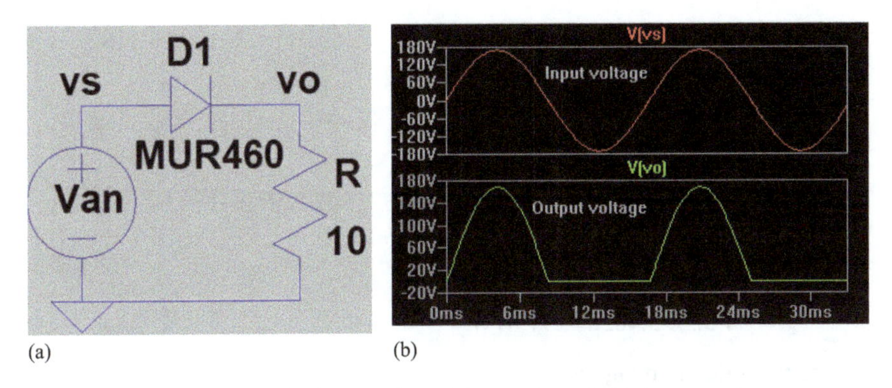

(a) (b)

Figure 3.2 Half-wave diode rectifier: (a) circuit and (b) input and output voltages

During the positive half-cycle in the input voltage, D_1 conducts, and the input voltage appears across the load. During the negative half-cycle in the input voltage, D_1 is reverse-biased, and the diode behaves as an open circuit. The voltage across the load is zero. The output voltage can be described as given by

$$\begin{aligned} v_o(t) &= V_m \sin(\omega t) \quad \text{for } \omega t \leq \pi \\ &= 0 \quad \text{for } \omega t > \pi \end{aligned} \tag{3.4}$$

The plots of the input and output voltages are shown in Figure 3.2(b). The single-phase half-wave rectifiers are normally used up to 100 W power level.

The performance parameters of diode rectifiers can be divided into three types:

- Output-side parameters,
- Rectifier parameters, and
- Input-side parameters.

Let us consider an input voltage of $V_s = 120$ (root-mean-square – rms), $V_m = V_s = 120 \times \sqrt{2} = 169.7$ V, and $R = 10\ \Omega$.

Note that unless specified, an AC voltage is generally quoted in rms values.

3.3.1 Output-side parameters

The average output voltage is

$$\begin{aligned} V_{o(av)} &= V_{dc} = \frac{1}{2\pi}\int_0^{\pi} V_m \sin\theta d\theta = \frac{V_m}{\pi} \\ &= \frac{169.7}{\pi} = 54.02\ \text{V} \end{aligned} \tag{3.5}$$

The average output current is

$$I_{o(av)} = I_{dc} = \frac{V_{o(av)}}{R}$$
$$= \frac{54.02}{10} = 5.402 \text{ A}$$

The DC output power is

$$P_{dc} = V_{dc} I_{dc}$$
$$= 54.02 \times 5.402 = 291.81\text{W}$$

The rms output voltage is

$$V_{o(rms)} = V_{rms} = \sqrt{\frac{1}{2\pi}\int_0^{\pi} V_m^2 \sin^2\theta d\theta} = \frac{V_m}{2}$$
$$= \frac{169.7}{2} = 84.853 \text{ V} \tag{3.6}$$

The rms output current is

$$I_{o(rms)} = I_{rms} = \frac{V_m}{2R}$$
$$= \frac{I_{o(rms)}}{R} = \frac{84.853}{10} = 8.4853 \text{ A}$$

The AC output power is

$$P_{ac} = V_{rms} I_{rms}$$
$$= 84.853 \times 8.4853 = 720 \text{ W}$$

The output AC power

$$P_{out} = I_{rms}^2 R$$
$$= 8.4853^2 \times 10 = 720 \text{ W}$$

Note: For a resistive load, $P_{out} = P_{ac}$ and $I_{rms}^2 R$ is the average or the real power.

The rms ripple content of the output voltage

$$V_{ac} = \sqrt{V_{rms}^2 - V_{dc}^2}$$
$$= \sqrt{84.853^2 - 54.02^2} = 65.437 \text{ V} \tag{3.7}$$

Note: We can add or subtract mean-square values, $V_{rms}^2 = V_{ac}^2 + V_{dc}^2$, but not the rms values.

The rectification *efficiency or ratio*

$$\eta = \frac{P_{dc}}{P_{ac}}$$

$$= \frac{291.81}{720} = 40.53\%$$

Note: It is a figure of merit and measures the quality of the conversion from AC to DC.

The ripple *form factor of the output voltage*

$$RF_{ov} = \frac{V_{ac}}{V_{dc}}$$

$$= \frac{84.853}{54.02} = 121.136\%$$

The *form factor of the output voltage*

$$FF_{ov} = \frac{V_{rms}}{V_{dc}}$$

$$= \frac{84.853}{54.02} = 157.08\%$$

3.3.2 Rectifier parameters

The peak diode current

$$I_{p(diode)} = \frac{V_m}{R}$$

$$= \frac{169.7}{10} = 16.97 \text{ A}$$

The average diode current is the same as the average value of the load current

$$I_{D(av)} = I_{o(av)}$$

$$= 5.402 \text{ A}$$

The rms diode current is the same as the rms value of the load current

$$I_{D(rms)} = I_{rms}$$

$$= 8.4853 \text{ A}$$

3.3.3 Input-side parameters

The rms input current is

$$I_s = I_{rms}$$
$$= 8.4853\ \text{A}$$

The input power

$$P_{in} = P_{out}$$
$$= 720\ \text{W}$$

Input power factor

$$PF_i = \frac{P_{in}}{V_s I_s}$$
$$= \frac{720}{120 \times 8.4853} = 0.707\ (lagging)$$

The *transformer utilization factor*

$$TUF = \frac{P_{dc}}{V_s I_s}$$
$$= \frac{291.81}{120 \times 8.4853} = 0.287$$

Note: The inverse of TUF is a measure of the transformer volt-amp rating to deliver the desired DC output power.

3.3.4 Output LC-filter

An LC-filter as shown in Figure 3.3 is often connected to the output side to limit the ripple contents of the load resistor, R. The capacitor C provides a low impedance path for the ripple currents to flow through it and the inductor L provides a high impedance for the ripple currents. The fundamental frequency f_o of output voltage ripples is the same as the supply frequency $f = 1/T$, where $T = 1/f$ is the period of the input supply voltage. We should note that although the inductor L offers a high impedance to the ripple currents, and the capacitor C offers a low impedance path

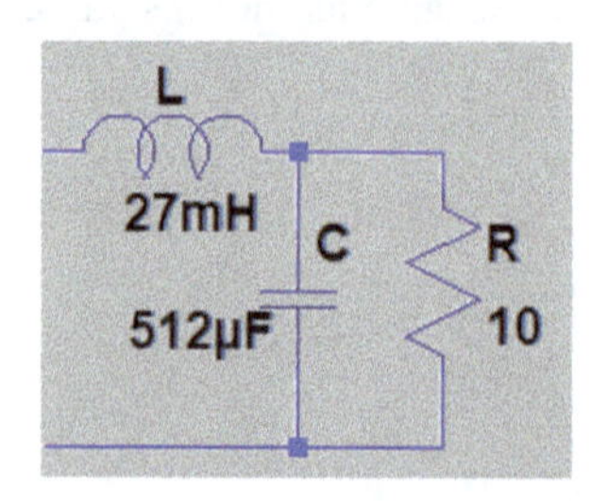

Figure 3.3 LC-filter

for the ripple currents, the average voltage at the output of the rectifier and the voltage across the load resistance R remains the same.

For a desired value of the ripple factor (RF) of the output voltage $RF_v = 5\% = 0.05$, we can find the ripple voltage as

$$\begin{aligned}\Delta V &= RF_v\, V_{dc} \\ &= 0.05 \times 54.02 = 3.241\ \text{V}\end{aligned}$$

Similarly, for a desired value of the RF of the output current $RF_i = 6\% = 0.06$, we can find the ripple current as

$$\begin{aligned}\Delta I &= RF_i\, I_{dc} \\ &= 0.06 \times 5.402 = 0.27\text{A}\end{aligned}$$

We can find the inductor L from the ripple voltage ΔV and the ripple current ΔI as

$$L\frac{\Delta I}{\Delta T} = \Delta V$$

which gives the inductor value L as

$$\begin{aligned}L &= \frac{\Delta V \times \Delta T}{\Delta I} = \frac{\Delta V}{\Delta I} \times \frac{T}{2} \\ &= \frac{3.241}{0.27} \times \frac{16.67 \times 10^{-3}}{2} = 100\ \text{mH}\end{aligned} \tag{3.8}$$

Note: The ripple content varies from a minimum value to a maximum value around the average value over a period of T. Thus, $\Delta T = T/2$ for a half-wave rectifier.

We can find the capacitor C from the ripple voltage ΔV and the ripple current ΔI as

$$C\frac{\Delta V}{\Delta T} = \Delta I$$

which gives the capacitor value C as

$$\begin{aligned}C &= \frac{\Delta I \times \Delta T}{\Delta V} = \frac{\Delta I}{\Delta V} \times \frac{T}{2} \\ &= \frac{0.27}{3.241} \times \frac{16.67 \times 10^{-3}}{2} = 694.4\ \mu\text{F}\end{aligned} \tag{3.9}$$

Figure 3.4(a) shows the LTspice schematic of the single-phase half-wave rectifier with an LC-filter of $C = 694\ \mu$F and $L = 100$ mH. The LTspice plots of the inductor current and the current through the load resistor R are shown in Figure 3.4(b). We can notice that the inductor current is discontinuous, but the load current is continuous with a ripple. Increasing the filter inductor value should make the inductor current continuous. The values of L and C can be adjusted to get the desired ripple content of the load current.

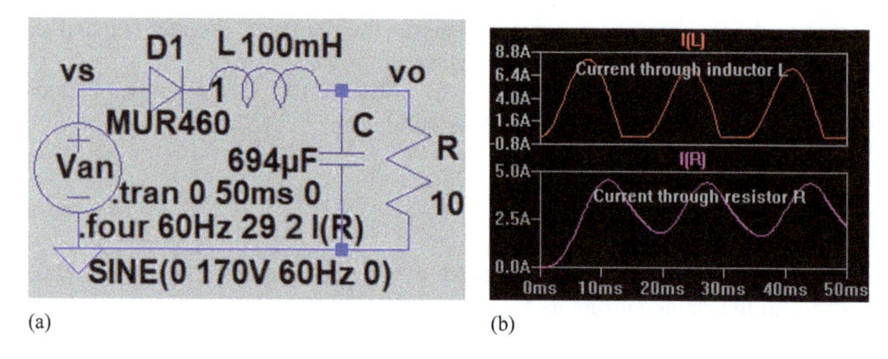

(a) (b)

Figure 3.4 Inductor and load currents: (a) schematic with LC-filter and (b) currents for $C = 694\ \mu\text{F}$, $L = 100$ mH *[4,5]*

Fourier analysis: The LTspice command (.four 60 Hz 29 2 $I(R)$) for Fourier Analysis gives

Total harmonic distortion (THD): 43.782373% (43.783700%) for $C = 694$ pF and $L = 100$ nH

THD: 10.528147% (11.433739%) for $C = 694\ \mu\text{F}$ and $L = 100$ mH

which indicates a significant reduction of the THD with an output filter. LTspice gives the Fourier components as follows:

Fourier components of $I(r)$

DC component: 2.99417

Harmonic number	Frequency (Hz)	Fourier component	Normalized component	Phase (degrees)	Normalized phase (degrees)
1	6.000e+01	1.323e+00	1.000e+00	−140.40	0.00
2	1.200e+02	1.325e−01	1.002e−01	23.48	163.88
3	1.800e+02	3.955e−02	2.990e−02	65.07	205.47
4	2.400e+02	7.063e−03	5.340e−03	105.30	245.69
5	3.000e+02	5.249e−03	3.969e−03	−43.98	96.42
6	3.600e+02	7.906e−03	5.978e−03	−9.49	130.90
7	4.200e+02	6.086e−03	4.601e−03	17.73	158.12
8	4.800e+02	3.505e−03	2.651e−03	22.47	162.87
9	5.400e+02	2.537e−03	1.918e−03	−10.48	129.92
10	6.000e+02	3.225e−03	2.439e−03	−4.01	136.39
11	6.600e+02	3.561e−03	2.693e−03	1.77	142.16
12	7.200e+02	2.270e−03	1.717e−03	9.18	149.57
13	7.800e+02	2.080e−03	1.573e−03	8.07	148.47
14	8.400e+02	2.337e−03	1.767e−03	−8.84	131.55
15	9.000e+02	2.047e−03	1.548e−03	0.92	141.32
16	9.600e+02	2.028e−03	1.533e−03	4.98	145.37

(Continues)

(*Continued*)

Fourier components of $I(r)$					
DC component: 2.99417					
Harmonic number	**Frequency (Hz)**	**Fourier component**	**Normalized component**	**Phase (degrees)**	**Normalized phase (degrees)**
17	1.020e+03	1.591e−03	1.203e−03	4.66	145.05
18	1.080e+03	1.677e−03	1.268e−03	−2.42	137.97
19	1.140e+03	1.632e−03	1.234e−03	−0.94	139.46
20	1.200e+03	1.560e−03	1.179e−03	3.07	143.46
21	1.260e+03	1.430e−03	1.081e−03	3.67	144.06
22	1.320e+03	1.331e−03	1.006e−03	−0.85	139.55
23	1.380e+03	1.308e−03	9.894e−04	−2.15	138.25
24	1.440e+03	1.290e−03	9.752e−04	2.71	143.10
25	1.500e+03	1.226e−03	9.273e−04	0.57	140.96
26	1.560e+03	1.092e−03	8.258e−04	3.10	143.50
27	1.620e+03	1.132e−03	8.559e−04	−1.19	139.20
28	1.680e+03	1.092e−03	8.259e−04	0.40	140.80
29	1.740e+03	1.073e−03	8.114e−04	2.01	142.41
THD: 10.528147% (11.433739%)					

L-Filter: Only the inductor L is connected in series with the load resistance R, and it limits the amount of ripple current ΔI of the inductor around the average inductor current. This inductor ripple current also reduces the ripple voltage of the load resistance R. We can apply (3.8) to determine the inductor value to limit the amount of ripple contents.

C-Filter: Only the capacitor C is connected across the load resistance, and it offers a low impedance to the ripple contents. Most of the ripple currents flow through the capacitor C, rather than the load resistance R. The capacitor C acts as a bypass circuit for the ripple contents. As a result, the current through the load resistance R should contain less ripple currents. We can apply (3.9) to determine the capacitor value to limit the amount of ripple contents through the load resistance.

The impedance of the capacitor forms a parallel circuit with the load resistance R. Thus, selecting the capacitance impedance much smaller than R should make the result parallel impedance tending to R. That is,

$$R \gg \frac{1}{2\pi f C} \tag{3.10}$$

where f is the frequency of the lowest order harmonic, supply frequency for a single-phase half-wave rectifier.

A ratio of 10:1 generally satisfies the condition of (3.10) and gives the relation as given by

$$R = \frac{10}{2\pi fC} \tag{3.11}$$

which gives the value of capacitance C as

$$C = \frac{10}{2\pi fR} \tag{3.12}$$

The designer needs to consider the cost–benefit in selecting the value of C. For example, a higher value of capacitance C would increase the cost versus the reduction in the ripple content. A ratio of 5:1 rather than 10:1 may be a compromise in many applications.

3.3.5 THD versus RF

The THD, which is a measure of closeness in shape between a waveform and its average (DC) component, is defined as

$$THD = \frac{V_{ac}}{V_{dc}} = \frac{1}{V_{dc}} \sum_{n=1}^{\infty} V_{on}^2 \tag{3.13}$$

where V_{on} is the rms value of the nth harmonic component of the output voltage.

Ripple factor: The *ripple factor* of a rectifier is a measure of the ripple content in a waveform as defined by

$$RF = \frac{V_{ac}}{V_{dc}} \tag{3.14}$$

where V_{ac} is the rms value of the ripple content and V_{dc} is the average value of the output waveform.

Harmonic factor (HF) is a measure of the distortion of a waveform and is also known as *total harmonic distortion* (THD). *Ripple Factor* (RF) is also called the harmonic factor (HF), which is a measure of the rms value of the harmonic content in a waveform.

3.3.6 Circuit model

$R = 10\ \Omega$, $C = 694\ \mu F$ and $L = 100$ mH, $f = 60$ Hz

$$\omega_o = 2\pi f_o = 2\pi \times 60 = 377\ \text{rad/s}$$

The Fourier components of the output voltage are given by [1]

$$V_{dc} = a_o = \frac{1}{2\pi}\int_o^{\pi} V_m \sin(\omega t) d(\omega t) = \frac{V_m}{\pi} \tag{3.15}$$

$$\begin{aligned} a(n) &= \frac{1}{2\pi}\int_o^{\pi} V_m \sin(\omega t)\sin(n\omega t) d(\omega t) = \frac{-2V_m}{(n-1)(n+1)} \quad \text{for } n = 2, 4, \ldots \\ &= 0 \quad \text{for } n = 3, 5, \ldots \end{aligned} \tag{3.16}$$

$$b(n) = \frac{1}{2\pi}\int_o^{\pi} V_m \sin(\omega t)\cos(n\omega t) d(\omega t) = 0 \quad \text{for } n = 2, 3, 4, \ldots \tag{3.17}$$

$$b(1) = \frac{1}{2\pi}\int_o^{\pi} V_m \sin(\omega t)\cos(\omega t) d(\omega t) = \frac{V_m}{2} \quad \text{for } n = 1 \tag{3.18}$$

Therefore, the Fourier series of the output voltage is given by

$$\begin{aligned} v_o(t) &= V_{dc} + b(1)\sin(\omega t) + \sum_{n=2,4\ldots}^{\infty} b(n)\sin(n\omega t) \\ &= \frac{V_m}{\pi} + \frac{V_m}{2}\sin(\omega t) + \sum_{n=2,4\ldots}^{\infty} \frac{2V_m}{n\pi(n-1)(n+1)}\sin(n\omega t) \end{aligned} \tag{3.19}$$

The nth harmonic impedance of the load with the LC-filter as shown in Figure 3.3 is given by

$$Z(n) = jn\omega_o L + \frac{R \| \frac{1}{jn\omega_o C}}{1 + jn\omega_o RC} = \frac{R + jn\omega_o L - (n\omega_o)RLC}{1 + jn\omega_o RC} \tag{3.20}$$

Dividing the voltage expression in (3.19) by *Z(n)* in (3.20) gives the Fourier components of the load currents as [1]

$$i_o(t) = \frac{V_m}{\pi R} + \frac{V_m}{2 \times Z(1)}\sin(\omega t) + \sum_{n=2,4\ldots}^{\infty} \frac{2V_m}{n\pi(n-1)(n+1)Z(n)}\sin(n\omega t) \tag{3.21}$$

which gives the peak magnitude of the fundamental harmonic current for $n = 1$ as

$$I_m(1) = \frac{V_m}{2 \times Z(1)} \quad \text{for } n = 1 \tag{3.22}$$

which gives the peak magnitude of the nth harmonic currents

$$I_m(n) = \frac{2V_m}{n\pi(n-1)(n+1)Z(n)} \quad \text{for } n = 2, 4, \ldots \tag{3.23}$$

which for $n = 1$, $n = 2$, $n = 4$, and $n = 6$ gives the peak values of the harmonic currents as

$$I_m(1) = \frac{169.7}{2 \times Z(1)} = 2.467\angle - 87.878^\circ$$

$$I_m(2) = \frac{2 \times 169.7}{3 \times \pi \times Z(2)} = 0.49\angle - 89.726^\circ$$

$$I_m(4) = \frac{2 \times 169.7}{15 \times \pi \times Z(4)} = 0.048\angle - 89.965^\circ$$

$$I_m(6) = \frac{2 \times 169.7}{35 \times \pi \times Z(4)} = 0.015\angle - 89.965^\circ$$

Dividing the peak values by $\sqrt{2}$ gives the rms values for $n = 1$, $n = 2$, $n = 4$, and $n = 6$ as

$$I_1 = \frac{I_m(1)}{\sqrt{2}} = \frac{169.7}{\sqrt{2} \times 2 \times Z(1)} = 1.745\angle - 87.878^\circ$$

$$I_2 = \frac{I_m(2)}{\sqrt{2}} = \frac{2 \times 169.7}{\sqrt{2} \times 3 \times \pi \times Z(2)} = 0.346\angle - 9.726^\circ$$

$$I_4 = \frac{I_m(4)}{\sqrt{2}} = \frac{2 \times 169.7}{\sqrt{2} \times 15 \times \pi \times Z(4)} = 0.034\angle - 89.965^\circ$$

$$I_6 = \frac{I_m(6)}{\sqrt{2}} = \frac{2 \times 169.7}{\sqrt{2} \times 35 \times \pi \times Z(4)} = 0.015\angle - 89.965^\circ$$

The total rms value of the load ripple current can be determined by adding the square of the individual rms values and then taking the square root. That is,

$$I_{ripple} = \sqrt{\sum_{n=1,2,4,6}^{6} \left(|I_1|^2 + |I_2|^2 + |I_4|^2 + |I_6|^2\right)} \tag{3.24}$$
$$= 1.779 \text{ A}$$

The average output current $I_{o(av)}$ of a half-wave rectifier is

$$I_{o(av)} = I_{dc} = \frac{a_o}{R} = \frac{V_m}{\pi R}$$
$$= \frac{169.7}{\pi \times 10} = 5.402 \text{ A}$$

The rms load current can be found from

$$I_{rms} = \sqrt{I_{dc}^2 + I_{ripple}^2}$$
$$= 5.687 \text{ A}$$

Thus, the ripple factor of the load current can be found from

$$RF_i = \frac{I_{ripple}}{I_{dc}}$$
$$= \frac{1.779}{5.402} = 32.935\%$$

3.4 Single-phase full-wave rectifiers

We can connect two single-phase half-wave rectifiers to make a single-phase full-wave rectifier as shown.

In the LTspice schematic, Figure 3.5(a) for a resistive load and an AC input voltage as given by

$$v_s(t) = V_m \sin(\omega t) = V_m \sin(2\pi f t) \tag{3.25}$$

where V_m is the peak value of the AC input voltage, V; f is the supply frequency, Hz; and $\omega = 2\pi f$ is the supply frequency, rad/s.

During the positive half-cycle in the input voltage, diode D_1 conducts, and the positive half-cycle of the input voltage appears across the load. During the negative half-cycle in the input voltage, diode D_2 conducts, and the negative half-cycle of the input voltage appears across the load. The output voltage can be described as

$$\begin{aligned} v_o(t) &= V_m \sin(\omega t) \quad \text{for } 0 < \omega t \le \pi \\ &= V_m \sin(\omega t) \quad \text{for } \pi < \omega t \le 2\pi \\ &= |V_m \sin(\omega t)| \end{aligned} \tag{3.26}$$

The turn ratio of the primary voltage to the secondary voltage of the input transformer can be found the primary and secondary inductances as given by

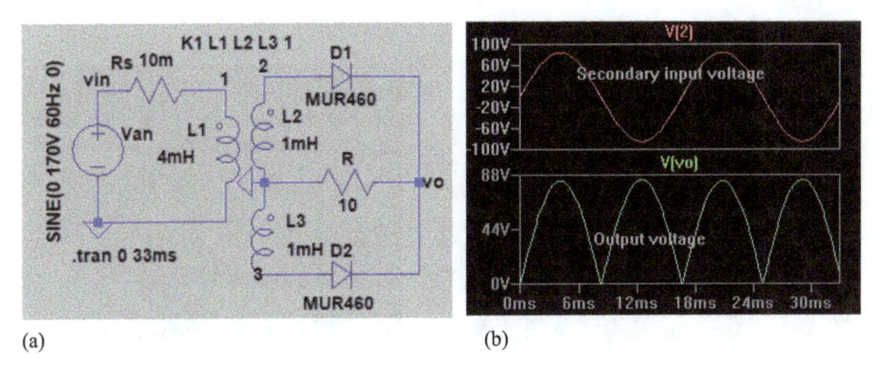

(a) (b)

Figure 3.5 Single-phase full-wave diode rectifier: (a) circuit and (b) input and output voltages

$$TR = \frac{N_{primary}}{N_{secondary}} = \sqrt{\frac{L_1}{L_2}}$$
$$= \sqrt{\frac{4 \times 10^{-3}}{1 \times 10^{-3}}} = 2 \tag{3.27}$$

The plots of the secondary input voltage and the output voltage are shown in Figure 3.4(b) for a transformer of turn ratio of 2:1. There are two pulses on the output voltage. That is, the frequency of the output pulses is twice the supply frequency, $f_o = 2f$.

The single-phase full-wave rectifiers are normally used up to 1 kW power level. The performance parameters of a full-wave rectifier are the same as those of a full-wave bridge rectifier in Section 3.5.

3.5 Single-phase bridge rectifiers

A full-wave bridge rectifier uses four diodes as shown in the LTspice schematic, Figure 3.6(a) with a resistive load for an AC input voltage as

$$v_s(t) = V_m \sin(\omega t) = V_m \sin(2\pi f t) \tag{3.28}$$

where V_m is the peak value of the AC input voltage, V; f is the supply frequency, Hz; and $\omega = 2\pi f$ is the supply frequency, rad/s.

During the positive half-cycle in the input voltage, diodes D_1 and D_2 conduct, and the positive half-cycle of the input voltage appears across the load. During the negative half-cycle in the input voltage, D_3 and D_4 conduct, and the negative half-cycle of the input voltage appears across the load. The output voltage can be described as

$$\begin{aligned} v_o(t) &= V_m \sin(\omega t) \quad \text{for } 0 < \omega t \le \pi \\ &= V_m \sin(\omega t) \quad \text{for } \pi < \omega t \le 2\pi \\ &= |V_m \sin(\omega t)| \end{aligned} \tag{3.29}$$

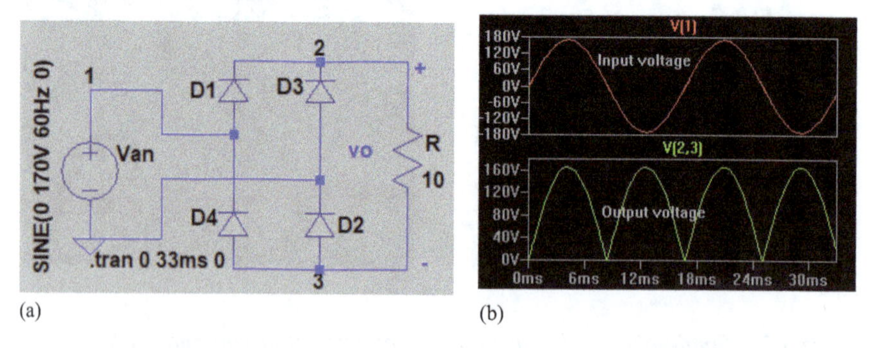

(a) (b)

Figure 3.6 Single-phase bridge-diode rectifier: (a) circuit and (b) input and output voltages

The plots of the input and output voltages are shown in Figure 3.6(b). As expected, there are two pulses per cycle of the input voltage. That is, $f_o = 2f$.

Note: The diodes are numbered in the order of conducting.

The performance parameters of diode rectifiers can be divided into three types:

- Output-side parameters,
- Rectifier parameters, and
- Input-side parameters.

Let us consider an input voltage of $V_s = 120$ rms and $V_m = V_s = 120 \times \sqrt{2} = 169.7$ V.

3.5.1 Output-side parameters

The average output voltage is

$$\begin{aligned} V_{o(av)} &= V_{dc} = \frac{1}{2\pi}\int_0^{\pi} V_m \sin\,\theta d\theta = \frac{2V_m}{\pi} \\ &= \frac{2 \times 169.7}{\pi} = 108.038 \text{ V} \end{aligned} \tag{3.30}$$

Note: The rms value of the output voltage is the same as the rms value of the input voltage. Because the rms value of the positive half-cycle is the rms of the negative half-cycle.

The average output current is

$$\begin{aligned} I_{o(av)} &= I_{dc} = \frac{V_{o(av)}}{R} \\ &= \frac{108.038}{10} = 10.804 \text{ A} \end{aligned}$$

The DC output power is

$$\begin{aligned} P_{dc} &= V_{dc} I_{dc} \\ &= 108.038 \times 10.804 = 1.167 \text{ kW} \end{aligned}$$

The rms output voltage is

$$\begin{aligned} V_{o(rms)} &= V_{rms} = \sqrt{\frac{2}{2\pi}\int_0^{\pi} V_m^2 \sin^2\theta d\theta} = \frac{V_m}{\sqrt{2}} \\ &= \frac{169.7}{\sqrt{2}} = 120 \text{ V} \end{aligned} \tag{3.31}$$

The rms output current is

$$I_{o(rms)} = I_{rms} = \frac{V_m}{\sqrt{2}R}$$
$$= \frac{I_{o(rms)}}{R} = \frac{120}{10} = 12.0 \text{ A}$$

The AC output power is

$$P_{ac} = V_{rms} I_{rms}$$
$$= 120 \times 12 = 1.44 \text{ kW}$$

The output AC power

$$P_{out} = I_{rms}^2 R$$
$$= 12^2 \times 10 = 1.44 \text{ kW}$$

The rms ripple content of the output voltage

$$V_{ac} = \sqrt{V_{rms}^2 - V_{dc}^2}$$
$$= \sqrt{120^2 - 108.038^2} = 52.228 \text{ V} \tag{3.32}$$

The rectification *efficiency or ratio*

$$\eta = \frac{P_{dc}}{P_{ac}}$$
$$= \frac{1.167 \text{ k}}{1.44 \text{ k}} = 81.057\%$$

The *ripple factor* of the output voltage

$$RF_{ov} = \frac{V_{ac}}{V_{dc}}$$
$$= \frac{52.228}{108.038} = 48.343\%$$

The *form factor of the output voltage*

$$FF_{ov} = \frac{V_{rms}}{V_{dc}}$$
$$= \frac{120}{108.038} = 111.072\%$$

3.5.2 Rectifier parameters

The peak diode current

$$I_{p(diode)} = \frac{2V_m}{R}$$
$$= \frac{2 \times 169.7}{10} = 33.941 \text{ A}$$

Since the average load current is shared by two diodes, the average diode current is

$$I_{D(av)} = \frac{I_{o(av)}}{2}$$
$$= \frac{10.804}{2} = 5.402 \text{ A}$$

Since the rms load current is shared by two diodes, the rms diode current is

$$I_{D(rms)} = \frac{I_{rms}}{\sqrt{2}}$$
$$= \frac{12}{\sqrt{2}} = 8.4853 \text{ A}$$

3.5.3 *Input-side parameters*

The rms input current drawn by two diodes is

$$I_s = I_{rms}$$
$$= \sqrt{2} \times 8.4853 = 12 \text{ A}$$

The input power

$$P_{in} = P_{out}$$
$$= 1.44 \text{ kW}$$

Input power factor

$$PF_i = \frac{P_{in}}{V_s I_s}$$
$$= \frac{1440}{120 \times 12} = 1.0 \ (lagging)$$

The *transformer utilization factor*

$$TUF = \frac{P_{dc}}{V_s I_s}$$
$$= \frac{1167}{120 \times 120} = 0.811$$

3.5.4 *Output LC-filter*

An LC-filter as shown in Figure 3.3 is often connected to the output side to limit the ripple contents of the load resistor, *R*. The capacitor *C* provides a low impedance path for the ripple currents to flow through it, and the inductor *L* provides a high impedance for the ripple currents. The fundamental frequency f_o of output voltage ripples is twice the supply frequency f. That is, $f_o = 2f$.

For a desired value of the ripple factor of the output voltage $RF_v = 6\% = 0.06$, we can find the ripple voltage as

$$\begin{aligned}\Delta V &= RF_v\, V_{dc} \\ &= 0.06 \times 108.038 = 6.482\ \text{V}\end{aligned}$$

Similarly, for a desired value of the ripple factor of the output current $RF_i = 6\% = 0.06$, we can find the ripple current as

$$\begin{aligned}\Delta I &= RF_i\, I_{dc} \\ &= 0.06 \times 10.804 = 0.648\ \text{A}\end{aligned}$$

We can find the inductor L from the ripple voltage ΔV and the ripple current ΔI as

$$L\frac{\Delta I}{\Delta T} = \Delta V$$

which gives the inductor value L as

$$\begin{aligned}L &= \frac{\Delta V \times \Delta T}{\Delta I} = \frac{\Delta V}{\Delta I} \times \frac{T}{4} \\ &= \frac{6.482}{0.648} \times \frac{16.67 \times 10^{-3}}{4} = 41.667\ \text{mH}\end{aligned} \tag{3.33}$$

Note: The ripple content varies from a minimum value to a maximum value around the average value over a period of $T/2$. Thus, $\Delta T = T/4$ for a full-wave rectifier.

We can find the capacitor C from the ripple voltage ΔV and the ripple current ΔI as

$$C\frac{\Delta V}{\Delta T} = \Delta I$$

which gives the capacitor value C as

$$\begin{aligned}L &= \frac{\Delta I \times \Delta T}{\Delta V} = \frac{\Delta I}{\Delta V} \times \frac{T}{4} \\ &= \frac{0.648}{6.482} \times \frac{16.67 \times 10^{-3}}{4} = 416.667\ \mu\text{F}\end{aligned} \tag{3.34}$$

Figure 3.7(a) shows the LTspice schematic for the single-phase bridge rectifier with an LC-filter of $C = 694\ \mu\text{F}$ and $L = 100$ mH. The LTspice lots of the inductor current and the current through the load resistor R are shown in Figure 3.7(b). We can notice that both the inductor current and the load current are continuous with a ripple. Increasing the filter inductor value should reduce the ripple contents. The values of L and C can be adjusted to get the desired ripple content of the load current.

Fourier analysis: The LTspice command (.four 60 Hz 29 2 *I*(*R*)) for Fourier Analysis gives

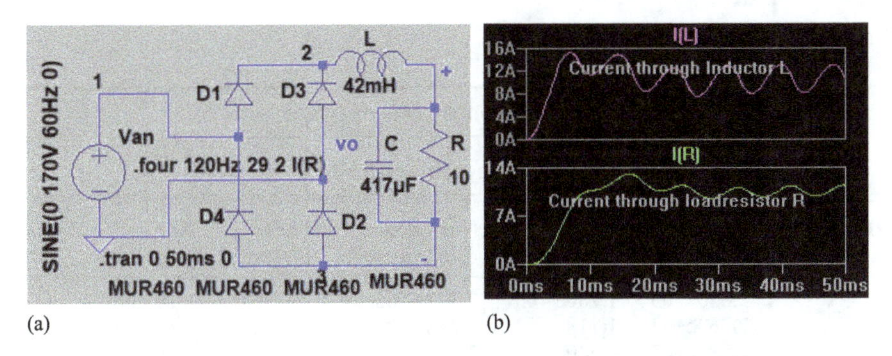

(a) (b)

Figure 3.7 Inductor and load currents: (a) schematic with LC-filter and (b) currents for C = 417 μF L = 42 mH [6].

THD: 22.650358% (22.651220%) for $C=417$ pF and $L = 42$ nH
THD: 6.483097% (14.613308%) for $C=417$ μF and $L = 42$ mH

which indicates a significant reduction of the THD with an output filter. LTspice gives the Fourier components as follows:

Fourier components of *I(r)*

DC component: 10.5331

Harmonic number	Frequency (Hz)	Fourier component	Normalized component	Phase (degrees)	Normalized phase (degrees)
1	1.200e+02	7.640e−01	1.000e+00	112.73	0.00
2	2.400e+02	4.265e−02	5.583e−02	126.87	14.14
3	3.600e+02	1.545e−02	2.022e−02	156.77	44.04
4	4.800e+02	1.024e−02	1.341e−02	171.94	59.20
5	6.000e+02	7.714e−03	1.010e−02	175.86	63.12
6	7.200e+02	6.693e−03	8.760e−03	176.11	63.38
7	8.400e+02	5.783e−03	7.570e−03	−179.62	−292.36
8	9.600e+02	4.846e−03	6.343e−03	−178.20	−290.93
9	1.080e+03	4.308e−03	5.639e−03	179.48	66.74
10	1.200e+03	3.920e−03	5.130e−03	−179.69	−292.42
11	1.320e+03	3.654e−03	4.782e−03	179.82	67.09
12	1.440e+03	3.300e−03	4.319e−03	−178.61	−291.34
13	1.560e+03	2.997e−03	3.923e−03	−178.81	−291.54
14	1.680e+03	2.781e−03	3.640e−03	−179.71	−292.45
15	1.800e+03	2.686e−03	3.516e−03	178.31	65.57
16	1.920e+03	2.486e−03	3.253e−03	−177.67	−290.40
17	2.040e+03	2.248e−03	2.942e−03	−179.08	−291.82
18	2.160e+03	2.206e−03	2.888e−03	178.93	66.20
19	2.280e+03	2.052e−03	2.686e−03	−179.79	−292.52
20	2.400e+03	1.978e−03	2.588e−03	179.93	67.19
21	2.520e+03	1.893e−03	2.478e−03	−179.53	−292.27

(Continues)

(*Continued*)

Fourier components of *I(r)*					
DC component: 10.5331					
Harmonic number	**Frequency (Hz)**	**Fourier component**	**Normalized component**	**Phase (degrees)**	**Normalized phase (degrees)**
22	2.640e+03	1.786e−03	2.337e−03	−179.33	−292.06
23	2.760e+03	1.718e−03	2.248e−03	179.67	66.94
24	2.880e+03	1.640e−03	2.147e−03	−179.58	−292.31
25	3.000e+03	1.586e−03	2.075e−03	−179.81	−292.54
26	3.120e+03	1.514e−03	1.982e−03	−178.96	−291.69
27	3.240e+03	1.437e−03	1.881e−03	−179.75	−292.48
28	3.360e+03	1.376e−03	1.801e−03	179.20	66.46
29	3.480e+03	1.370e−03	1.793e−03	179.90	67.17
THD: 6.483097% (14.613308%)					

L-Filter: Only the inductor L is connected in series with the load resistance R, and it limits the amount of ripple current ΔI of the inductor around the average inductor current. This inductor ripple current also reduces the ripple voltage of the load resistance R. We can apply (3.33) to determine the inductor value to limit the amount of ripple contents.

C-Filter: Only the capacitor C is connected across the load resistance, and it offers a low impedance to the ripple contents. Most of the ripple currents flow through the capacitor C, rather than the load resistance R. The capacitor C acts as a bypass circuit for the ripple contents. As a result, the current through the load resistance R should contain less ripple currents. We can apply (3.34) to determine the capacitor value to limit the amount of ripple contents through the load resistance.

The impedance of the capacitor forms a parallel circuit with the load resistance R. Thus, selecting the capacitance impedance much smaller than R should make the result parallel impedance tending to R. That is,

$$R \gg \frac{1}{2\pi \times 2fC} \tag{3.35}$$

where f is the frequency of the lowest order harmonic, supply frequency for a single-phase half-wave rectifier. A ratio of 10:1 generally satisfies the condition of (3.35) and gives the relation as given by

$$R = \frac{10}{2\pi \times 2 \times fC} \tag{3.36}$$

which gives the value of capacitance C as

$$C = \frac{10}{2\pi \times 2fR} \tag{3.37}$$

The designer needs to consider the cost–benefit in selecting the value of C. For example, a higher value of capacitance C would increase the cost versus the reduction in the ripple content. A ratio of 5:1 rather than 10:1 may be a compromise in many applications.

3.5.5 Circuit model

$R = 10\ \Omega$, $C = 417\ \mu\text{F}$ and $L = 42$ mH, $f = 60$ Hz

$$\omega_o = 2\pi f_o = 2\pi \times 60 = 377\ \text{rad/s}$$

The Fourier components of the output voltage are given by [1]

$$V_{dc} = a_o = \frac{2}{2\pi}\int_o^{\pi} V_m \sin(\omega t)d(\omega t) = \frac{2V_m}{\pi} \tag{3.38}$$

$$\begin{aligned} a(n) &= \frac{2}{2\pi}\int_o^{\pi} V_m \sin(\omega t)\sin(n\omega t)d(\omega t) = \frac{-4V_m}{(n-1)(n+1)} \quad \text{for } n = 2,4,\ldots \\ &= 0 \quad \text{for } n = 3,5,\ldots \end{aligned} \tag{3.39}$$

$$b(n) = \frac{2}{2\pi}\int_o^{\pi} V_m \sin(\omega t)\cos(n\omega t)d(\omega t) = 0 \tag{3.40}$$

Therefore, the Fourier series of the output voltage is given by

$$\begin{aligned} v_o(t) &= V_{dc} + \sum_{n=2,4\ldots}^{\infty} b(n)\sin(n\omega t) \\ &= \frac{2V_m}{\pi} + \sum_{n=2,4\ldots}^{\infty} \frac{4V_m}{n\pi(n-1)(n+1)}\sin(n\omega t - \theta_n) \end{aligned} \tag{3.41}$$

The nth harmonic impedance of the load with the LC-filter is given by

$$Z(n) = jn\omega_o L + \frac{R\|\frac{1}{jn\omega_o C}}{1 + jn\omega_o RC} = \frac{R + jn\omega_o L - (n\omega_o)RLC}{1 + jn\omega_o RC} \tag{3.42}$$

Dividing the voltage expression in (3.41) by *Z*(*n*) in (3.42) gives the Fourier components of the load currents as [1]

$$i_o(t) = \frac{2V_m}{\pi R} + \sin(\omega t) + \sum_{n=2,4\ldots}^{\infty} \frac{4V_m}{n\pi(n-1)(n+1)Z(n)}\sin(n\omega t - \theta_n) \tag{3.43}$$

where θ_n is the impedance angle of the nth harmonic component, rad, which gives the peak magnitude of the nth harmonic currents

$$I_m(n) = \frac{4V_m}{n\pi(n-1)(n+1)Z(n)} \quad \text{for } n = 2,4,\ldots \tag{3.44}$$

which for $n=2$, $n=4$, and $n=6$ gives

$$I_m(2) = \frac{4 \times 169.7}{3 \times \pi \times Z(2)} = 2.524\angle - 88.153^\circ$$

$$I_m(4) = \frac{4 \times 169.7}{15 \times \pi \times Z(4)} = 0.235\angle - 89.769^\circ$$

$$I_m(6) = \frac{4 \times 169.7}{35 \times \pi \times Z(6)} = 0.066\angle - 89.932^\circ$$

Dividing the peak value by $\sqrt{2}$ gives the rms values for $n=2$, $n=4$, and $n=6$ gives

$$I_2 = \frac{I_m(2)}{\sqrt{2}} = \frac{4 \times 169.7}{\sqrt{2} \times 3 \times \pi \times Z(2)} = 1.784\angle - 88.153^\circ$$

$$I_4 = \frac{I_m(4)}{\sqrt{2}} = \frac{4 \times 169.7}{\sqrt{2} \times 15 \times \pi \times Z(4)} = 0.166\angle - 89.769^\circ$$

$$I_6 = \frac{I_m(6)}{\sqrt{2}} = \frac{4 \times 169.7}{\sqrt{2} \times 35 \times \pi \times Z(6)} = 0.047\angle - 89.932^\circ$$

The total rms value of the load ripple current can be determined by adding the square of the individual rms values up to sixth harmonics and then taking the square root

$$\begin{aligned} I_{ripple} &= \sqrt{\sum_{n=1,2,4,6}^{6} \left(|I_2|^2 + |I_4|^2 + |I_6|^2\right)} \\ &= 1.793 \text{ A} \end{aligned} \tag{3.45}$$

The average output current $I_{o(av)}$ of a full-wave rectifier is

$$\begin{aligned} I_{dc} &= I_{o(av)} = \frac{a_o}{R} = \frac{2V_m}{\pi R} \\ &= \frac{2 \times 169.7}{\pi \times 10} = 10.804 \text{ A} \end{aligned} \tag{3.46}$$

The rms load current can be found from

$$\begin{aligned} I_{rms} &= \sqrt{I_{dc}^2 + I_{ripple}^2} \\ &= 10.952 \text{ A} \end{aligned} \tag{3.47}$$

The ripple factor of the load current can be found from

$$\begin{aligned} RF_i &= \frac{I_{ripple}}{I_{dc}} \\ &= \frac{1.793}{10.804} = 16.594\% \end{aligned} \tag{3.48}$$

3.6 Three-phase bridge rectifiers

A three-phase bridge rectifier is commonly used in high-power applications, and it is shown in the LTspice schematic, Figure 3.8(a). An LC-filter with values of $C = 140$ pF and $L = 14$ nH is connected to the load circuit to make the load generalized. Although the input supply voltages can be connected to either Wye or Delta, the input voltages in Figure 3.8(a) are shown in Wye connected. The phase voltages can be expressed as given by [1]

$$\begin{aligned} v_{an}(t) &= V_m \sin(\omega t) = V_m \sin(2\pi f t) \\ v_{bn}(t) &= V_m \sin(\omega t - 2\pi/3) = V_m \sin(2\pi f t - 2\pi/3) \\ v_{cn}(t) &= V_m \sin(\omega t - 4\pi/3) = V_m \sin(2\pi f t - 4\pi/3) \end{aligned} \tag{3.49}$$

where V_m = peak value of the AC input phase voltage, V; f = supply frequency, Hz; and $\omega = 2\pi f$ = supply frequency, rad/s.

Since the line-to-line voltages is $\sqrt{3}$ times the phase voltage of a three-phase Wye-connected source with a phase shift of $+\pi/6$, we can express the line-to-line voltages as given by

$$\begin{aligned} v_{ab}(t) &= \sqrt{3}V_m \sin(\omega t + \pi/6) = \sqrt{3}V_m \sin(2\pi f t + \pi/6) \\ v_{bc}(t) &= \sqrt{3}V_m \sin(\omega t - 2\pi/3 + \pi/6) = \sqrt{3}V_m \sin(2\pi f t - \pi/2) \\ v_{ca}(t) &= \sqrt{3}V_m \sin(\omega t - 4\pi/3 + \pi/6) = \sqrt{3}V_m \sin(2\pi f t - 7\pi/6) \end{aligned} \tag{3.50}$$

The rectifier can operate with or without a transformer. There are six-pulse ripples on the output voltage. The diodes are numbered in order of conduction sequences and each one conducts 120°. The conduction sequence for diodes is D_1–D_2, D_2–D_3, D_3–D_4, D_4–D_5, D_5–D_6, D_6–D_1, and the sequence repeats. The pair of diodes that are connected between that pair of supply lines having the highest amount of instantaneous line-to-line voltage will conduct.

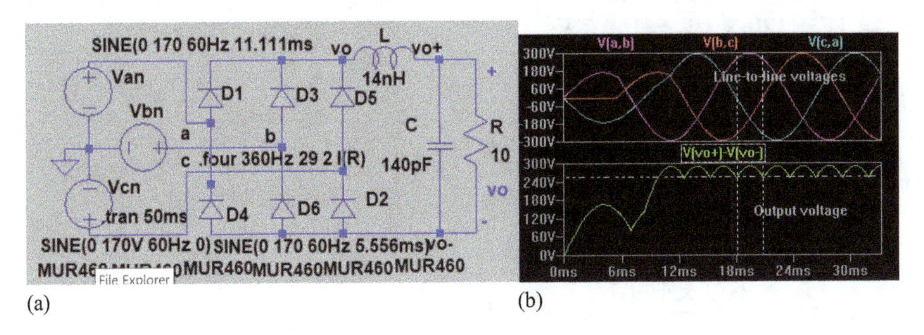

Figure 3.8 Three-phase bridge-diode rectifier: (a) circuit and (b) input and output voltages [2,7]

The waveforms of the line-to-line voltages and the current through load resistor R are shown in Figure 3.8(b). There are six ripple pulses on the output voltage. The frequency of the out-ripple pulses is six times the supply frequency, f. The waveform of the output voltage is identical to the ripple of the load current for a resistive load.

The average output voltage contributed by one pair of diodes can be determined by integrating sine-function from $\pi/6$ to $\pi/2$ or integrating cosine-function from $-\pi/6$ to $+\pi/6$. Using a cosine function gives the average output voltage as given by

$$V_{o(avg)} = V_{dc} = \frac{2}{2\pi/6}\int_{0}^{\pi/6} \sqrt{3}V_m \cos\,\theta d\theta = \sqrt{3}V_m\frac{6}{\pi}\,\sin\,\frac{\pi}{6} = \frac{3\sqrt{3}}{\pi}V_m$$
$$= 1.654V_m \tag{3.51}$$

Using a cosine function gives the rms output voltage as given by

$$V_{o(rms)} = V_{rms} = \sqrt{\frac{2}{2\pi/6}\int_{0}^{\pi/6} 3V_m^2\cos^2\theta d\theta} = \sqrt{3}V_m\sqrt{\frac{6}{2\pi}\left(\frac{\pi}{6}+\frac{1}{2}\,\sin\,\frac{2\pi}{6}\right)}$$
$$= V_m\sqrt{\left(\frac{3}{2}+\frac{9\sqrt{3}}{4\pi}\right)} = 1.6554V_m \tag{3.52}$$

where V_m is the peak phase voltage is, not the peak line-to-line voltage.

The performance parameters of diode rectifiers can be divided into three types:

- Output-side parameters,
- Rectifier parameters, and
- Input-side parameters.

Let us consider an input phase voltage of $V_s = 120$ rms and $V_m = V_s = 120 \times \sqrt{2} = 169.7$ V.

3.6.1 Output-side parameters

From (3.51), the average output voltage is

$$V_{o(av)} = V_{dc} = \frac{3\sqrt{3}}{\pi}V_m$$
$$= \frac{3\sqrt{3}\times 169.7}{\pi} = 280.691 \text{ V}$$

The average output current is

$$I_{o(av)} = I_{dc} = \frac{V_{o(av)}}{R}$$
$$= \frac{280.691}{10} = 28.069 \text{ A}$$

The DC output power is

$$P_{dc} = V_{dc} I_{dc}$$
$$= 280.691 \times 28.069 = 7.879 \text{ kW}$$

From (3.52), the rms output voltage is

$$V_{o(rms)} = V_{rms} == V_m \sqrt{\left(\frac{3}{2} + \frac{9\sqrt{3}}{4\pi}\right)}$$
$$= 169.7 \times \sqrt{\left(\frac{3}{2} + \frac{9\sqrt{3}}{4\pi}\right)} = 280.938 \text{ V}$$

The rms output current is

$$I_{o(rms)} = I_{rms} = \frac{V_{o(rms)}}{R}$$
$$= \frac{280.938}{10} = 28.094 \text{ A}$$

The AC output power is

$$P_{ac} = V_{rms} I_{rms}$$
$$= 280.938 \times 28.094 = 7.893 \text{ kW}$$

The output power

$$P_{out} = I_{rms}^2 R$$
$$= 28.094^2 \times 10 = 7.893 \text{ kW}$$

The rms ripple content of the output voltage

$$V_{ac} = \sqrt{V_{rms}^2 - V_{dc}^2}$$
$$= \sqrt{280.938^2 - 280.691^2} = 11.78 \text{ V}$$

The rectification *efficiency or ratio*

$$\eta = \frac{P_{dc}}{P_{ac}}$$
$$= \frac{7.879 \text{ k}}{7.893 \text{ k}} = 99.824\%$$

The *ripple factor of* output voltage

$$RF = \frac{V_{ac}}{V_{dc}} = \frac{11.78}{280.691} = 4.197\%$$

The form factor *of the output voltage*

$$FF_{ov} = \frac{V_{rms}}{V_{dc}} = \frac{280.938}{280.691} = 100.1\%$$

3.6.2 Rectifier parameters

The peak diode current due to the peak-line to line voltage

$$I_{p(diode)} = \frac{\sqrt{3}V_m}{R} = \frac{\sqrt{3} \times 169.7}{10} = 29.394 \text{ A}$$

The load current is shared by three diodes, the average diode current

$$I_{D(av)} = \frac{I_{o(av)}}{3} = \frac{28.069}{3} = 9.356 \text{ A} \tag{3.53}$$

The rms load current is shared by three diodes, the rms diode current

$$I_{D(rms)} = \frac{I_{rms}}{\sqrt{3}} = \frac{28.094}{\sqrt{3}} = 16.22 \text{ A} \tag{3.54}$$

3.6.3 Input-side parameters

The rms input phase current is continued by two diodes

$$I_s = I_{rms} = \sqrt{2}I_{D(rms)} = \sqrt{\frac{2}{3}} \times 16.22 = 22.938 \text{ A} \tag{3.55}$$

The input power

$$P_{in} = P_{out} = 7.893 \text{ kW}$$

Input power factor for three phases,

$$PF_i = \frac{P_{in}}{3V_sI_s} = \frac{7.893\text{ kW}}{3 \times 120 \times 22.938} = 0.956\ (lagging) \tag{3.56}$$

The *transformer utilization factor,*

$$TUF = \frac{P_{dc}}{3V_sI_s} = \frac{7.879\text{ kW}}{3 \times 120 \times 22.938} = 0.954 \tag{3.57}$$

Fourier analysis: The LTspice command (.four 360 Hz 29 2 $I(R)$) for Fourier Analysis for $C = 140$ pF and $L = 14$ nH gives

Fourier components of *I*(*r*)

DC component: 27.7019

Harmonic number	Frequency (Hz)	Fourier component	Normalized component	Phase (degrees)	Normalized phase (degrees)
1	3.600e+02	1.592e+00	1.000e+00	89.91	0.00
2	7.200e+02	3.874e−01	2.432e−01	−89.87	−179.78
3	1.080e+03	1.726e−01	1.084e−01	89.95	0.04
4	1.440e+03	9.656e−02	6.064e−02	−90.59	−180.50
5	1.800e+03	6.060e−02	3.806e−02	90.40	0.49
6	2.160e+03	4.226e−02	2.654e−02	−89.86	−179.77
7	2.520e+03	3.070e−02	1.928e−02	88.52	−1.39
8	2.880e+03	2.331e−02	1.464e−02	−89.56	−179.47
9	3.240e+03	1.845e−02	1.159e−02	89.97	0.06
10	3.600e+03	1.461e−02	9.173e−03	−90.33	−180.24
11	3.960e+03	1.221e−02	7.667e−03	89.02	−0.89
12	4.320e+03	9.854e−03	6.188e−03	−91.91	−181.82
13	4.680e+03	8.217e−03	5.160e−03	89.80	−0.12
14	5.040e+03	7.496e−03	4.707e−03	−89.56	−179.47
15	5.400e+03	6.307e−03	3.960e−03	87.28	−2.64
16	5.760e+03	5.192e−03	3.260e−03	−91.81	−181.72
17	6.120e+03	4.725e−03	2.967e−03	86.67	−3.24
18	6.480e+03	4.161e−03	2.613e−03	−91.68	−181.59
19	6.840e+03	3.771e−03	2.368e−03	88.99	−0.92
20	7.200e+03	3.302e−03	2.073e−03	−93.43	−183.34
21	7.560e+03	2.897e−03	1.819e−03	84.79	−5.12
22	7.920e+03	2.642e−03	1.659e−03	−94.42	−184.33
23	8.280e+03	2.414e−03	1.516e−03	84.91	−5.00
24	8.640e+03	2.155e−03	1.354e−03	−91.22	−181.13
25	9.000e+03	1.986e−03	1.247e−03	85.55	−4.36

(Continues)

(*Continued*)

Fourier components of *I*(*r*)					
DC component: 27.7019					
Harmonic number	**Frequency (Hz)**	**Fourier component**	**Normalized component**	**Phase (degrees)**	**Normalized phase (degrees)**
26	9.360e+03	1.792e−03	1.125e−03	−98.82	−188.73
27	9.720e+03	1.686e−03	1.059e−03	80.93	−8.98
28	1.008e+04	1.449e−03	9.100e−04	−92.32	−182.23
29	1.044e+04	1.344e−03	8.442e−04	86.79	−3.12
THD: 27.885835% (27.887313%)					

Note: The output harmonic frequency is $6 \times 60 = 360$ Hz.

3.6.4 Output LC-filter

An LC-filter as shown in Figure 3.3 is often connected to the output side to limit the ripple contents of the load resistor, R. The capacitor C provides a low impedance path for the ripple currents to flow through it, and the inductor L provides a high impedance for the ripple currents. The fundamental frequency f_o of output voltage ripples is six times the supply frequency f. That is, $f_o = 6f$

For a desired value of the ripple factor of the output voltage $RF_v = 6\% = 0.06$, we can find the ripple voltage as

$$\Delta V = RF_v\, V_{dc}$$
$$= 0.06 \times 280.691 = 16.482 \text{ V}$$

Similarly, for a desired value of the ripple factor of the output current $RF_i = 6\% = 0.06$, we can find the ripple current as

$$\Delta I = RF_i\, I_{dc}$$
$$= 0.06 \times 28.069 = 1.684 \text{ A}$$

We can find the inductor L from the ripple voltage ΔV and the ripple current ΔI as

$$L\frac{\Delta I}{\Delta T} = \Delta V$$

which gives the inductor value L for six pulses as

$$\begin{aligned} L &= \frac{\Delta V \times \Delta T}{\Delta I} = \frac{\Delta V}{\Delta I} \times \frac{T}{2 \times 6} \\ &= \frac{16.482}{1.684} \times \frac{16.67 \times 10^{-3}}{2 \times 6} = 13.889 \text{ mH} \end{aligned} \tag{3.58}$$

Note: The ripple content varies from a minimum value to a maximum value around the average value over a period of *T*/6. Thus, $\Delta T = T/12$ for a full-wave rectifier.

We can find the capacitor *C* from the ripple voltage ΔV and the ripple current ΔI as

$$C\frac{\Delta V}{\Delta T} = \Delta I$$

which gives the capacitor value *C* as

$$\begin{aligned} C &= \frac{\Delta I \times \Delta T}{\Delta V} = \frac{\Delta I}{\Delta V} \times \frac{T}{2 \times 6} \\ &= \frac{1.684}{16.482} \times \frac{16.67 \times 10^{-3}}{2 \times 6} = 138.889\ \mu\text{F} \end{aligned} \tag{3.59}$$

Let us choose $L = 14$ mH and $C = 140\ \mu$F.

Fourier analysis: The LTspice command (.four 360 Hz 29 2 *I*(*R*)) for Fourier Analysis gives

THD: 27.885835% (27.887313%) for $C = 140$ pF and $L = 14$ nH
THD: 5.971740% (6.008168%) for $C = 140\ \mu$F and $L = 14$ mH

which indicates a significant reduction of the THD with an output filter. LTspice gives the Fourier components as follows:

Fourier components of *I(r)*

DC component: 27.7054

Harmonic number	Frequency (Hz)	Fourier component	Normalized component	Phase (degrees)	Normalized phase (degrees)
1	3.600e+02	1.672e−01	1.000e+00	−70.63	0.00+
2	7.200e+02	9.763e−03	5.841e−02	99.74	170.37
3	1.080e+03	1.948e−03	1.165e−02	−86.54	−15.90
4	1.440e+03	5.946e−04	3.557e−03	98.14	168.77
5	1.800e+03	3.242e−04	1.939e−03	−92.40	−21.76
6	2.160e+03	1.388e−04	8.301e−04	87.78	158.41
7	2.520e+03	1.166e−04	6.972e−04	−89.14	−18.50
8	2.880e+03	9.891e−05	5.917e−04	81.18	151.81
9	3.240e+03	5.161e−05	3.088e−04	−114.14	−43.50
10	3.600e+03	7.892e−05	4.721e−04	100.26	170.89
11	3.960e+03	6.660e−05	3.984e−04	−136.09	−65.46
12	4.320e+03	4.015e−05	2.402e−04	124.33	194.96
13	4.680e+03	7.329e−05	4.384e−04	−117.94	−47.31
14	5.040e+03	8.936e−06	5.346e−05	54.31	124.95
15	5.400e+03	5.101e−05	3.051e−04	−90.79	−20.15
16	5.760e+03	3.288e−05	1.967e−04	40.12	110.75
17	6.120e+03	1.525e−05	9.121e−05	−74.19	−3.56

(Continues)

(*Continued*)

Fourier components of *I(r)*					
DC component: 27.7054					
Harmonic number	**Frequency (Hz)**	**Fourier component**	**Normalized component**	**Phase (degrees)**	**Normalized phase (degrees)**
18	6.480e+03	3.793e−05	2.269e−04	77.93	148.57
19	6.840e+03	1.816e−05	1.086e−04	−176.58	−105.94
20	7.200e+03	2.715e−05	1.624e−04	127.60	198.23
21	7.560e+03	3.132e−05	1.874e−04	−146.85	−76.21
22	7.920e+03	1.398e−05	8.363e−05	−158.93	−88.30
23	8.280e+03	2.760e−05	1.651e−04	−103.76	−33.13
24	8.640e+03	1.037e−05	6.201e−05	−50.50	20.14
25	9.000e+03	1.519e−05	9.086e−05	−46.82	23.81
26	9.360e+03	1.214e−05	7.262e−05	40.08	110.71
27	9.720e+03	7.588e−06	4.539e−05	59.18	129.81
28	1.008e+04	1.399e−05	8.368e−05	115.81	186.45
29	1.044e+04	1.025e−05	6.129e−05	157.96	228.59
THD: 5.971740% (6.008168%)					

Note: The output harmonic frequency is $6 \times 60 = 360$ Hz.

L-Filter: Only the inductor L is connected in series with the load resistance R, and it limits the amount of ripple current ΔI of the inductor around the average inductor current. This inductor ripple current also reduces the ripple voltage of the load resistance R. We can apply (3.58) to determine the inductor value to limit the amount of ripple contents.

C-Filter: Only the capacitor C is connected across the load resistance, and it offers a low impedance to the ripple contents. Most of the ripple currents flow through the capacitor C, rather than the load resistance R. The capacitor C acts as a bypass circuit for the ripple contents. As a result, the current through the load resistance R should contain less ripple currents. We can apply (3.59) to determine the capacitor value to limit the amount of ripple contents through the load resistance.

The impedance of the capacitor forms a parallel circuit with the load resistance R. Thus, selecting the capacitance impedance much smaller than R should make the result parallel impedance tending to R. That is,

$$R \gg \frac{1}{2\pi \times 6fC} \tag{3.60}$$

where f is the frequency of the lowest order harmonic, supply frequency for a single-phase half-wave rectifier. A ratio of 10:1 generally satisfies the condition of (3.35) and gives the relation as given by

$$R = \frac{10}{2\pi \times 6 \times fC} \tag{3.61}$$

which gives the value of capacitance C as

$$C = \frac{10}{2\pi \times 6fR} \tag{3.62}$$

The designer needs to consider the cost-benefit in selecting the value of C. For example, a higher value of capacitance C would increase the cost versus the reduction in the ripple content. A ratio of 5:1 rather than 10:1 may be a compromise in many applications.

3.6.5 Circuit model

$R = 10\ \Omega$, $C = 140\ \mu\text{F}$ and $L = 14$ mH, $f = 60$ Hz

$$\omega_o = 2\pi f_o = 2\pi \times 6 \times 60 = 2262\ \text{rad/s}$$

As shown in Figure 3.8(b), the output waveforms have six pulses per cycle of the input supply frequency. The frequency of the output is six times the fundamental component ($6f$). To find the constants of the Fourier series of the output voltage, we integrate a cosine function of the line-line voltage from $-\pi/6$ to $\pi/6$, and the constants are given by [1]

$$b(n) = \frac{1}{\pi/6}\int_{-\pi/6}^{\pi/6} \sqrt{3}V_m \cos(\omega t)\sin(n\omega t)d(\omega t) = 0 \tag{3.63}$$

$$\begin{aligned}
a(n) &= \frac{1}{\pi/6}\int_{-\pi/6}^{\pi/6} \sqrt{3}V_m \cos(\omega t)\cos(n\omega t)d(\omega t) \\
&= \frac{6 \times \sqrt{3}V_m}{\pi}\left[\frac{\sin(n-1)\pi/6}{n-1} + \frac{\sin(n+1)\pi/6}{n+1}\right] \quad \text{for } n = 0, 2, 4, \ldots \\
&= \frac{6 \times \sqrt{3}V_m}{\pi}\left[\frac{(n+1)\sin(n-1)\pi/6 + (n-1)\sin(n+1)\pi/6}{n^2-1}\right] \\
&= 0 \quad \text{for } n = 3, 5, \ldots
\end{aligned} \tag{3.64}$$

Using the following trigonometrical relations:

$$\sin(A+B) = \sin A \cos B + \cos A \sin B$$

and

$$\sin(A-B) = \sin A \cos B - \cos A \sin B$$

Equation (3.46) can be simplified to

$$a(n) = \frac{2 \times 6 \times \sqrt{3}V_m}{\pi(n^2-1)}\left[n \sin\left(\frac{n\pi}{6}\right)\cos\left(\frac{\pi}{6}\right) - \cos\left(\frac{n\pi}{6}\right)\sin\left(\frac{\pi}{6}\right)\right] \tag{3.64}$$

Since the term $\sin(n\pi/6) = 0$, (3.64) can be simplified to

$$a(n) = -\frac{2 \times 6 \times \sqrt{3}V_m}{\pi(n^2-1)} \cos\left(\frac{n\pi}{6}\right) \sin\left(\frac{\pi}{6}\right) \tag{3.66}$$

The DC component is found by letting $n = 0$ and is given by

$$V_{dc} = \frac{a(0)}{2} = \frac{6 \times \sqrt{3}V_m}{\pi} \sin\left(\frac{\pi}{6}\right) \tag{3.67}$$

Therefore, the Fourier series of the output voltage is given by

$$\begin{aligned} v_o(t) &= \frac{6 \times \sqrt{3}V_m}{\pi} \sin\left(\frac{\pi}{6}\right) + \sum_{n=6,12\ldots}^{\infty} -\frac{2 \times 6 \times \sqrt{3}V_m}{\pi(n^2-1)} \cos\left(\frac{n\pi}{6}\right) \sin\left(\frac{\pi}{6}\right) \cos(n\omega t) \\ &= \frac{6 \times \sqrt{3}V_m}{\pi} \sin\left(\frac{\pi}{6}\right) \left[1 - \sum_{n=6,12\ldots}^{\infty} -\frac{2}{(n^2-1)} \cos\left(\frac{n\pi}{6}\right) \cos(n\omega t)\right] \end{aligned} \tag{3.68}$$

which can be expressed as given by

$$\begin{aligned} v_o(t) &= \frac{6 \times \sqrt{3}V_m}{\pi} \sin\left(\frac{\pi}{6}\right) + \sum_{n=6,12\ldots}^{\infty} -\frac{2 \times 6 \times \sqrt{3}V_m}{\pi(n^2-1)} \cos\left(\frac{n\pi}{6}\right) \sin\left(\frac{\pi}{6}\right) \cos(n\omega t) \\ &= \frac{6 \times \sqrt{3}V_m}{\pi} \sin\left(\frac{\pi}{6}\right) \left[1 - \sum_{n=6,12\ldots}^{\infty} -\frac{2}{(n^2-1)} \cos\left(\frac{n\pi}{6}\right) \cos(n\omega t)\right] \\ &= 0.9549 \times \sqrt{3}V_m\left(1 + \frac{2}{35}\cos(6\omega t) - \frac{2}{143}\cos(12\omega t) + \frac{2}{323}\cos(18\omega t) - \ldots\right. \end{aligned} \tag{3.69}$$

The nth harmonic impedance of the load with the LC-filter is given by

$$Z(n) = jn\omega_o L + \frac{R \| \frac{1}{jn\omega_o C}}{1 + jn\omega_o RC} = \frac{R + jn\omega_o L - (n\omega_o)RLC}{1 + jn\omega_o RC} \tag{3.70}$$

Dividing the voltage expression in (3.69) by $Z(n)$ in (3.70) gives the Fourier components of the load currents as [1]

$$i_o(t) = \frac{6 \times \sqrt{3}V_m}{\pi} \sin\left(\frac{\pi}{6}\right) \left[\frac{1}{R} - \sum_{n=6,12\ldots}^{\infty} -\frac{2}{(n^2-1)Z(n)} \cos\left(\frac{n\pi}{6}\right) \cos(n\omega t)\right] \tag{3.71}$$

which gives the peak magnitude of the nth harmonic currents

$$I_m(n) = \frac{6 \times \sqrt{3}V_m}{\pi} \sin\left(\frac{\pi}{6}\right) \times \frac{2}{(n^2-1)Z(n)} \cos\left(\frac{n\pi}{6}\right) \text{ for } n = 6, 12, 18, \ldots \tag{3.72}$$

which for $n=6$, $n=12$, and $n=8$ gives

$$I_m(6) = \frac{6 \times \sqrt{3}V_m}{\pi} \sin\left(\frac{\pi}{6}\right) \times \frac{2}{(n^2-1)Z(6)} \cos\left(\frac{6\pi}{6}\right) = 0.562\angle -88.153^\circ$$

$$I_m(12) = \frac{6 \times \sqrt{3}V_m}{\pi} \sin\left(\frac{\pi}{6}\right) \times \frac{2}{(n^2-1)Z(12)} \cos\left(\frac{12\pi}{6}\right) = 0.064\angle 90.231^\circ$$

$$I_m(18) = \frac{6 \times \sqrt{3}V_m}{\pi} \sin\left(\frac{\pi}{6}\right) \times \frac{2}{(n^2-1)Z(18)} \cos\left(\frac{18\pi}{6}\right)$$
$$= 0.019\angle -89.932^\circ$$

Dividing the peak value by $\sqrt{2}$ gives the rms values for $n=6$, $n=12$, and $n=18$ gives

$$I_6 = \frac{I_m(6)}{\sqrt{2}} = \frac{0.562\angle -88.153^\circ}{\sqrt{2}} = 0.397\angle -88.153^\circ$$

$$I_{12} = \frac{I_m(12)}{\sqrt{2}} = \frac{0.064\angle 90.231^\circ}{\sqrt{2}} = 0.045\angle 90.231^\circ$$

$$I_{18} = \frac{I_m(18)}{\sqrt{2}} = \frac{0.019\angle -89.932^\circ}{\sqrt{2}} = 0.013\angle -89.932^\circ$$

The total rms value of the load ripple current can be determined by adding the square of the individual rms values and then taking the square root

$$\begin{aligned} I_{ripple} &= \sqrt{\sum_{n=6,12,\ldots}^{6} \left(|I_6|^2 + |I_{12}|^2 + |I_{18}|^2\right)} \\ &= 0.4 \text{ A} \end{aligned} \tag{3.73}$$

The average output current $I_{o(av)}$ of a three-phase rectifier is

$$\begin{aligned} I_{dc} = I_{o(av)} &= \frac{V_{dc}}{R} = \frac{6 \times \sqrt{3}V_m}{\pi R} \sin\left(\frac{\pi}{6}\right) \\ &= \frac{6 \times \sqrt{3} \times 169.7}{\pi \times 10} \sin\left(\frac{\pi}{6}\right) = 28.069 \text{ A} \end{aligned} \tag{3.74}$$

The rms load current can be found from

$$\begin{aligned} I_{rms} &= \sqrt{I_{dc}^2 + I_{ripple}^2} \\ &= 28.072 \text{ A} \end{aligned} \tag{3.75}$$

The ripple factor of the load current can be found from

$$RF_i = \frac{I_{ripple}}{I_{dc}} = \frac{0.4}{28.072} = 1.426\% \tag{3.76}$$

3.7 Summary

A diode rectifier converts a fixed AC voltage to a variable DC voltage and uses diodes as switching devices. It uses one or more diodes to make a unidirectional current flow during the positive half-cycle of the AC input voltage and a unidirectional current flow during the negative half-cycle of the AC input voltage. Thus, the voltage and the current on the input side are of AC types and the voltage and current on the output side are of DC types. Ideally, one or more diodes conduct during the positive half-cycle of the input voltage, and one or more diodes conduct during the negative half-cycle of the input voltage. The average output voltage depends on the peak value of the input AC voltage and thus gives a fixed output voltage. Since the peak voltages of a three-phase supply are higher than a single-phase AC source, a three-phase rectifier gives a higher average output voltage than that of a single-phase rectifier. *L*, *C*, and LC filters are often connected to a resistive load to limit the ripple contents of the output voltage and current. For a resistive load, the load current can be discontinuous and falls to zero. For an inductive load, the load current becomes continuous, and the conducting pair of diodes continues to conduct until the next pair of diodes is turned on.

Problems

1. The single-phase half-wave rectifier as shown in Figure 3.2(a) has a purely resistive load of $R = 5\ \Omega$, a supply frequency of $f = 60$ Hz, and the peak voltage input voltage of $V_m = 170$ V. Determine (a) the average value of the output voltage, $V_{o(av)}$, (b) the rms value of the output voltage, $V_{o(rms)}$, (c) the DC output power is P_{dc}, (d) the output AC power, P_{out}, and (e) input power factor PF_i. Assume ideal diode switch.
2. The single-phase half-wave rectifier as shown in Figure 3.2(a) has a purely resistive load of $R = 5\ \Omega$, a supply frequency of $f = 60$ Hz, and the peak voltage input voltage $V_m = 170$ V. Calculate the performance parameters – (a) Output-side parameters, (b) controller parameters, and (c) input-side parameters. Assume ideal thyristor switches. Assume ideal diode switch.
3. The single-phase half-wave rectifier as shown in Figure 3.4(a) has a purely resistive load of $R = 5\ \Omega$, a supply frequency of $f = 60$ Hz, and the peak voltage input voltage $V_m = 170$ V. Determine the values of the filter inductor L and capacitor C to limit the ripple factor of the output voltage

$RF_v = 6\% = 0.06$ and the ripple factor of the output current $RF_i = 5\% = 0.05$. Assume ideal diode switch.

4. The single-phase half-wave rectifier as shown in Figure P3.4 is connected to the input supply through a transformer of turn ratio $n = 10{:}1$. The load resistive is $R = 5\ \Omega$, a supply frequency of $f = 60$ Hz, and the peak voltage input voltage $V_m = 170$ V at the primary side of the transformer. Determine (a) the average value of the output voltage, $V_{o(av)}$, (b) the rms value of the output voltage, $V_{o(rms)}$, (c) the DC output power is P_{dc}, (d) the output AC power, P_{out}, and (e) input power factor PF_i Assume ideal diode switch.
5. The single-phase half-wave rectifier as shown in Figure P3.4 is connected to the input supply through a transformer of turn ratio $n = 10{:}1$. The load resistive is $R = 5\ \Omega$, a supply frequency of $f = 60$ Hz, and the peak voltage input voltage $V_m = 170$ V at the primary side of the transformer. Calculate the performance parameters – (a) Output-side parameters, (b) controller parameters, and (c) input-side parameters. Assume ideal thyristor switches. Assume ideal diode switch,
6. The single-phase half-wave rectifier as shown in Figure P3.4 is connected to the input supply through a transformer of turn ratio $n = 10{:}1$. The load resistive is $R = 5\ \Omega$, a supply frequency of $f = 60$ Hz, and the peak voltage input voltage $V_m = 170$ at the primary side of the transformer. Determine the values of the filter inductor L and capacitor C to limit the ripple factor of the output voltage $RF_v = 6\% = 0.06$ and the ripple factor of the output current $RF_i = 5\% = 0.05$. Assume ideal diode switch.

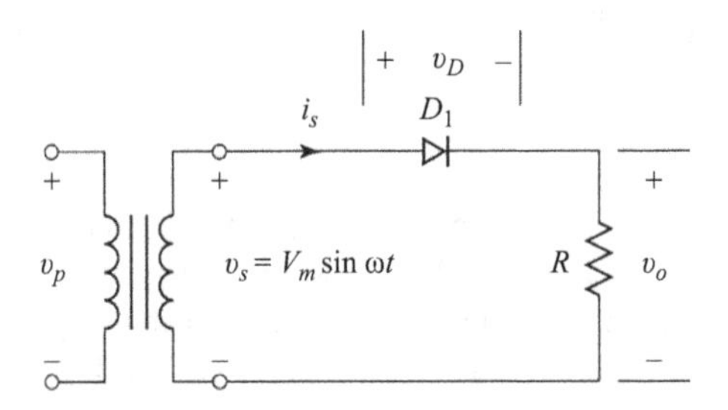

Figure P3.4 Single-phase half-wave rectifier

7. The single-phase full-wave rectifier as shown in Figure 3.4(a) is connected to the input supply through a transformer of turn ratio $n = 10{:}1$. The load resistive is $R = 5\ \Omega$, a supply frequency of $f = 60$ Hz, and the peak voltage input voltage $V_m = 170$ V. Determine (a) the average value of the output voltage, $V_{o(av)}$, (b) the rms value of the output voltage, $V_{o(rms)}$, (c) the DC output power is P_{dc}, (d) he output AC power, P_{out}, and (e) input power factor PF_i
8. The single-phase full-wave rectifier as shown in Figure 3.4(a) is connected to the input supply through a transformer of turn ratio $n = 10{:}1$. The load

resistive is $R = 5\ \Omega$, a supply frequency of $f = 60$ Hz, and the peak voltage input voltage $V_m = 170$ V. Calculate the performance parameters – (a) Output-side parameters, (b) controller parameters, and (c) input-side parameters. Assume ideal thyristor switches. Assume ideal diode switch,

9. The single-phase full-wave rectifier as shown in Figure 3.4(a) is connected to the input supply through a transformer of turn ratio $n = 10{:}1$. The load resistive is $R = 5\ \Omega$, a supply frequency of $f = 60$ Hz, and the peak voltage input voltage $V_m = 170$ V. Determine the values of the filter inductor L and capacitor C to limit the ripple factor of the output voltage $RF_v = 6\% = 0.06$ and the ripple factor of the output current $RF_i = 5\% = 0.05$.
10. The single-phase bridge rectifier as shown in Figure 3.5(a) has a purely resistive load of $R = 5\ \Omega$, a supply frequency of $f = 60$ Hz, and the peak voltage input voltage $V_m = 170$ V. Determine (a) the average value of the output voltage, $V_{o(av)}$, (b) the rms value of the output voltage, $V_{o(rms)}$, (c) the DC output power is P_{dc}, (d) he output AC power, P_{out}, and (e) input power factor PF_i
11. The single-phase bridge rectifier as shown in Figure 3.5(a) has a purely resistive load of $R = 5\ \Omega$, a supply frequency of $f = 60$ Hz, and the peak voltage input voltage $V_m = 170$ V. Calculate the performance parameters: (a) output-side parameters, (b) controller parameters, and (c) input-side parameters. Assume ideal thyristor switches. Assume ideal diode switch.
12. The single-phase bridge rectifier as shown in Figure 3.5(a) has a purely resistive load of $R = 5\ \Omega$, a supply frequency of $f = 60$ Hz, and the peak voltage input voltage $V_m = 170$ V. Determine the values of the filter inductor L and capacitor C to limit the ripple factor of the output voltage $RF_v = 6\% = 0.06$ and the ripple factor of the output current $RF_i = 5\% = 0.05$.
13. The three-phase bridge rectifier as shown in Figure 3.8(a) is connected to a three-phase Wye-connected input voltage source. The load resistive is $R = 5\ \Omega$, a supply frequency of $f = 60$ Hz, and the peak voltage input phase voltage $V_m = 170$ V. Determine (a) the average value of the output voltage, $V_{o(av)}$, (b) the rms value of the output voltage, $V_{o(rms)}$, (c) the DC output power is P_{dc}, (d) he output AC power, P_{out} and (e) input power factor PF_i.
14. The three-phase bridge rectifier as shown in Figure 3.8(a) is connected to a three-phase wye-connected input voltage source. The load resistive is $R = 5\ \Omega$, a supply frequency of $f = 60$ Hz, and the peak voltage input phase voltage of $V_m = 170$ V. Calculate the performance parameters – (a) Output-side parameters, (b) controller parameters, and (c) input-side parameters. Assume ideal thyristor switches. Assume ideal diode switches.
15. The three-phase bridge rectifier as shown in Figure 3.8(a) is connected to a three-phase Wye-connected input voltage source. The load resistive is $R = 5\ \Omega$, a supply frequency of $f = 60$ Hz, and the peak voltage input phase voltage $V_m = 170$ V. Determine the values of the filter inductor L and capacitor C to limit the ripple factor of the output voltage $RF_v = 6\% = 0.06$ and the ripple factor of the output current $RF_i = 5\% = 0.05$.

References

[1] Rashid, M.H. *Power Electronics: Circuits, Devices, and Applications*, 4th ed., Englewood Cliffs, NJ: Prentice-Hall, 2014, Chapter 3.
[2] M. H. Rashid, *Introduction to PSpice Using OrCAD for Circuits and Electronics*. Englewood, NJ: Prentice-Hall, Inc., 2003 (3/e).
[3] M. H. Rashid, *Electronics Analysis and Design Using Electronics Workbench*. Boston, MA: PWS Publishing, 1997.
[4] M. H. Rashid, *SPICE and LTspice for Power Electronics and Electric Power*. Boca Raton, FL: CRC Press, 2024.
[5] M. H. Rashid, *Microelectronic Circuits: Analysis and Design*. Boston, MA: Cengage Publishing, 2017.
[6] *LTspice IV Manual*, Milpitas, CA: Linear Technology Corporation, 2013.
[7] R. Jacob Baker, *SPICE Software, MOSFET Models, and MOSIS Information*. *CMOS Circuit Design, Layout, and Simulation*, 3rd edn. New York, NY: Wiley-IEEE Press, 2010.

Chapter 4
DC–DC converters

4.1 Introduction

A DC-to-DC converter converts a direct current (DC) input voltage to a direct current (DC) output voltage. The output voltage can be higher than or lower than the input voltage. It normally uses one transistor acting as a switch and one or more diodes to transfer energy from the source to the load. The amount of energy transfer depends on the switching on-time of the transistor commonly known as the *duty cycle* of the transistor switch. The converter is normally operated under closed-loop control to vary the duty cycle to give the desired output voltage. Thus, the voltage and the current on the input side and the output side are of DC types.

4.2 Transistor switches

A DC-to-DC converter uses a power semiconductor device as a switch to turn on and off the DC supply to the load. The switching action can be implemented by a bipolar junction transistor (BJT), a metal–oxide-semiconductor field-effect-transistor (MOSFET), or an insulated-gate bipolar-transistor (IGBT). A DC–DC converter with only one switch is often known as a DC *chopper*.

4.2.1 BJT switch

A BJT is a semiconductor device and has three terminals: base, collector, and emitter. There are two types of transistors: NPN and PNP. An NPN transistor has a p-type semiconductor layer between two n-type layers, whereas a PNP transistor has an n-type layer between two p-type layers. NPN BJTs work from a positive supply voltage, whereas PNP BJTs work from a negative supply voltage. NPN BJTs are generally used for power electronic applications. If a base–current is caused to flow through the base–emitter junction, it causes a current flow through the collector terminal. The base–emitter junction follows the diode characteristics, and the collector current i_C is related to the base–emitter voltage v_{BE} by the Shockley equation as given by

$$i_C = I_S\left(e^{\frac{-v_{BE}}{V_T}} - 1\right) \tag{4.1}$$

where

$$V_T = \frac{kT_K}{q} \tag{4.2}$$

V_T = thermal voltage, V;
i_C = collector current, A;
v_{BE} = base–emitter voltage with the base terminal positive with respect to the emitter, V;
I_S = leakage (or reverse saturation) current, typically in the range to 10^{-6} A to 10^{-15} A;
η = empirical constant known as *emission constant*, or the ideality factor whose value varies from 1 to 2;
q = electron charge = 1.6022×10^{-19} C;
T_k = absolute temperature in kelvins = $273 + T_{Celsius}$; and
k = Boltzmann's constant = 1.3806×10^{-23} J/K.
At a junction temperature of 25 °C, we can obtain the value of V_T as

$$V_T = \frac{kT_K}{q} = \frac{(1.3806 \times 10^{-23})(273 + 25)}{1.6022 \times 10^{-19}} = \frac{T_K}{11,605.1} \approx 25.7 \text{ mV} \tag{4.3}$$

The collector current i_C is also normally related to the corresponding base current i_B by a factor known as the *forward current gain* β_F as given by

$$\beta_F = \frac{i_C}{i_B} \tag{4.4}$$

The emitter current i_E is the sum of the base current i_B and collector current i_C as given by

$$\begin{aligned} i_E &= i_C + i_B \\ i_E &= i_C + i_B = (\beta_F i_B + i_B) = (1 + \beta_F) i_B \\ i_E &= i_C + i_B = i_C + \frac{i_C}{\beta_F} = \left(\frac{1 + \beta_F}{\beta_F}\right) i_C \end{aligned} \tag{4.5}$$

The ratio of the collector current to the emitter is also as the *current ratio* as given by

$$\alpha_F = \frac{i_C}{i_E} = \frac{\beta_F}{1 + \beta_F} \tag{4.6}$$

The LTspice schematic for plotting the output characteristics of a BJT is shown in Figure 4.1(a) for BJT parameters of $I_S = 2.37 \times 10^{-8}, \beta_F = 73$ A, and $\eta = 1.26$. A BJT is a current dependent device, and its collector current depends on the base current. As the base–current is increased, the collector current is also increased. As the collector–emitter voltage increases, the collector current increases until the collector current reaches the saturation region and remains practically

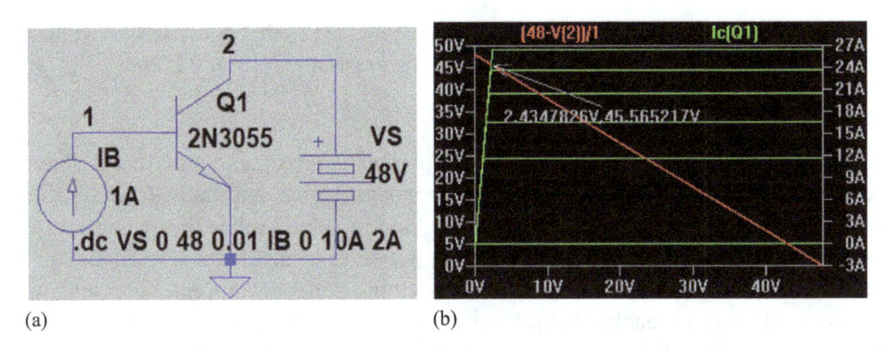

(a) (b)

Figure 4.1 LTspice schematic BJT characteristics: (a) BJT biasing and (b) BJT output characteristics

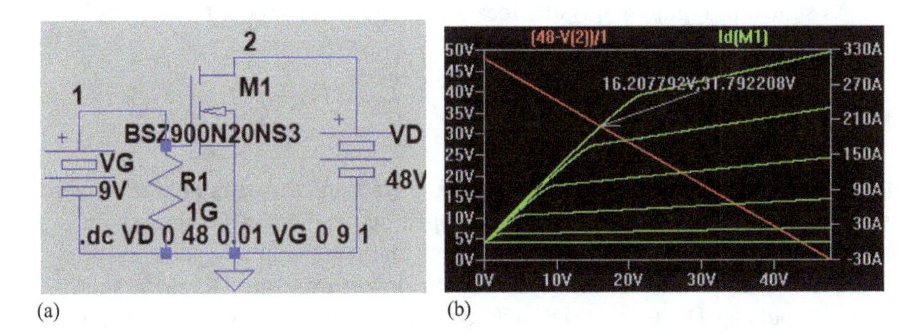

(a) (b)

Figure 4.2 LTspice schematic MOSFET characteristics: (a) MOSFET biasing and (b) MOSFET output characteristics

constant. For operating the BJT as a switching device, its base current must be high enough to keep the collector–emitter low for a specific load line as shown in Figure 4.1(b) for a load resistance of 1 Ω, the collector emitter voltage, $V_{CE} = 2.43$ V, and the voltage across the load resistance, $V_L = 45.56$ V.

Note: For the analysis of DC–DC converters for power electronics, we will assume that the transistors are operated as switches and ideal BJTs, that is, the voltage drop across the BJTs is zero, $v_{CE} = 0$.

4.2.2 MOSFET switch

The LTspice schematic for plotting the output characteristics of a MOSFET is shown in Figure 4.2(a). A MOSFET is a voltage-dependent device, and its drain current depends on the gate–source voltage. As the gate–source voltage is increased, the drain current is also increased. As the drain–source voltage increases, the drain current increases until the drain current reaches the saturation region and remains practically constant. For operating the MOSFET as a switching device, its gate–source voltage must be high enough to keep the drain–source voltage low for a specific load line as shown in Figure 4.2(b) for a load resistance of 1 Ω, the drain–source voltage, $V_{DS} = 16.207$ V, and the voltage across the load resistance, $V_L = 31.79$ V.

4.2.3 IGBT switch

The LTspice schematic for plotting the output characteristics of an IGBT is shown in Figure 4.3(a). An IGBT is a voltage-dependent device, and its collector current depends on the gate–emitter voltage. From the input side, it has the characteristics of a MOSFET; while looking from the output side, it has the characteristics of a BJT. As the gate–emitter voltage is increased, the collector current increases until the collector current reaches the saturation region and remains practically constant. For operating the IGBT as a switching device, its gate–emitter voltage must be high enough to keep the collector–emitter low for a specific load line as shown in Figure 4.3(b) for a load resistance of 1 Ω, the collector–emitter source voltage, $V_{CE} = 8.247$ V, and the voltage across the load resistance, $V_L = 39.75$ V.*

As shown in Figure 4.3(a), the location of the IGBT model for FGW50XS65. Lib is included by the .lib command.

4.3 DC–DC converters

The input voltage to a DC–DC converter is DC type, and the output of the converter is also DC. The output voltage can be stepped up or down. The output voltage is normally a pulsing DC with a ripple content. The input source voltage to the converter is normally DC and the input current drawn from the DC input source is generally pulsing DC type. A DC–DC converter in general uses a transistor as a switching element, a diode to control the direction of current to the load, an inductor as an energy storage element to main the ripple content within a specified limit ripple content, and a capacitor to maintain the load voltage within a specified

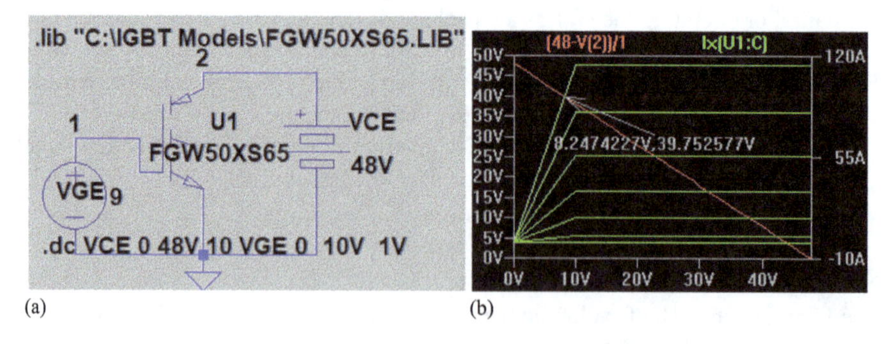

Figure 4.3 Characteristics of Fuji Electric IGBT FGW50XS6: (a) IGBT test circuit and (b) V–I output characteristics [5,6]*

*Discrete IGBT P-Spice Models, https://www.fujielectric.com/products/semiconductor/model/igbt/technical/design_tool.html [1–6].

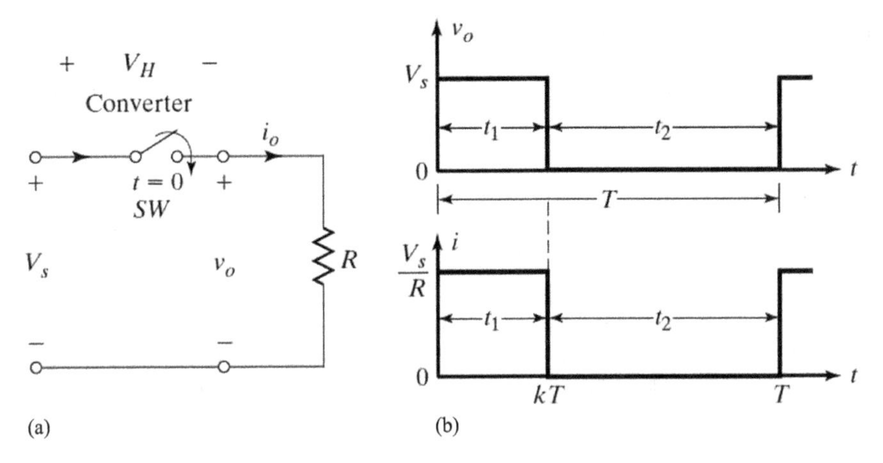

Figure 4.4 Simple switched DC–DC converter

ripple content. Figure 4.4(a) shows that the load is connected to the DC source through a switch. When the switch is turned on, the input DC voltage is connected to the load, the output voltage is $v_o = V_S$, and when the switched is turned off, the output voltage is zero. $v_o = 0$ as shown in Figure 4.4(b). The average output voltage can be found from the switch on-time t_1 and off-time t_2 as given by

$$V_{dc} = \frac{t_1}{t_1 + t_2} V_S = \frac{t_1}{T} V_S = t_1 f V_S = k V_S \tag{4.7}$$

where k is the ratio of the on-time to the switching period T. commonly known as the *duty-cycle*

$f = 1/T$ is the switching frequency, Hz.

Therefore, by varying the duty cycle k from 0 to 1, we can vary the average output voltage V_{dc} in (4.7) from 0 to V_S. The output voltage as shown in Figure 4.4(b) is a pulsed wave. Inductors and capacitors are generally connected to limit the variations of the load voltage and the current. Since the action of the switch makes the output voltage chopped DC voltage, the switch is often known as a *chopper*.

A DC–DC converter generally consists of a power switching device, an inductor, a capacitor, and a diode. Depending on the layout of these elements, the DC–DC converters can be classified into six types as follows:

1. Chopper with an inductive load,
2. Buck converter (step-down),
3. Boost converter (step-up),
4. Buck–Boost converter (step up and step-down),
5. Single-ended primary inductance converter (SEPIC) – Buck–Boost converter,
6. Cúk converter (Buck–Boost converter).

Note: The performances of the Ćuk converter are like those of the SEPIC converter, not included.

4.4 Chopper with an inductive load

The LTspice schematic of a MOSFET chopper with an inductive load is shown in Figure 4.5(a). The plots of the gating signals, the output voltage, and the load current are shown in Figure 4.5(b). The waveforms of the gating pulses and the output voltage are identical, except the gating pulses are pulsed waveforms of 0–5 V and the output waveforms are also of pulsed waveforms of 0–98 V. However, the load current is continuous due to the load inductance. The gating pulses are connected between the gate and the source terminals of the MOSFET through a voltage-controlled voltage-source to isolate the gating circuit from the power circuit. A high resistance of $R_G = 1\ \text{M}\Omega$ is also connected between the gate and the source terminals of the MOSFET to provide a finite impedance between the gate and source terminals of the MOSFET.

The performance parameters of the DC–DC converters can be divided into three types:

- Output-side parameters,
- Converter parameters, and
- Input-side parameters.

Let us consider an input voltage of $V_S = 98$ V, $E = 0$ V, $L = 450\ \mu$H, and $R = 4\ \Omega$. The chopper is operating at a switching frequency of $f_s = 10$ kHz, a switching period of $T = 1/f = 100\ \mu$s, and a duty cycle of $k = 60\%$. Thus, the on-time of the switch is $t_1 = kT = 0.6 \times 100\ \mu\text{s} = 60\ \mu\text{s}$, and the off-time of the switch is $t_2 = (1 - k)T = (1 - 0.6) \times 100\ \mu\text{s} = 40\ \mu\text{s}$.

Depending on whether the switch is turned on or off, there are two modes of operation. During mode 1, the switch is turned on; and during mode 2, the switch is turned off [7,8].

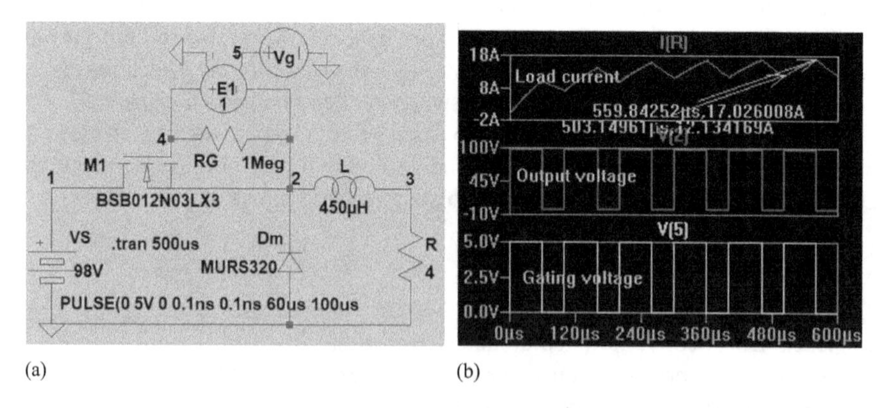

Figure 4.5 A MOSFET chopper with an inductive load: (a) schematic and (b) converter waveforms

Mode 1: When the switch is turned on, the DC supply voltage V_S is connected to the load, and the load current i_1 can be described by

$$V_S = L\frac{di_1}{dt} + Ri_1 + E \tag{4.8}$$

which for the initial current $i_1(t = 0) = I_1$ gives the load current as given by

$$i_1(t) = I_1 e^{-tR/L} + \frac{V_S - E}{R}\left(1 - e^{-tR/L}\right) \tag{4.9}$$

Note: The voltage source of E in (4.8) is included in series with R and L to derive generalized expressions of the converter performance parameters. E can be an external battery to be charged or the back emf of a DC motor.

At the end of this mode at $t = kT$, we get the load current in (4.9) as given by

$$i_1(t = t_1 = kT) = I_2 = I_1 e^{-kTR/L} + \frac{V_S - E}{R}\left(1 - e^{-kTR/L}\right) \tag{4.10}$$

Mode 2: When the switch s turned off, the DC supply voltage V_S is disconnected from the load, and the load current tends to fall. As a result, the inductor L induces a voltage in the opposite direction and turns on diode D_m, known as the free-wheeling diode. Thus, the load current i_2 during this mode can be described by

$$0 = L\frac{di_2}{dt} + Ri_2 + E \tag{4.11}$$

which for the initial current $i_2(t = 0) = I_2$ gives the load current as

$$i_2(t) = I_2 e^{-tR/L} - \frac{E}{R}\left(1 - e^{-tR/L}\right) \tag{4.12}$$

At the end of this mode at $t = (1 - k)T$, we get the load current in (4.12) as given by

$$i_2(t = t_2 = (1 - k)T) = I_3 \tag{4.13}$$

Under steady-state conditions, $I_3 = I_1$, and (4.10) and (4.13) gives

$$I_1 = I_3 = I_2 e^{-(1-kT)R/L} - \frac{E}{R}\left(1 - e^{-(1-kT)R/L}\right) \tag{4.14}$$

$$I_2 = I_1 e^{-kTR/L} + \frac{V_S - E}{R}\left(1 - e^{-kTR/L}\right) \tag{4.15}$$

Solving for the steady-state values of I_1 and I_2 from (4.14) and (4.15), we obtain

$$I_1 = \frac{V_S}{R}\left(\frac{e^{kz} - 1}{e^{z} - 1}\right) - \frac{E}{R} \tag{4.16}$$

$$I_2 = \frac{V_S}{R}\left(\frac{e^{-kz} - 1}{e^{-z} - 1}\right) - \frac{E}{R} \tag{4.17}$$

where $z = \frac{TR}{L} = \frac{R}{fL}$ is the ratio of the switching period (T) to the load time constant $\tau = L/R$.

From (4.16) and (4.17), we can find the peak to peal ripple current as

$$\Delta I = I_2 - I_1 = \frac{V_S}{R}\left(\frac{e^{-kz}-1}{e^{-z}-1}\right) - \frac{V_S}{R}\left(\frac{e^{kz}-1}{e^{z}-1}\right)$$

$$= \frac{V_S}{R}\left(\frac{1-e^{-kz}+e^{-z}-e^{-(1-k)z}}{1-e^{-z}}\right) \tag{4.18}$$

which gives the condition for maximum ripple as

$$\frac{d(\Delta I)}{dk} = 0 \tag{4.19}$$

which gives $e^{-kz} - e^{-(1-k)z} = 0$ or $-k = -(1-k)$, or $k = 0.5$. Thus, the maximum peak to peak ripple current at $k = 0.5$ is

$$\Delta I_{\max} == \frac{V_S}{R}\tanh\left(\frac{R}{4fL}\right) \tag{4.20}$$

For a small value of θ, tanh $(\theta) \simeq \theta$. For $4fL \gg R$, the maximum ripple current in (4.20) can be approximated to

$$\Delta I_{\max} == \frac{V_S}{R} \times \frac{R}{4fL} = \frac{V_S}{4fL} \tag{4.21}$$

4.4.1 Output-side parameters

The parameter, $z = \frac{R}{fL} = \frac{4}{10\times10^3\times450\times10^{-6}} = 0.889$

(a) The average output voltage

$$V_{dc} = \frac{1}{T}\int_0^{kT} V_S dt = kV_s$$
$$= 0.6 \times 98 = 58.8 \text{ V}$$

(b) The rms output voltage

$$V_{rms} = \sqrt{\frac{1}{T}\int_0^{kT} V_S^2 dt} = \sqrt{k}V_s$$
$$= \sqrt{0.6} \times 98 = 75.91 \text{ V}$$

(c) From (4.16), the steady-state minimum load current

$$I_1 = \frac{V_S}{R}\left(\frac{e^{kz}-1}{e^{z}-1}\right) - \frac{E}{R} = \frac{98}{4} \times \left(\frac{e^{0.6\times0.889}-1}{e^{0.889}-1}\right) - \frac{0}{4} = 12.051 \text{ A}$$

(d) From (4.17), the steady-state maximum load current

$$I_2 = \frac{V_S}{R}\left(\frac{e^{-kz}-1}{e^{-z}-1}\right) - \frac{E}{R} = \frac{98}{4} \times \left(\frac{e^{-0.6\times0.889}-1}{e^{-0.889}-1}\right) - \frac{0}{4}$$

$$= 17.197 \text{ A } (17.03 \text{ A})$$

(e) The peak-to-peak load ripple current $\Delta I_{peak} = I_2 - I_1 = 17.197-12.051 = 5.146$ A

(f) The maximum possible ripple current at $k = 0.5$, $\Delta I_{\max} = \frac{V_S}{4fl} = \frac{98}{4\times10\times10^3\times450\times10^{-6}} = 5.444$ A

(g) Assuming triangular form, the average load current $I_{av} = \frac{I_2+I_1}{2} = \frac{17.197+12.051}{2} = 14.624$ A

Alternately, $I_a = I_1 + \frac{I_2-I_1}{2} = 12.051 + \frac{17.197-12.051}{2} = 14.624$

(h) Assuming triangular form, the rms load current

$$I_{rms} = \sqrt{I_1{}^2 + \frac{I_1{}^2 + I_1I_2 + I_2{}^2}{3}}$$

$$= \sqrt{\frac{12.051^2 + 12.051 \times 17.197 + 17.197^2}{3}} = 14.7 \text{ A}$$

Alternately,

$$I_{rms} = \sqrt{I_1{}^2 + \frac{(I_2-I_1)^2}{3} + I_1(I_2-I_1)}$$

$$= 12.051^2 + \frac{(17.197-12.051)^2}{3} + 12.051\times(17.197-12.051) = 14.7 \text{ A}$$

(i) Power delivered to the load $P_{out} = I_{rms}{}^2R = 14.7^2 \times 4 = 864.305$ W

4.4.2 Converter parameters

(a) The peak transistor current, $I_p = I_2 = 17.197$ A

(b) The average transistor current $I_{av} = kI_a = 0.6 \times 14.624 = 8.775$ A

(c) The rms transistor current $I_r = \sqrt{k}I_{rms} = \sqrt{0.6} \times 14.7 = 11.386$ A

(d) The peak transistor voltage $V_{peak} = V_S = 98$ V

4.4.3 Input-side parameters

(a) The average input current is

$$I_s = I_{av}$$
$$= 8.775 \text{ A}$$

(b) The rms input current is

$$I_i = I_{rms}$$
$$= 14.7 \text{ A}$$

(c) The input power

$$P_{in} = P_{out} \\ = 864.305 \text{ W}$$

(d) The ripple content of the input current

$$I_{ripple} = \sqrt{I_r^2 - I_s^2} \\ = \sqrt{11.386^2 - 8.775^2} = 7.256 \text{ A}$$

(e) The ripple factor of the input current

$$RF_i = \frac{I_{ripple}}{I_s} \\ = \frac{7.256}{8.775} \times 100 = 82.696\%$$

4.4.4 Output L-filter

The inductor L in the load shown in Figure 4.5(a) as the filter to limit the ripple current of the load resistor, R. The inductor L provides a high impedance for the ripple currents. The fundamental frequency f_o of output voltage ripples is the same as the switching frequency f.

For a desired value of the ripple factor of the output voltage $RF_v = 6\% = 0.06$, we can find the ripple voltage as

$$\Delta V = RF_v\, V_{dc} \\ = 0.05 \times 58.8 = 3.528 \text{ V}$$

Similarly, for a desired value of the ripple factor of the output current $RF_i = 6\% = 0.06$, we can find the ripple current as

$$\Delta I = RF_i\, I_a \\ = 0.06 \times 14.624 = 0.877 \text{ A}$$

We can find the inductor L from the ripple voltage ΔV and the ripple current ΔI as

$$L\frac{\Delta I}{\Delta T} = \Delta V$$

which gives the inductor value L as

$$L = \frac{\Delta V \times kT}{\Delta I} = \frac{\Delta V}{\Delta I} \times kT \\ = \frac{3.528}{0.877} \times 0.6 \times 100 \times 10^{-6} = 241.243 \text{ μH}$$

Note: The inductor current rises from a minimum value to a maximum value during the on-time of the chopper switch kT and $\Delta T = kT$.

4.5 Buck converter

The LTspice schematic of a MOSFET Buck converter is shown In Figure 4.6(a). The plots of the gating signals, the output voltage and the load current are shown in Figure 4.6(b). The performance parameters of the DC–DC converters can be divided into three types.

- Output-side parameters,
- Converter parameters, and
- Input-side parameters.

Let us consider an input voltage of $V_S = 98$ V, $E = 0$ V, $L = 450$ μH, $C = 4.7$ μF, and $R = 4$ Ω. The chopper is operating at a switching frequency of $f_s = 10$ kHz, a switching period of $T = 1/f = 100$ μs, and a duty cycle of $k = 60\%$. Thus, the on-time of the switch is $t_1 = kT = 0.6 \times 100$ μs $= 60$ μs and the off-time of the switch is $t_2 = (1 - k)T = (1 - 0.6) \times 100$ μs $= 40$ μs.

Depending on whether the switch is turned on to off, there are two modes of operation. During mode 1 the switch is turned on and during mode 2 the switch is turned off.

Mode 1: When the switch is turned on, the DC supply voltage V_S is connected to the load and the inductor current rises exponentially from I_1 at $t = 0$ and reaches to I_2 at the end of this mode at $t_1 = kT$ under-steady state conditions. Assuming that V_a is the average load current across the capacitor C, the voltage across the inductor L is given by

$$V_S - V_a = L\frac{I_2 - I_1}{t_1} = L\frac{\Delta I}{t_1} \tag{4.8}$$

which gives

$$t_1 = L\frac{\Delta I}{V_S - V_a} \tag{4.9}$$

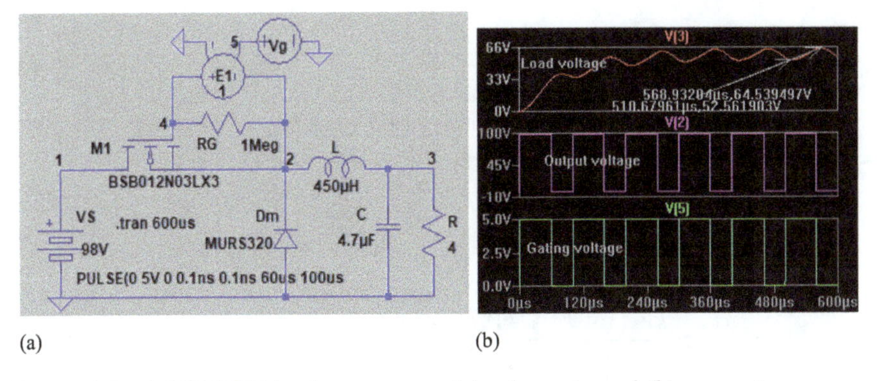

Figure 4.6 A MOSFET Buck converter: (a) schematic and (b) converter waveforms [8]

Mode 2: When the switch is turned off, the DC supply voltage V_S is disconnected and free-wheeling diode D_m is turned on due to the current in the inductor. The inductor current falls from I_2 to I_1 at the end of this mode at $t_2 = (1-k)T$ under the steady-state condition. The voltage across the inductor L is given by

$$-V_a = L\frac{I_1 - I_2}{t_2} = -L\frac{\Delta I}{t_2} \tag{4.9}$$

which gives

$$t_2 = L\frac{\Delta I}{V_a} \tag{4.10}$$

The switching period T is related to t_1 and t_2 by

$$T = \frac{1}{f} = t_1 + t_2 = L\frac{\Delta I}{V_S - V_a} + L\frac{\Delta I}{V_a} = \frac{V_S L \Delta I}{V_a(V_S - V_a)}$$

which gives the peak-to-peak ripple current of the inductor as

$$\Delta I = \frac{V_a(V_S - V_a)}{f L V_S} \tag{4.11}$$

The average voltage across the load as shown in Figure 4.6(b) is given by

$$V_a = kV_S \tag{4.12}$$

Substituting $V_a = kV_S$ into (4.11), we obtain

$$\Delta I = \frac{kV_S(1-k)V_S}{fLV_S} = \frac{kV_S(1-k)}{fL} \tag{4.13}$$

Thus, the ripple current ΔI depends on the duty cycle, k, the inductance L, and the switching frequency, f. As the inductor current varies from I_1 to I_2, the current through the filter capacitor varies from $(I_1 - I_a)$ to $(I_2 - I_a)$. The current flows through the capacitor for a time interval of $T/2$. Thus, the average capacitor current is approximately

$$I_c = \frac{\Delta I}{4} \tag{4.14}$$

Thus, the peak-to-peak ripple voltage of the capacitor is given by

$$\Delta V_c = \frac{1}{C}\int_0^{T/2} I_c dt = \frac{1}{C}\int_0^{T/2} \frac{\Delta I}{4} dt = \frac{\Delta I T}{8C} = \frac{\Delta I}{8fC} \tag{4.15}$$

Substituting ΔI from (4.13) gives

$$\Delta V_c = \frac{1}{8fC} \times \frac{kV_S(1-k)}{fL} = \frac{kV_S(1-k)}{8LCf^2} \tag{4.16}$$

Condition for continuous inductor current and capacitor voltage: Assuming I_L is the average inductor current, $\Delta I \geq 2I_L$, (4.13) gives

$$\frac{kV_S(1-k)}{fL} \geq 2I_L = 2I_a = \frac{2kV_S}{R} \tag{4.17}$$

which gives the critical value of the inductor as

$$L_c = L = \frac{(1-k)R}{2f} \tag{4.18}$$

Similarly, V_c is the average capacitor voltage, $\Delta V_c \geq 2V_a$. Thus, (4.16) gives

$$\frac{kV_S(1-k)}{8LCf^2} \geq 2V_a = 2kV_S \tag{4.19}$$

which gives the critical value of the capacitor as

$$C_c = C = \frac{1-k}{16Lf^2} \tag{4.20}$$

4.5.1 Output-side parameters

(a) The average output voltage, $V_a = kV_s = 0.6 \times 98 = 58.8$ V
(b) The average load current $I_a = \frac{V_a}{R} = \frac{58.8}{4} = 14.7$ A
(c) The rms output voltage

$$V_{rms} = \sqrt{\frac{1}{T}\int_0^{(1-k)T} V_a^2 dt} = \sqrt{(1-k)}V_a$$
$$= \sqrt{(1-0.6)} \times 58.8 = 75.91 \text{ V}$$

(d) Equation (4.13) gives the peak-to-peak load ripple current

$$\Delta I = \frac{0.6 \times 98 \times (1-0.6)}{10 \times 10^3 \times 450 \times 10^{-6}} = 2.613 \text{ A}$$

(e) The steady-state minimum inductor current

$$I_1 = I_a - \frac{\Delta I}{2} = 14.7 - \frac{2.613}{2} = 13.393 \text{ A}$$

(f) The steady-state maximum inductor current

$$I_2 = I_a + \frac{\Delta I}{2} = 14.7 + \frac{2.613}{2} = 16.007 \text{ A}$$

(g) The rms value of the triangular inductor current

$$I_{rms} = \sqrt{\frac{{I_1}^2 + I_1 I_2 + {I_2}^2}{3}}$$
$$= \sqrt{\frac{13.393^2 + 13.393 \times 16.007 + 16.007^2}{3}}$$
$$= 9.401 \text{ A}$$

(h) Equation (4.16) gives the peak-to-peak capacitor voltage

$$\Delta V_c = \frac{kV_S(1-k)}{8LCf^2} = \frac{0.6 \times 98 \times (1-0.6)}{8 \times 450 \times 10^3 \times 4.7 \times 10^{-6} \times \left(10 \times 10^3\right)^2} = 3.63 \text{ V}$$

(i) The steady-state minimum capacitor voltage

$$V_{C1} = V_a - \frac{\Delta V_c}{2} = 58.8 - \frac{3.63}{2} = 56.985 \text{ V}$$

(j) The steady-state maximum capacitor voltage

$$V_{C2} = V_a + \frac{\Delta V_c}{2} = 58.8 + \frac{3.63}{2} = 60.615 \text{ V}$$

(k) The rms value of the triangular capacitor voltage

$$V_{rms} = \sqrt{\frac{{V_{c1}}^2 + V_{c1} V_{c2} + {V_{c2}}^2}{3}}$$
$$= \sqrt{\frac{56.985^2 + 56.985 \times 60.615 + 60.615^2}{3}}$$
$$= 58.809 \text{ V}$$

(l) Power delivered to the load

$$P_{out} = \frac{{V_{rms}}^2}{R} = \frac{58.809^2}{4} = 14.7^2 \times 4 = 864.634 \text{ W}$$

(m) DC power delivered to the load

$$P_{dc} = \frac{{V_a}^2}{R} = \frac{58.8^2}{4} = 864.36 \text{ W}$$

4.5.2 Converter parameters

(a) The peak transistor current $I_p = I_2 = 16.007$ A
(b) The average transistor current $I_{av} = kI_a = 0.6 \times 14.7 = 8.82$ A
(c) For a duty cycle of k, the rms transistor current $I_r = \sqrt{k}I_{rms} = \sqrt{0.6} \times 14.719 = 11.402$ A
(d) The peak transistor voltage $V_{peak} = V_S = 98$ V

4.5.3 Input-side parameters

(f) The average input current is

$$I_s = I_{av}$$
$$= 8.82 \text{ A}$$

(g) The rms input current is

$$I_i = I_r$$
$$= 11.402 \text{ A}$$

(h) The input power

$$P_{in} = P_{out}$$
$$== 864.634 \text{ W}$$

(i) The ripple content of the input current

$$I_{ripple} = \sqrt{I_i^2 - I_s^2}$$
$$= \sqrt{11.402^2 - 8.82^2} = 7.225 \text{ A}$$

(j) The ripple factor of the input current

$$RF_i = \frac{I_{ripple}}{I_s}$$
$$= \frac{7.225}{8.82} \times 100 = 81.918\%$$

4.5.4 Condition for continuous inductor current and capacitor voltage

Equation (4.18) gives the critical value of the inductor

$$L_c = \frac{(1-k)R}{2f}$$
$$= \frac{(1-0.6)\times 4}{2\times 10\times 10^3} = 40\ \mu\text{H}$$

Equation (4.20) gives the critical value of the capacitor as

$$C_c = \frac{1-k}{16L_c f^2}$$
$$= \frac{1-0.6}{16\times 40\times 10^{-6}\times \left(10\times 10^3\right)^2} = 1.563\ \mu\text{F}$$

4.5.5 Output L-filter

The inductor L as shown in Figure 4.6(a) provides a high impedance for the ripple currents and limits the ripple current of the load resistor, R. The capacitor

C provides a low impedance path for the ripple currents and limits the ripple voltage of the load resistor, R.

For a desired value of the ripple factor of the output voltage $RF_v = 5\% = 0.05$, we can find the ripple voltage as

$$\Delta V = RF_v\, V_a$$
$$= 0.05 \times 58.8 = 2.94 \text{ V}$$

Similarly, for a desired value of the ripple factor of the output current $RF_i = 10\% = 0.1$, we can find the ripple current as

$$\Delta I = RF_i\, I_a$$
$$= 0.1 \times 14.7 = 1.47 \text{ A}$$

Using (4.13), we can find the inductor

$$L = \frac{kV_S(1-k)}{f\Delta I}$$
$$= \frac{0.6 \times 98 \times (1-0.6)}{10 \times 10^3 \times 1.47} = 800\ \mu\text{H}$$

Using (4.16), we can find the inductor

$$C = \frac{kV_S(1-k)}{8Lf^2\Delta V_c}$$
$$= \frac{0.6 \times 98 \times (1-0.6)}{8 \times 800 \times 10^{-6} \times \left(10 \times 10^3\right)^2 \times 2.94} = 3.125\ \mu\text{F}$$

4.6 Boost converter

The LTspice schematic of a MOSFET Buck converter is shown in Figure 4.7(a). The plots of the inductor current, the switch voltage, and the voltage across the transistor are shown in Figure 4.7(b). The performance parameters of the DC–DC converters can be divided into three types [7].

- Output-side parameters,
- Converter parameters, and
- Input-side parameters.

Let us consider an input voltage of $V_S = 6$ V, $L = 150\ \mu$H, $C = 47\ \mu$F, and $R = 4\ \Omega$. The transistor is operating at a switching frequency of $f_s = 10$ kHz, a switching period of $T = 1/f = 100\ \mu$s, and a duty cycle of $k = 60\%$. Thus, the on-time of the switch is $t_1 = kT = 0.6 \times 100\ \mu\text{s} = 60\ \mu$s and the off-time of the switch is $t_2 = (1-k)T = (1-0.6) \times 100\ \mu\text{s} = 40\ \mu$s.

Depending on whether the switch is turned on or off, there are two modes of operation. During mode 1, the switch is turned on, and during mode 2, the switch is turned off.

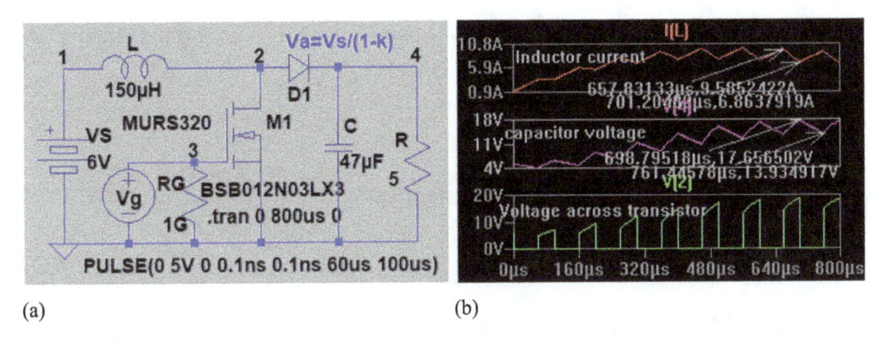

(a) (b)

Figure 4.7 A MOSFET Boost converter: (a) schematic and (b) converter waveforms [8]

Mode 1: When the switch is turned on, the DC supply voltage V_S is connected across the inductor L and the inductor current rises from I_1 to I_2 at the end of this mode at $t_1 = kT$ under the steady-state condition. That is,

$$V_S = L\frac{I_2 - I_1}{t_1} = L\frac{\Delta I}{t_1} \tag{4.21}$$

which gives

$$t_1 = L\frac{\Delta I}{V_S} \tag{4.22}$$

Mode 2: When the switch is turned off, diode D_m is turned on due to the current in the inductor and charges the capacitor. The inductor current falls from I_2 to I_1 at the end of this mode at $t_2 = (1 - k)T$ under the steady-state condition. That is, the inductor voltage is given by

$$V_S - V_a = L\frac{I_1 - I_2}{t_2} = -L\frac{\Delta I}{t_2} \tag{4.23}$$

which gives

$$t_2 = -L\frac{\Delta I}{V_S - V_a} = L\frac{\Delta I}{V_a - V_S} \tag{4.24}$$

Equating ΔI from (4.22) to ΔI from (4.24), we obtain

$$\Delta I = \frac{t_1 V_S}{L} = \frac{(V_a - V_S)t_2}{L} \tag{4.25}$$

Substituting $t_1 = kT$ and $t_2 = (1 - k)T$, (4.25) can be simplified to give the average output voltage as

$$V_a = V_S\frac{T}{t_2} = \frac{V_S}{1 - k} \tag{4.26}$$

Assuming a lossless circuit, the input power from the source must equal to the output power. That is,

$$V_S I_S = V_a I_a = \frac{V_S}{1-k} I_a$$

which gives the average input supply current as

$$I_S = \frac{I_a}{1-k} \tag{4.27}$$

Using (4.22) and (4.24), the switching period T is related to t_1 and t_2 by

$$T = \frac{1}{f} = t_1 + t_2 = L\frac{\Delta I}{V_S} + L\frac{\Delta I}{V_a - V_S} = \frac{V_a L \Delta I}{V_S(V_a - V_S)} \tag{4.28}$$

which gives the peak-to-peak current of the inductor as

$$\Delta I = \frac{V_S(V_a - V_S)}{fLV_a} \tag{4.29}$$

which substituting V_a from (4.26) gives

$$\Delta I = \frac{V_S\left(\frac{V_S}{1-k} - V_S\right)}{fL\frac{V_S}{1-k}} = \frac{V_S k}{fL} \tag{4.30}$$

When the transistor is turned on, the capacitor supplies the load current I_a for time $t = t_1$ and the average capacitor current is $I_c = I_a$. Thus, the peak-to-peak ripple of the capacitor is given by

$$\Delta V_c = \frac{1}{C}\int_0^{t_1} I_a dt = \frac{I_a t_1}{C} = \frac{I_a}{C} \times L\frac{\Delta I}{V_S} \tag{4.31}$$

Substituting ΔI from (4.30) gives

$$\Delta V_c = \frac{I_a}{C} L \frac{1}{V_S} \times \frac{V_S k}{fL} = \frac{I_a k}{fC} \tag{4.32}$$

Condition for continuous inductor current and capacitor voltage: Assuming I_L is the average inductor current, $\Delta I \geq 21_L$ which gives

$$\begin{aligned} \frac{V_S k}{fL} &\geq 21_L = 21_s = 2\frac{I_a}{1-k} = 2\frac{V_a}{R(1-k)} \\ &= 2\frac{1}{R(1-k)}\frac{V_S}{(1-k)} = \frac{2V_S}{R(1-k)^2} \end{aligned} \tag{4.33}$$

which gives the critical value of the inductor as

$$L_c = L = \frac{k(1-k)^2 R}{2f} \tag{4.34}$$

Similarly, V_c is the average capacitor voltage, $\Delta V_c \geq 2V_a$ which gives

$$\frac{I_a k}{fC} \geq 2V_a = kI_a R \tag{4.35}$$

which gives the critical value of the capacitor as

$$C_c = C = \frac{k}{2fR} \tag{4.36}$$

4.6.1 Output-side parameters

(a) From (4.26), the average output voltage,

$$V_a = \frac{V_s}{1-k} = \frac{6}{1-0.6} = 15 \text{ V}$$

(b) The average load current

$$I_a = \frac{V_a}{R} = \frac{15}{4} = 3.75 \text{ A}$$

(c) Equation (4.30) gives the peak-to-peak load ripple current

$$\Delta I = \frac{V_S k}{fL}$$
$$= \frac{6 \times 0.6}{10 \times 10^3 \times 150 \times 10^{-6}} = 2.4 \text{ A}$$

(d) The steady-state minimum inductor current

$$I_1 = I_S - \frac{\Delta I}{2}$$
$$= 9.375 - \frac{2.4}{2} = 8.175 \text{ A}$$

(e) The steady-state maximum inductor current

$$I_2 = I_S + \frac{\Delta I}{2}$$
$$= 9.375 + \frac{2.4}{2} = 10.575 \text{ A}$$

(f) The rms value of the triangular inductor current

$$I_{rms} = \sqrt{\frac{I_1{}^2 + I_1 I_2 + I_2{}^2}{3}}$$
$$= \sqrt{\frac{8.175^2 + 8.175 \times 10.575 + 10.575^2}{3}}$$
$$= 9.401 \text{ A}$$

(g) Equation (4.27) gives the average current dorm the supply

$$I_S = \frac{I_a}{1-k} = \frac{3.75}{1-0.6} = 9.375 \text{ A}$$

(h) Equation (4.16) gives the peak-to-peak capacitor voltage

$$\Delta V_c = \frac{I_a k}{fC} = \frac{3.75 \times 0.6}{10 \times 10^3 \times 47 \times 10^{-6}} = 4.787 \text{ V}$$

(i) The steady-state minimum capacitor voltage

$$V_{C1} = V_a - \frac{\Delta V_c}{2} = 15 - \frac{4.787}{2} = 12.606 \text{ V}$$

(j) The steady-state maximum capacitor voltage

$$V_{C2} = V_a + \frac{\Delta V_c}{2} = 15 + \frac{4.787}{2} = 17.394 \text{ V}$$

(k) The rms value of the triangular capacitor voltage

$$V_{rms} = \sqrt{\frac{{V_{c1}}^2 + V_{c1}V_{c2} + {V_{c2}}^2}{3}} = \sqrt{\frac{12.606^2 + 12.606 \times 17.394 + 17.394^2}{3}} = 15.064 \text{ V}$$

(l) Power delivered to the load

$$P_{out} = \frac{{V_{rms}}^2}{R} = \frac{15.064^2}{4} = 56.727 \text{ W}$$

(m) DC power delivered to the load

$$P_{dc} = \frac{{V_a}^2}{R} = \frac{15^2}{4} = 56.25 \text{ W}$$

4.6.2 Converter parameters

(a) The peak transistor current $I_p = I_2 = 10.575$ A
(b) The average transistor current $I_{av} = kI_S = 0.6 \times 9.375 = 5.625$ A
(c) The rms transistor current $I_r = \sqrt{k} I_{rms} = \sqrt{0.6} \times 9.401 = 7.282$ A
(d) The peak transistor voltage $V_{peak} = \frac{V_S}{1-k} = \frac{6}{1-0.98} = 300\text{V}$ for $k = 0.98$

4.6.3 Input-side parameters

(a) The average input current is $I_s = 9.375$ A
(b) The rms input current is $I_i = I_{rms} = 9.401$ A

(c) The input power, $P_{in} = P_{out} = 56.727$ W
(d) The ripple content of the input current

$$I_{ripple} = \sqrt{I_i^2 - I_s^2}$$
$$= \sqrt{9.401^2 - 9.375^2} = 0.693 \text{ A}$$

(e) The ripple factor of the input current

$$RF_i = \frac{I_{ripple}}{I_s}$$
$$= \frac{0.693}{9.375} \times 100 = 7.39\%$$

4.6.4 Condition for continuous inductor current and capacitor voltage

Equation (4.34) gives the critical value of the inductor

$$L_c = \frac{k(1-k)^2 R}{2f} = \frac{0.6 \times (1-0.6)^2 \times 4}{2 \times 10 \times 10^3} = 19.2 \text{ μH}$$

Equation (4.36) gives the critical value of the capacitor as

$$C_c = \frac{k}{2fR} = \frac{0.6}{2 \times 10 \times 10^3 \times 4} = 7.5 \text{ μF}$$

4.6.5 Output LC-filter

The inductor L as shown in Figure 4.7(a) provides a high impedance for the inductor ripple current and limits the ripple current of the inductor L. The capacitor C provides a low impedance path for the load ripple currents and limits the ripple voltage of the load resistor, R.

For a desired value of the ripple factor of the output voltage $RF_v = 5\% = 0.05$, we can find the ripple voltage as

$$\Delta V = RF_v V_a$$
$$= 0.05 \times 15 = 0.75 \text{ V}$$

Similarly, for a desired value of the ripple factor of the inductor current $RF_i = 10\% = 0.1$, we can find the ripple current as

$$\Delta I = RF_i I_S$$
$$= 0.1 \times 9.375 = 0.9375 \text{ A}$$

Using (4.30), we can find the inductor

$$L == \frac{V_S k}{f \Delta I} = \frac{6 \times 0.6}{10 \times 10^3 \times 0.9375} = 150 \text{ μH}$$

Using (4.32), we can find the capacitor

$$C = \frac{I_a k}{f} = \frac{3.75 \times 0.6}{10 \times 10^3} = 300\ \mu\text{F}$$

4.7 Buck–Boost converter

The LTspice schematic of a MOSFET Buck–Boost converter is shown In Figure 4.8(a), the plots of the output of the switch a node 2, the inductor current $I(L)$ and the load voltage $V(3)$ at node 3 are shown in Figure 4.8(b). Note that the location of the diode and the inductor are interchanged if we compare the Buck converter as shown in Figure 4.6(a). The performance parameters of the DC–DC converters can be divided into three types.

- Output-side parameters,
- Converter parameters, and
- Input-side parameters.

Let us consider an input voltage of $V_S = 6$ V, $L = 150\ \mu$H, $C = 47\ \mu$F, and $R = 4\ \Omega$. The chopper is operating at a switching frequency of $f_s = 20$ kHz, a switching period of $T = 1/f = 50\ \mu$s and a duty cycle of $k = 90\%$. Thus, the on-time of the switch is $t_1 = kT = 0.9 \times 50\ \mu\text{s} = 50\ \mu$s and the off-time of the switch is $t_2 = (1 - k)T = (1 - 0.9) \times 50\ \mu\text{s} = 5\ \mu$s.

Depending on whether the switch is turned on or off, there are two modes of operation. During mode 1, the switch is turned on and during mode 2 the switch is turned off. The switch disconnected the load from the DC source.

Mode 1: When the switch is turned on, the DC supply voltage V_S is connected across the inductor L and the inductor current rises from I_1 to I_2 at the end of this mode at $t_1 = kT$ under the steady-state condition. Thus, the voltage across the

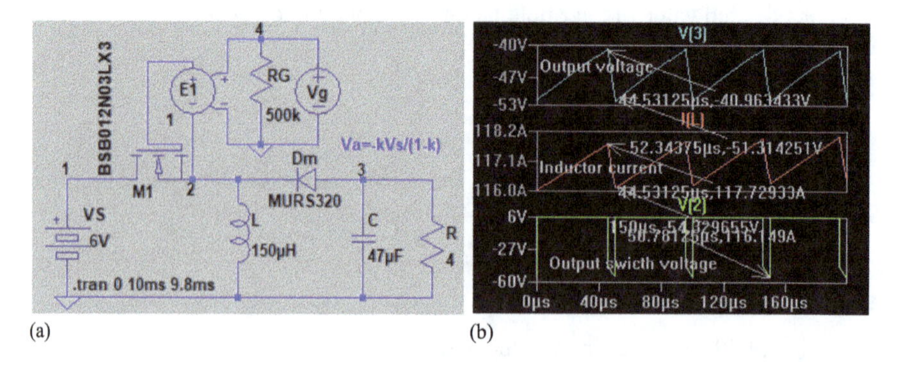

(a) (b)

Figure 4.8 A MOSFET Buck–Boost converter: (a) schematic and (b) converter waveforms inductor voltage is

$$V_S = L\frac{I_2 - I_1}{t_1} = L\frac{\Delta I}{t_1} \tag{4.37}$$

which gives

$$t_1 = L\frac{\Delta I}{V_S} \tag{4.38}$$

Mode 2: When the switch is turned off, diode D_m is turned on due to the current in the inductor and charges the capacitor C in the opposite direction. The inductor current falls from I_2 to I_1 at the end of this mode at $t_2 = (1 - k)T$ under the steady-state condition. Thus, the voltage across the inductor voltage is

$$V_a = L\frac{I_1 - I_2}{t_2} = -L\frac{\Delta I}{t_2} \tag{4.39}$$

which gives

$$t_2 = -L\frac{\Delta I}{V_a} \tag{4.40}$$

Equating ΔI from (4.38) to ΔI from (4.40), we obtain

$$\Delta I = \frac{t_1 V_S}{L} = -\frac{V_a t_2}{L} \tag{4.41}$$

Substituting $t_1 = kT$ and $t_2 = (1 - k)T$, (4.41) can be simplified to give the average output voltage as

$$V_a = -V_S\frac{t_1}{t_2} = -\frac{V_S k}{1 - k} \tag{4.42}$$

Note: The polarity of the output voltage is negative.

Assuming a lossless circuit, the input power from the source must equal to the output power. That is,

$$V_S I_S = -V_a I_a = \frac{V_S k}{1 - k} I_a$$

which gives the average input supply current as

$$I_S = \frac{I_a k}{1 - k} \tag{4.43}$$

The switching period T is related to t_1 and t_2 by

$$T = \frac{1}{f} = t_1 + t_2 = L\frac{\Delta I}{V_S} - L\frac{\Delta I}{V_a} = \frac{L\Delta I(V_a - V_S)}{V_S V_a} \tag{4.44}$$

which gives the peak-to-peak inductor current as

$$\Delta I = \frac{V_S V_a}{fL(V_a - V_S)} \tag{4.45}$$

which substituting V_a from (4.42) gives

$$\Delta I = \frac{V_S k}{fL} \tag{4.46}$$

When the transistor is turned on, diode D_m isolates the input from the load and the capacitor supplies the load current I_a for time $t = t_1$ and the average capacitor current is $-I_c = I_a$. Thus, the peak-to-peak ripple voltage of the capacitor is given by

$$\Delta V_c = \frac{1}{C}\int_0^{t_1} I_a dt = \frac{I_a t_1}{C} = \frac{I_a}{C} \times L\frac{\Delta I}{V_S} \tag{4.47}$$

Substituting ΔI from (4.46) gives

$$\Delta V_c = \frac{I_a}{C} L \frac{1}{V_S} \times \frac{V_S k}{fL} = \frac{I_a k}{fC} \tag{4.48}$$

Condition for continuous inductor current and capacitor voltage: Assuming I_L is the average inductor current, $\Delta I \geq 21_L$ and using (4.43) and (4.46), we obtain

$$\Delta I = \frac{V_S k}{fL} \geq 21_L = 21_s = 2\frac{I_a k}{1-k} = \frac{2kV_S}{(1-k)R} \tag{4.49}$$

which gives the critical value of the inductor as

$$L_c = L = \frac{(1-k)R}{2f} \tag{4.50}$$

Similarly, V_c is the average capacitor voltage, $\Delta V_c \geq 2V_a$ and (4.48) gives.

$$\frac{I_a k}{fC} \geq 2V_a = kI_a R \tag{4.51}$$

which gives the critical value of the capacitor as

$$C_c = C = \frac{k}{2fR} \tag{4.52}$$

4.7.1 Output-side parameters

(a) Equation (4.42) gives the average output voltage

$$V_a = -\frac{V_S k}{1-k} = -\frac{6 \times 0.9}{1-0.9} = -54 \text{ V}$$

(b) The average load current

$$I_a = \frac{V_a}{R} = \frac{-45.107}{4} = -13.5 \text{ A}$$

(c) Equation (4.43) gives the average current drawn the supply

$$I_S = \frac{I_a k}{1-k} = \frac{13.5 \times 0.9}{1-0.9} = 121.5 \text{ A}$$

(d) Equation (4.46) gives the peak-to-peak load ripple current

$$\Delta I = \frac{V_S k}{fL} = \frac{6 \times 0.9}{20 \times 10^3 \times 150 \times 10^{-6}} = 1.8 \text{ A}$$

(e) Since the switch conducts for a duty cycle of k, the steady-state average supply current is related to the average inductor current I_L by $I_s = kI_L$ which gives

$$I_L = \frac{I_S}{k} = \frac{121.5}{0.9} = 135 \text{ A}$$

(f) The steady-state minimum inductor current is

$$I_1 = I_L - \frac{\Delta I}{2} = \frac{I_S}{k} - \frac{\Delta I}{2} = \frac{121.5}{0.9} - \frac{1.8}{2} = 134.1 \text{ A}$$

(g) The steady-state maximum inductor current

$$I_2 = I_L + \frac{\Delta I}{2} = \frac{I_S}{k} + \frac{\Delta I}{2} = \frac{121.5}{0.6} + \frac{1.8}{2} = 135.9 \text{ A}$$

(h) The rms value of the triangular inductor current

$$I_{rms} = \sqrt{\frac{I_1{}^2 + I_1 I_2 + I_2{}^2}{3}}$$
$$= \sqrt{\frac{134.1^2 + 134.1 \times 135.9 + 135.9^2}{3}} = 135.001 \text{ A}$$

(i) Equation (4.48) gives the peak-to-peak capacitor voltage

$$\Delta V_c = \frac{I_a k}{fC} = \frac{13.5 \times 0.9}{20 \times 10^3 \times 47 \times 10^{-6}} = 12.926 \text{ V}$$

(j) The steady-state minimum capacitor voltage

$$V_{C1} = V_a - \frac{\Delta V_c}{2} = -54 - \frac{12.926}{2} = -47.537 \text{ V}$$

(k) The steady-state maximum capacitor voltage

$$V_{C2} = V_a + \frac{\Delta V_c}{2} = -54 + \frac{12.926}{2} = -60.463 \text{ V}$$

(l) The rms value of the triangular capacitor voltage

$$V_{rms} = \sqrt{\frac{{V_{c1}}^2 + V_{c1}V_{c2} + {V_{c2}}^2}{3}}$$

$$= \sqrt{\frac{47.537^2 + 47.537 \times 60.463 + 60.463^2}{3}}$$

$$= 54.129 \text{ V}$$

(m) Power delivered to the load

$$P_{out} = \frac{{V_{rms}}^2}{R} = \frac{54.129^2}{4} = 732.481 \text{ W}$$

(n) DC power delivered to the load

$$P_{dc} = \frac{V_a^2}{R} = \frac{54^2}{4} = 729 \text{ W}$$

4.7.2 *Converter parameters*

(a) The peak transistor current, $I_p = I_2 = 135.9$ A
(b) The average transistor current $I_{av} = I_S = 121.5$ A
(c) The rms transistor current $I_r = \sqrt{k}I_{rms} = \sqrt{0.9} \times 135.001 = 128.073$ A
(d) The peak transistor voltage $V_{peak} = -V_S + V_{C2} = -6 - 60.463 \text{ V} = -66.463$ V

4.7.3 *Input-side parameters*

(a) The average input current is $I_s = 121.5$ A
(b) The rms input current is $I_i = I_r = \sqrt{k}I_{rms} = \sqrt{0.9} \times 135.001 = 128.073$ A
(c) The input power, $P_{in} = P_{out} = 732.481$ W
(d) The ripple content of the input current

$$I_{ripple} = \sqrt{I_i^2 - I_s^2}$$

$$= \sqrt{128.073^2 - 121.5^2} = 40.503 \text{ A}$$

(e) The ripple factor of the input current

$$RF_i = \frac{I_{ripple}}{I_s}$$

$$= \frac{40.503}{121.5} \times 100 = 33.336\%$$

4.7.4 *Condition for continuous inductor current and capacitor voltage*

Equation (4.50) gives the critical value of the inductor

$$L_c = \frac{(1-k)R}{2f} = \frac{(1-0.9) \times 4}{2 \times 20 \times 10^3} = 10 \text{ μH}$$

Equation (4.52) gives the critical value of the capacitor as

$$C_c = \frac{k}{2fR} = \frac{0.9}{2 \times 20 \times 10^3 \times 4} = 5.625\ \mu\text{F}$$

4.7.5 Output LC-filter

The inductor L as shown in Figure 4.8(a) provides a high impedance for the inductor ripple current and limits the ripple current of the inductor L. The capacitor C provides a low impedance path for the load ripple currents and limits the ripple voltage of the load resistor, R.

For a desired value of the ripple factor of the output voltage $RF_v = 5\% = 0.05$, we can find the ripple voltage as

$$\begin{aligned}\Delta V_d &= RF_v\, V_a \\ &= 0.05 \times 54 = 2.7\ \text{V}\end{aligned}$$

Similarly, for a desired value of the ripple factor of the inductor current $RF_i = 6\% = 0.06$, we can find the ripple current as

$$\begin{aligned}\Delta I_d &= RF_i\, I_L \\ &= 0.06 \times 135 = 8.1\ \text{A}\end{aligned}$$

Using (4.46), we can find the inductor

$$L = \frac{V_S k}{f \Delta I_d} = \frac{6 \times 0.9}{20 \times 10^3 \times 8.1} = 33.333\ \mu\text{H}$$

Using (4.48), we can find the capacitor

$$C = \frac{I_a k}{f \Delta V_d} = \frac{13.5 \times 0.9}{20 \times 10^3 \times 2.7} = 225\ \mu\text{F}$$

4.8 SEPIC converter

The Cúk converter is an inverting Buck–Boost characteristic and does not exhibit pulsating input and output currents. However the output voltage of the Cúk converter is inverted. An single-ended primary inductance converter (SEPIC) is a non-inverting Cúk converter and gives positive output voltage. The LTspice schematic of a MOSFET SEPIC converter is shown in Figure 4.9. The circuit is like the Boost converter with the addition of a capacitor C_1 and an inductor L_2. The plots of the gating signals, the output voltage and the load current are shown in Figure 4.10. The performance parameters of the DC–DC converters can be divided into three types [7,8]

- Output-side parameters,
- Converter parameters, and
- Input-side parameters.

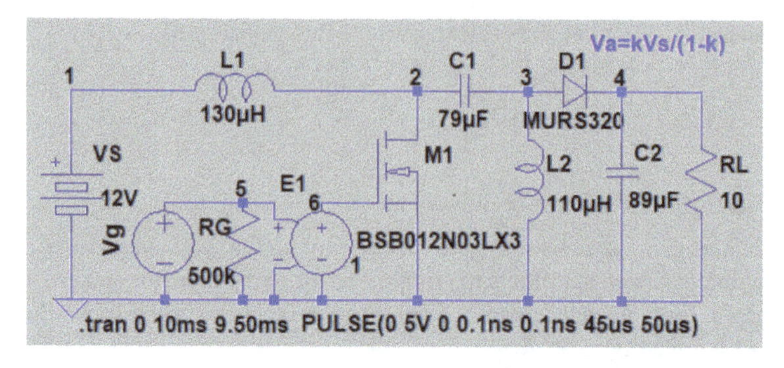

Figure 4.9 Schematic of a MOSFET SEPIC converter

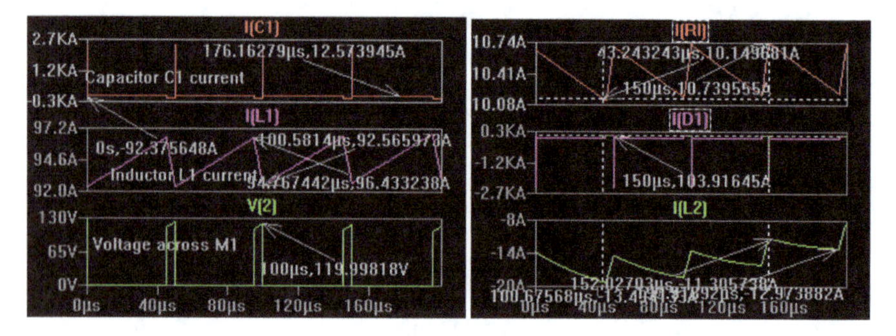

Figure 4.10 Waveforms of a MOSFET SEPIC converter

Let us consider an input voltage of $V_S = 12$ V, $L_1 = 130$ µH, $L_2 = 110$ µH, $C_1 = 79$ µF, $C_2 = 89$ µF, and $R = 10\ \Omega$. The switch is operating at a switching frequency of $f_s = 20$ kHz, a switching period of $T = 1/f = 50$ µs and a duty cycle of $k = 90\%$. Thus, the on-time of the switch is $t_1 = kT = 0.9 \times 50$ µs $= 45$ µs and the off-time of the switch is $t_2 = (1 - k)T = (1 - 0.9) \times 50$ µs $= 10$ µs.

When the DC supply voltage V_S is turned on, the capacitor C_1 changes toV_{C11} through inductor L_1 and L_2. Depending on whether the chopper switch is turned on or off, there are two modes of operation. During mode 1 the switch is turned on and during mode 2 the switch is turned off.

Mode 1: When the switch is turned on, the DC supply voltage V_S is connected across the inductor L_1 and the inductor current rises from I_{11} to I_{12} at the end of this mode at $t_1 = kT$ under the steady-state condition. The capacitor C_1 discharges through the switch and the inductor L_2. The energy stored in the capacitor C_1 is transferred to the inductor L_2. The current through inductor L_2 current rises from I_{21} toI_{22}. The voltage across the capacitor C_1 falls from V_{11} to V_{12}. The voltage across the capacitor C_2 supplied the load current and falls from V_{21} to V_{22}.

Mode 2: When the switch is turned off, the capacitor $C_1$is connected to the DC supply voltage V_S through the inductor L_1 and is recharged back to voltage, V_{11}.

The current through inductor L_1falls from I_{12} toI_{11}. When the voltage at terminal 3 becomes higher than the output voltage at terminal 4, the diode D_1 conducts and the current through inductor L_2 replenishes the voltage of the output capacitor C_2. The current through inductor L_2 current falls from I_{22} to I_{21} and the voltage across the capacitor C_2 rises from V_{22} to V_{21}.

The following equations describes the performances of the SEPIC converter [1]. The average output voltage is given by

$$V_a = \frac{kV_S}{1-k} \tag{4.53}$$

The average load current for a load resistance of R is given by

$$I_a = \frac{kV_S}{R(1-k)} \tag{4.54}$$

The average current supplied by the DC source is given by

$$I_S = \frac{I_a k}{1-k} \tag{4.55}$$

The peak-to-peak ripple current of inductor L_1 is given by

$$\Delta I_1 = I_{12} - I_{11} = \frac{kV_S}{fL_1} \tag{4.56}$$

Since the average current of the inductor L_1 is the same as the average supply current I_S, the minimum value of the inductor L_1 current is given by

$$I_{11} = I_S - \frac{\Delta I_1}{2} \tag{4.57}$$

The maximum value of the inductor L_1 current is given by

$$I_{12} = I_S + \frac{\Delta I_1}{2} \tag{4.58}$$

The peak-to-peak voltage of capacitor C_1 is given by

$$\Delta V_{C1} = V_{11} - V_{12} = \frac{I_S(1-k)}{fC_1} \tag{4.59}$$

The average voltage across capacitor C_1 is given by

$$V_{C1} = \frac{V_a}{k} \tag{4.60}$$

The minimum value of the capacitor C_1 voltage is given by

$$V_{11} = V_{C1} - \frac{\Delta V_{C1}}{2} \tag{4.61}$$

The maximum value of the capacitor C_1 voltage is given by

$$V_{12} = V_{C1} + \frac{\Delta V_{C1}}{2} \tag{4.62}$$

The peak-to-peak ripple current of inductor L_2 is given by

$$\Delta I_2 = I_{21} - I_{22} = \frac{kV_S}{fL_2} \tag{4.63}$$

Since the average current of the inductor L_2 is the same as the average load current I_a, the minimum value of the inductor L_2 current is given by

$$I_{21} = I_a - \frac{\Delta I_2}{2} \tag{4.64}$$

The maximum value of the inductor L_2 current is given by

$$I_{22} = I_a + \frac{\Delta I_2}{2} \tag{4.65}$$

The peak-to-peak voltage of capacitor C_2 is given by

$$\Delta V_{C2} = V_{21} - V_{22} = \frac{\Delta I_2}{8fC_2} \tag{4.66}$$

The minimum value of the capacitor C_2 voltage is given by

$$V_{22} = V_a - \frac{\Delta V_{C2}}{2} \tag{4.67}$$

The maximum value of the capacitor C_2 voltage is given by

$$V_{21} = V_a + \frac{\Delta V_{C2}}{2} \tag{4.68}$$

Condition for continuous inductor current and capacitor voltage: Assuming I_L is the average inductor current, $\Delta I \geq 21_L$.

The critical value of the inductor L_1 as

$$L_{c1} = L_1 = \frac{(1-k)^2 R}{2kf} \tag{4.69}$$

The critical value of the inductor L_2 as

$$L_{c2} = L_2 = \frac{(1-k)R}{2f} \tag{4.70}$$

The critical value of the capacitor C_1 as

$$C_{c1} = C_1 = \frac{k}{2fR} \tag{4.71}$$

The critical value of the capacitor C_2 as

$$C_{c2} = C_2 = \frac{1}{8fR} \tag{4.72}$$

4.8.1 Output-side parameters

(a) From (4.53), the average output voltage

$$V_a = \frac{kV_s}{1-k} = \frac{0.9 \times 12}{1-0.9} = 108 \text{ V}$$

(b) The average load current

$$I_a = \frac{V_a}{R} = \frac{108}{10} = 10.8 \text{ A}$$

(c) Equation (4.55) gives the average current from the supply

$$I_S = \frac{I_a k}{1-k} = \frac{10.8 \times 0.9}{1-0.9} = 97.2 \text{ A}$$

(d) Equation (4.56) gives the peak-to-peak load ripple current of inductor L_1

$$\Delta I_1 = \frac{V_S k}{fL_1} = \frac{12 \times 0.9}{20 \times 10^3 \times 130 \times 10^{-6}} = 4.154 \text{ A}$$

(e) The steady-state minimum inductor current

$$I_{11} = I_S - \frac{\Delta I_1}{2} = 97.2 - \frac{4.154}{2} = 95.123 \text{ A}$$

(f) The steady-state maximum inductor current

$$I_{12} = I_S + \frac{\Delta I}{2} = 97.2 + \frac{4.154}{2} = 99.277 \text{ A}$$

(g) The rms value of the triangular inductor L_1 current

$$\begin{aligned} I_{rms1} &= \sqrt{\frac{{I_{11}}^2 + I_{11}I_{12} + {I_{12}}^2}{3}} \\ &= \sqrt{\frac{95.123^2 + 95.123 \times 99.277 + 99.277^2}{3}} \\ &= 97.207 \text{ A} \end{aligned}$$

(h) Equation (5.59) gives the peak-to-peak voltage of capacitor C_1

$$\Delta V_{C1} = \frac{I_S(1-k)}{fC_1} = \frac{97.2 \times (1-0.9)}{20 \times 10^3 \times 47 \times 10^{-6}} = 6.152 \text{ V}$$

(i) Equation (4.60) gives the average voltage across capacitor C_1 as

$$V_{C1} = \frac{V_a}{k} = \frac{108}{0.9} = 120$$

(j) The minimum value of the capacitor C_1 voltage

$$V_{11} = V_{C1} - \frac{\Delta V_{C1}}{2} = 120 - \frac{6.152}{2} = 116.924 \text{ V}$$

(k) The maximum value of the capacitor C_1 voltage

$$V_{12} = V_{C1} + \frac{\Delta V_{C1}}{2} = 120 + \frac{6.152}{2} = 123.076 \text{ V}$$

(l) The rms capacitor C_2 triangular voltage

$$V_{rms1} = \sqrt{\frac{{V_{11}}^2 + V_{11}V_{12} + {V_{12}}^2}{3}}$$
$$= \sqrt{\frac{116.924^2 + 116.924 \times 123.076 + 123.076^2}{3}}$$
$$= 120.013 \text{ V}$$

(m) Equation (4.63) gives the peak-to-peak ripple current of inductor L_2,

$$\Delta I_2 = \frac{kV_S}{fL_2} = \frac{12 \times 0.9}{20 \times 10^3 \times 110 \times 10^{-6}} = 4.909 \text{ A}$$

(n) The minimum value of the inductor L_2 current

$$I_{21} = I_a - \frac{\Delta I_2}{2} = 10.8 - \frac{4.909}{2} = 8.345 \text{ A}$$

(o) The maximum value of the inductor L_2 current

$$I_{22} = I_a + \frac{\Delta I_2}{2} = 10.8 + \frac{4.909}{2} = 13.255 \text{ A}$$

(p) The rms value of the triangular inductor L_2 current

$$I_{rms1} = \sqrt{\frac{{I_{21}}^2 + I_{21}I_{22} + {I_{22}}^2}{3}}$$
$$= \sqrt{\frac{8.345^2 + 8.345 \times 13.255 + 13.255^2}{3}}$$
$$= 10.893 \text{ A}$$

(q) Equation (4.66) gives the peak-to-peak capacitor voltage of capacitor C_2

$$\Delta V_{C2} = \frac{\Delta I_2}{8fC_2} = \frac{4.909}{8 \times 20 \times 10^3 \times 89 \times 10^{-6}} = 0.345 \text{ V}$$

(r) The steady-state minimum capacitor voltage

$$V_{21} = V_a - \frac{\Delta V_{c2}}{2} = 108 - \frac{4.787}{2} = 107.828 \text{ V}$$

(s) The steady-state maximum capacitor voltage

$$V_{C2} = V_a + \frac{\Delta V_{c2}}{2} = 108 + \frac{4.787}{2} = 108.172$$

(t) The rms value of the triangular capacitor voltage

$$\begin{aligned} V_{rms2} &= \sqrt{\frac{V_{21}{}^2 + V_{21}V_{22} + V_{22}{}^2}{3}} \\ &= \sqrt{\frac{107.828^2 + 107.828 \times 108.172 + 108.172^2}{3}} \\ &= 108 \text{ V} \end{aligned}$$

(u) Power delivered to the load

$$P_{out} = \frac{V_{rms2}{}^2}{R} = \frac{108^2}{10} = 1.166 \text{kW}$$

(v) DC power delivered to the load

$$P_{dc} = \frac{V_a{}^2}{R} = \frac{108^2}{10} = 1.166 \text{ kW}$$

4.8.2 Converter parameters

(a) The peak transistor current,

$$I_p = I_{12} = 99.277 \text{ A}$$

(b) The average transistor current

$$I_{av} = I_S = 97.2\text{A}$$

(c) The rms transistor current

$$I_r = \frac{P_{out}}{V_S} = \frac{1166}{12} = 97.2 \text{ A}$$

(d) The peak transistor voltage

$$V_{peak} = V_{C1} = \frac{V_a}{k} = \frac{108}{0.9} = 120 \text{ V for } k = 0.98$$

4.8.3 Input-side parameters

(a) The rms transistor current $I_r = 97.2$ A
(b) The average input current is $I_s = 97.2$ A

(c) The rms input current is $I_i = 97.2$ A
(d) The input power $P_{in} = P_{out} = 1166$ W
(e) The ripple content of the input current

$$\begin{aligned} I_{ripple} &= \sqrt{I_i^2 - I_s^2} \\ &= \sqrt{97.2^2 - 97.2^2} = 0.127 \text{ A} \end{aligned}$$

(f) The ripple factor of the input current

$$\begin{aligned} RF_i &= \frac{I_{ripple}}{I_s} \\ &= \frac{0.127}{97.2} \times 100 = 0.13\% \end{aligned}$$

4.8.4 Condition for continuous inductor current and capacitor voltage

Equation (4.69) gives the critical value of the inductor

$$\begin{aligned} L_{c1} &= \frac{(1-k)^2 R}{2kf} \\ &= \frac{(1-0.9)^2 \times 10}{2 \times 0.9 \times 20 \times 10^3} = 2.778 \text{ µH} \end{aligned}$$

Equation (4.70) gives the critical value of the inductor

$$\begin{aligned} L_{c2} &= \frac{(1-k)R}{2f} \\ &= \frac{(1-0.9) \times 10}{2 \times 20 \times 10^3} = 25 \text{ µH} \end{aligned}$$

Equation (4.71) gives the critical value of the capacitor C_1 as

$$\begin{aligned} C_{c1} &= \frac{k}{2fR} \\ &= \frac{0.9}{2 \times 20 \times 10^3 \times 10} = 2.25 \text{ µH} \end{aligned}$$

Equation (4.71) gives the critical value of the capacitor C_2 as

$$\begin{aligned} C_{c2} &= \frac{1}{8fR} \\ &= \frac{1}{8 \times 20 \times 10^3 \times 10} = 0.625 \text{ µH} \end{aligned}$$

4.8.5 Output LC-filter

The inductors L_1 and L_2 as shown in Figure 4.9 provide a high impedance for the inductor ripple current and limit the ripple currents of the inductors $L_1$and L_2. The capacitor C_1 provides a low impedance path to charge and transfer energy to inductor L_2. The capacitor C_2 provides a low impedance path for the load ripple currents and limits the ripple voltage of the load resistor, R.

For a desired value of the ripple factor of the capacitor C_1 voltage $RF_{v1} = 5\% = 0.05$, we can find the ripple voltage as

$$\Delta V_{1d} = RF_{v1}\, V_{c1}$$
$$= 0.05 \times 120 = 6 \text{ V}$$

For a desired value of the ripple factor of the capacitor C_2 voltage $RF_{v2} = 6\% = 0.06$, we can find the ripple voltage as

$$\Delta V_{2d} = RF_{v2}\, V_a$$
$$= 0.06 \times 108 = 6.48 \text{ V}$$

Similarly, for a desired value of the ripple factor of the inductor L_1 current $RF_{i1} = 7\% = 0.07$, we can find the ripple current as

$$\Delta I_{1d} = RF_{i1} I_S$$
$$= 0.07 \times 97.2 = 6.804 \text{ A}$$

For a desired value of the ripple factor of the inductor L_2 current $RF_{i2} = 8\% = 0.08$, we can find the ripple current as

$$\Delta I_{2d} = RF_{i2} I_a$$
$$= 0.08 \times 10.8 = 0.864 \text{ A}$$

Using (4.56), we can find the inductor

$$L_1 = \frac{V_S k}{f \Delta I_{1d}}$$
$$= \frac{12 \times 0.9}{20 \times 10^3 \times 6.804} = 79.365 \ \mu\text{H}$$

Using (4.63), we can find the inductor

$$L_2 = \frac{V_S k}{f \Delta I_{2d}}$$
$$= \frac{12 \times 0.9}{20 \times 10^3 \times 0.864} = 625 \ \mu\text{H}$$

Using (4.59), we can find the capacitor

$$C_1 = \frac{I_S(1-k)}{f \Delta V_{1d}} = \frac{97.2 \times (1 - 0.9)}{20 \times 10^3 \times 6} = 9 \ \mu\text{F}$$

Using (5.59) gives the peak-to-peak voltage of capacitor C_1

$$\Delta V_{C1} = \frac{I_S(1-k)}{f C_1}$$
$$= \frac{97.2 \times (1 - 0.9)}{20 \times 10^3 \times 47 \times 10^{-6}} = 6.152 \text{ V}$$

Using (4.66), we can find the capacitor

$$C_2 = \frac{\Delta I_{2d}}{8f\Delta V_{2d}}$$
$$= \frac{0.864}{8 \times 2 \times 10 \times 10^3 \times 6.48} = 0.833\ \mu\text{F}$$

4.9 PMW signal generator

The transistor switch of the converters is turned on and off through a pulse-width modulation (PWM). The duty cycle k of the converter can be varied by varying the width of the pulse. The variable width pulse is generated by comparing a DC reference signal v_r with a triangular carrier waveform v_{cr} at the switching frequency. This is shown in Figure 4.11(a) for a DC supply voltage of $v_s = 5$ V and a switching frequency of $f = 10$ kHz and a period of $T = 1/f = 100$ μs.

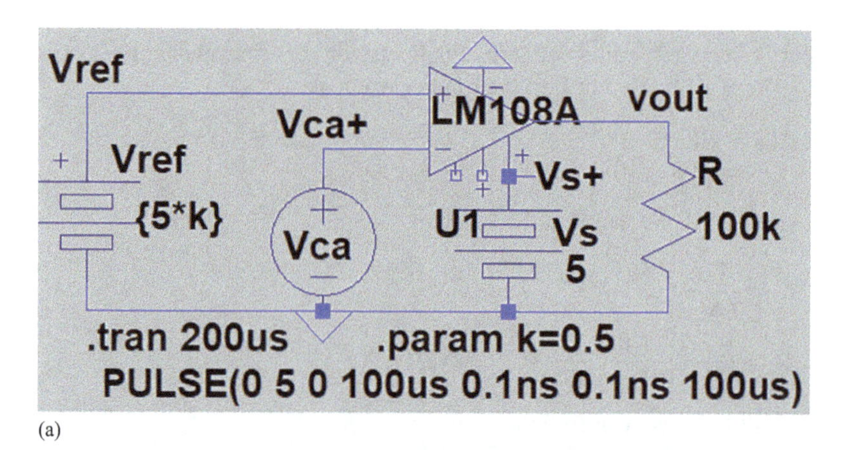

(a)

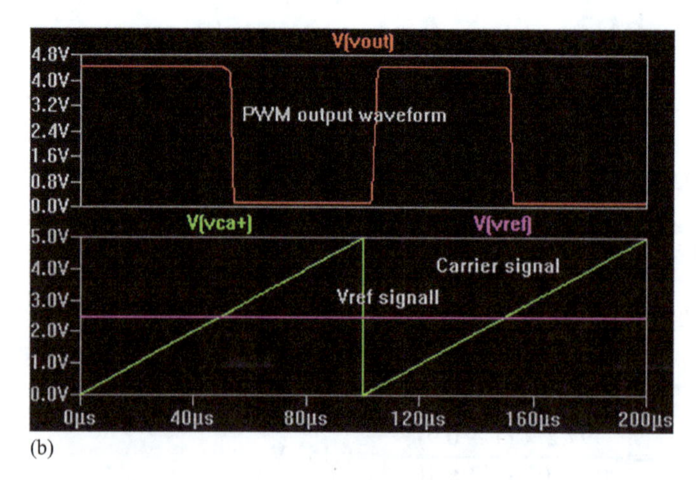

(b)

Figure 4.11 PWM generator: (a) circuit for a PWM generator and (b) PWM waveforms

The op-amp operates as a high-gain comparator. If the carrier signal is greater than the reference signal, the output becomes high otherwise the output is zero. For a duty cycle of $k = 0.4$, the reference voltage is $v_r = 5 \times k = 5 \times 0.4 = 2.0$ V and the carrier signal a triangular waveform of peak 5 V at a frequency of 10 kHz. The plots of the reference signal, carrier signal, and the PWM output gating signal are shown in Figure 4.11(b). For a hast transfer from a high to low and vice versa, the op-amp should be selected with a very high gain.

4.10 Summary

The DC–DC converters normally uses one controlled switch, one or two inductors, one or two capacitors, and one diode. Depending on the connection of these elements, the converter can be operated as a step down (known as Buck converter) or step-up (known as a Boost converter or step down and step-up known as Buck–Boost converter). Since the voltages and the currents are pulsating DC types, determination of the performances of the converters requires a clear understanding of the function of each element and the waveforms of their voltages and currents. The requirements of specifications (a) of an inductor includes specifying the average current, the rms current, the peak current, and peak-reverse voltage, (b) of a capacitor includes specifying the rms current, the rms voltage, the peak current, and peak-voltage, and (c) of a transistor or diode includes specifying the average current, the rms current, the peak current, and peak-reverse voltage.

Problems

1. The DC–DC converter circuit as shown in Figure 4.4 with a simple switch connected between the resistive load and the DC supply has $Vs = 110$ V, $R = 5$ Ω, and a duty cycle of $k = 50\%$. Determine (a) the average output voltage V_a, (b) the output DC power P_{dc}, (c) the average input current I_s and the input resistance R_i seen by the voltage source Vs. Assume an ideal switch.
2. The DC–DC converter circuit as shown in Figure 4.4 with a simple switch connected between the resistive load and the DC supply has $Vs = 110$ V, $R = 5$ Ω, and a duty cycle of $k = 80\%$. Determine (a) the average output voltage V_a, (b) the output DC power P_{dc}, (c) the average input current I_s and the input resistance R_i seen by the voltage source Vs. Assume an ideal switch.
3. The DC–DC converter circuit as shown in Figure 4.5(a) has $Vs = 98$ V, $E = 0$, $L = 4.5$ mH, $R = 4$ Ω, a duty cycle of $k = 50\%$, and a switching frequency of $f = 20$ kHz. Determine (a) the average output voltage V_a, (b) the output DC power P_{dc}, (c) the maximum load ripple current $\Delta I_{\max}$, and (d) the average input current I_s. Assume ideal transistor switch.
4. The DC–DC converter circuit as shown in Figure 4.5(a) has $Vs = 98$ V, $E = 0$, $L = 4.5$ mH, $R = 4$ Ω, a duty cycle of $k = 50\%$, and a switching frequency of $f = 20$ kHz. Calculate the performance parameters: (a) output-side parameters,

(b) controller parameters, and (c) input-side parameters. Assume ideal transistor switch.

5. The DC–DC converter circuit as shown in Figure 4.5(a) has $Vs = 98$ V, $E = 0$, $R = 4\ \Omega$, a duty cycle of $k = 50\%$, and a switching frequency of $f = 20$ kHz. Determine the value of the load inductor L to limit the ripple factor of the output voltage $RF_v = 6\% = 0.06$ and the ripple factor of the output current $RF_i = 5\% = 0.05$.
6. The DC–DC converter circuit as shown in Figure 4.5(a) has $Vs = 98$ V, $E = 0$, $L = 4.5$ mH, $R = 4\ \Omega$, a duty cycle of $k = 80\%$, and a switching frequency of $f = 20$ kHz. Determine (a) the average output voltage V_a, (b) the output DC power P_{dc}, (c) the maximum load ripple current ΔI_{max}, and (d) the average input current I_s. Assume ideal transistor switch.
7. The DC–DC converter circuit as shown in Figure 4.5(a) has $Vs = 98$ V, $E = 0$, $L = 4.5$ mH, $R = 4\ \Omega$, a duty cycle of $k = 80\%$, and a switching frequency of $f = 20$ kHz. Calculate the performance parameters: (a) output-side parameters, (b) controller parameters, and (c) input-side parameters. Assume ideal transistor switch.
8. The DC–DC converter circuit as shown in Figure 4.5(a) has $Vs = 98$ V, $E = 0$, $R = 4\ \Omega$, a duty cycle of $k = 80\%$, and a switching frequency of $f = 20$ kHz. Determine the value of the load inductor L to limit the ripple factor of the output voltage $RF_v = 6\% = 0.06$ and the ripple factor of the output current $RF_i = 5\% = 0.05$.
9. The Buck regulator as shown in Figure 5.6(a) has $Vs = 110$ V, $R = 5\ \Omega$, a duty cycle of $k = 50\%$, and the switching frequency is 25 kHz, $L = 220\ \mu$H, and $C = 20\ \mu$F. Determine (a) the average output voltage V_a, (b) the output DC power P_{dc}, (c) the average input current I_s, and (d) the input resistance R_i seen by the voltage source Vs.
10. The Buck regulator as shown in Figure 5.6(a) has $Vs = 110$ V, $R = 5\ \Omega$, a duty cycle of $k = 50\%$, and the switching frequency is 25 kHz, $L = 220\ \mu$H, and $C = 20\ \mu$F. Calculate the performance parameters: (a) output-side parameters, (b) controller parameters, and (c) input-side parameters. Assume ideal transistor switch.
11. The Buck regulator as shown in Figure 5.6(a) has $Vs = 110$ V, $R = 5\ \Omega$, a duty cycle of $k = 50\%$, and the switching frequency is 25 kHz. Determine the values of the filter inductor L and capacitor C to limit the ripple factor of the output voltage $RF_v = 6\% = 0.06$ and the ripple factor of the output current $RF_i = 5\% = 0.05$.
12. The Boost regulator as shown in Figure 5.7(a) has $Vs = 110$ V, $R = 5\ \Omega$ and a duty cycle of $k = 50\%$, and the switching frequency is 25 kHz, $L = 120\ \mu$H, and $C = 220\ \mu$F. Determine (a) the average output voltage V_a, (b) the output DC power P_{dc}, (c) the average input current I_s, and (d) the input resistance R_i seen by the voltage source Vs.
13. The Boost regulator as shown in Figure 5.7(a) has $Vs = 110$ V, $R = 5\ \Omega$, a duty cycle of $k = 50\%$, and the switching frequency is 25 kHz, $L = 120\ \mu$H, and $C = 220\ \mu$F. Calculate the performance parameters: (a) output-side

parameters, (b) controller parameters, and (c) input-side parameters. Assume ideal transistor switch.

14. The Boost regulator as shown in Figure 5.7(a) has $Vs = 110$ V, $R = 5\ \Omega$, a duty cycle of $k = 50\%$ and a switching frequency of $f = 20$ kHz. Determine the values of the filter inductor L and capacitor C to limit the ripple factor of the output voltage $RF_v = 6\% = 0.06$ and the ripple factor of the output current $RF_i = 5\% = 0.05$.
15. The Buck–Boost regulator as shown in Figure 5.8(a) has $Vs = 110$ V, $R = 5\ \Omega$ and a duty cycle of $k = 50\%$, the switching frequency is 25 kHz, $L = 120$ μH, and $C = 220$ μF. Determine (a) the average output voltage V_a, (b) the output DC power P_{dc}, (c) the average input current I_s, and (d) the input resistance R_i seen by the voltage source Vs.
16. The Buck–Boost regulator as shown in Figure 5.8(a) has $Vs = 110$ V, $R = 5\ \Omega$, duty cycle of $k = 50\%$, and a switching frequency is $f = 25$ kHz, $L = 120$ μH, and $C = 220$ μF. Calculate the performance parameters: (a) output-side parameters, (b) controller parameters, and (c) input-side parameters. Assume ideal transistor switch.
17. The Buck–Boost regulator as shown in Figure 5.8(a) has $Vs = 110$ V, $R = 5\ \Omega$, duty cycle of $k = 50\%$ and a switching frequency of $f = 25$ kHz. Determine the values of the filter inductor L and capacitor C to limit the ripple factor of the output voltage $RF_v = 6\% = 0.06$ and the ripple factor of the output current $RF_i = 5\% = 0.05$.
18. The SEPIC regulator as shown in Figure 5.9 has $Vs = 110$ V, $R = 5\ \Omega$, and duty cycle of $k = 50\%$, and a switching frequency of $f = 20$ kHz, $L_1 = 130$ μH, $L_2 = 110$ μH, $C_1 = 80$ μF, $C_1 = 10$ μF. Determine (a) the average output voltage V_a, (b) the output DC power P_{dc}, (c) the average input current I_s, and (d) the input resistance R_i seen by the voltage source Vs.
19. The SEPIC regulator as shown in Figure 5.9 has $Vs = 110$ V, $R = 5\ \Omega$, a duty cycle of $k = 50\%$, and a switching frequency of $f = 20$ kHz, $L_1 = 130$ μH, $L_2 = 110$ μH, $C_1 = 80$ μF, $C_1 = 10$ μF. Calculate the performance parameters: (a) output-side parameters, (b) controller parameters, and (c) input-side parameters. Assume ideal transistor switch.
20. The SEPIC regulator as shown in Figure 5.9 has $Vs = 110$ V, $R = 5\ \Omega$, duty cycle of $k = 50\%$ and a switching frequency of $f = 20$ kHz. Determine the values of the filter inductors L_1, L_2, and capacitors C_1, C_2 to limit the ripple factor of the output voltage $RF_v = 6\% = 0.06$ and the ripple factor of the output current $RF_i = 5\% = 0.05$.

References

[1] Rashid, M.H., *Power Electronics: Circuits, Devices, and Applications*, 4th ed., Englewood Cliffs, NJ: Prentice-Hall, 2014, Chapter 5.

[2] M. H. Rashid, *SPICE and LTspice for Power Electronics and Electric Power*, Boca Raton, FL: CRC Press, 2024

[3] Strollo, A.G.M., A new IGBT circuit model for SPICE simulation, *Power Electronics Specialists Conference*, June 1997, Vol. 1, pp. 133–138.

[4] Sheng, K., Finney, S.J., and Williams, B.W., A new analytical IGBT model with improved electrical characteristics, *IEEE Transactions on Power Electronics*, 14(1), 1999, 98–107.

[5] Sheng, K., Williams, B.W., and Finney, S.J., A review of IGBT models, *IEEE Transactions on Power Electronics*, 15(6), 2000, 1250–1266.

[6] Infineon-IKW30N65EL5-DataSheet-v02_02-EN. https://www.infineon.com/dgdl/Infineon-IGW30N65L5-DataSheet-v02_02-EN.pdf?fileId=5546d4624b0b249c014b11cd55583ac9

[7] IGBT Spice Model, https://www.globalspec.com/industrial-directory/igbt_spice_model

[8] Discrete IGBT P-Spice Models, https://www.fujielectric.com/products/semiconductor/model/igbt/technical/design_tool.html

Chapter 5
DC–AC converters

5.1 Introduction

A DC–AC converter is commonly known as an *inverter*. The input to an inverter is DC, and the output is AC. The power semiconductor devices perform the switching action, and the desired output is obtained by varying the turn-on and turn-off times of the switches. They must have controllable turn-on and turn-off characteristics. The commonly used devices are bipolar junction transistors (BJTs), metal-oxide field-effect-transistors (MOSFETs), and insulated-gate bipolar transistors (IGBTs) [1–6]. A pair of switches connect the DC input voltage to the load for a specified time duration t_1, and then another pair of switches connect the DC input voltage to the load in the opposite direction for an equal time duration $t_2 = t_1$. Each time interval may be followed by an off time t_{off} such that the total period T of the output waveform is $T = t_1 + t_{off} + t_2 + t_{off} = 2(t_1 + t_{off})$ and the frequency f of the output waveform is $f = 1/T$. We will use power MOSFETs as switches to obtain the voltage and current waveforms of the following types of voltage-source inverters:

- Single-phase voltage-source inverter,
- Single-phase pulse-width-modulated (PWM) voltage-source inverter,
- Single-phase sinusoidal PWM (SPWM) voltage-source inverter,
- Three-phase voltage-source inverter,
- Three-phase PWM voltage-source inverter, and
- Three-phase SPWM voltage-source inverter.

5.2 Single-phase voltage-source inverter

The input is a DC voltage V_S source, and the output voltage is an AC voltage of defined root-mean-square (rms) value V_o at a specified frequency f. The LTspice schematic of a single-phase inverter is shown in Figure 5.1(a). When n-channel MOSFETs (NMOS) transistors M_1 and M_2 are turned on by applying gating signals of 5 V for 50% of the period $T = 1/f$, while nMOSFET M_3 and M_3 remain turned off, the DC input voltage V_S appears across the output terminals a and b, and the output voltage is $v_o = V_S$. For a pair of switches, switching on one switch from the top left arm and one switch from the bottom right arm gives a positive output

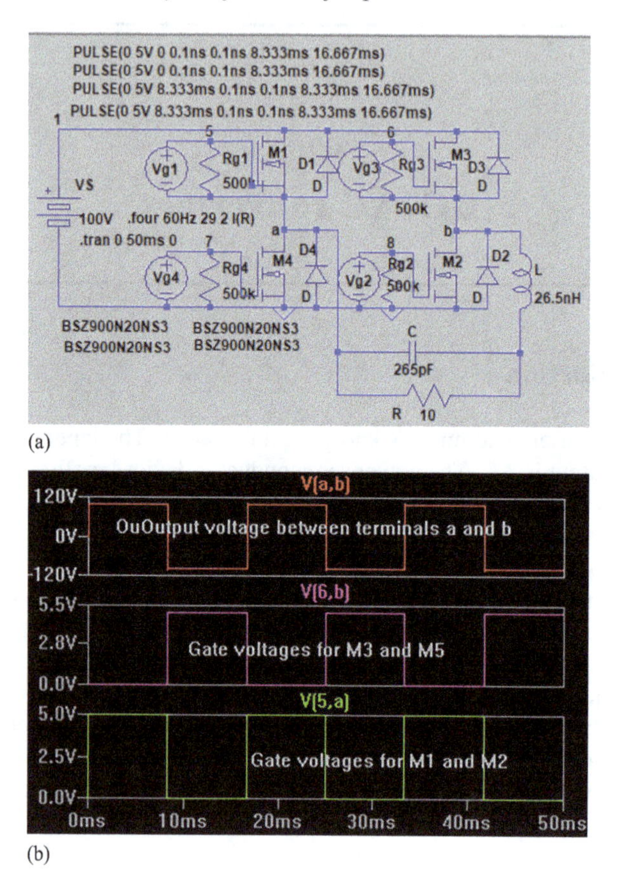

Figure 5.1 Single-phase inverter: (a) schematic and (b) waveforms

voltage. Whereas switching on one switch from the top right arm and one switch from the bottom left arm gives a negative output voltage. When nMOSFET M_3 and M_4 are turned on by applying gating signals of 5 V for 50% of the period $T = 1/f$, while nMOSFET M_1 and M_2 remain turned off, the DC input voltage V_S appears across the output terminals in the opposite direction and the output voltage is $v_o = -V_S$. The gating signals for M_1 and M_2, M_3 and M_3, and the output voltage v_o are shown in Figure 5.1(b). The addition of a small inductor $L = 100$ nH and a small capacitor $C = 442$ pF makes the circuit in a generalized form, and these elements are often connected to serve as a filter.

The rms value of the output voltage is given by

$$V_o = \sqrt{\frac{1}{T}\int_0^{T/2} v_o^2 dt} = \sqrt{\frac{1}{T}\int_0^{T/2} V_S^2 dt} = V_S \tag{5.1}$$

The output voltage v_o can be expanded to express the instantaneous output voltage in a Fourier series as [7]

$$v_o(t) = \sum_{n=1,3...}^{\infty} \frac{4V_S}{n\pi} \sin(n\omega t) \tag{5.2}$$

where angular frequency $\omega = 2\pi f$.

Equation (5.2) gives the fundamental component of the output voltage for $n = 1$ as

$$v_{o1}(t) = \frac{4V_S}{\pi} \sin(\omega t) \tag{5.3}$$

Dividing the peak value by $\sqrt{2}$, Equation (5.3) gives the rms value of the fundamental component as

$$V_{o1} = \frac{4V_S}{\sqrt{2}\pi} = 0.9V_S \tag{5.4}$$

The performance parameters of inverters can be divided into three types:

- Output-side parameters,
- Input-side parameters, and
- Inverter parameters.

Let us consider a DC input voltage of $V_S = 100$ V. The output frequency is 60 Hz with a load resistance of $R = 10\ \Omega$.

5.2.1 Output-side parameters

(a) Equation (5.1) gives the rms output voltage $V_o = V_S = 100$ V

(b) The rms output current is

$$I_o = \frac{V_o}{R} = \frac{100}{10} = 10 \text{ A}$$

(c) The output power is

$$P_{out} = \frac{V_o^2}{R} = \frac{100^2}{10} = 1 \text{ kW}$$

(d) Equation (5.4) gives the rms fundamental output voltage $V_{o1} = 0.9V_S = 0.9 \times 100 = 90$ V

(e) The rms fundamental output current is

$$I_{o1} = \frac{V_{o1}}{R} = \frac{90}{10} = 9 \text{ A}$$

(f) The fundamental output power is

$$P_{o1} = \frac{V_{o1}^2}{R} = \frac{90^2}{10} = 810 \text{ W}$$

(g) The rms ripple content of the output voltage

$$V_{ripple} = \sqrt{V_o^2 - V_{o1}^2}$$
$$= \sqrt{100^2 - 90^2} = 43.589 \text{ V}$$

(h) The total harmonic distortion (THD) of the output voltage

$$THD = \frac{V_{ripple}}{V_{o1}} = \frac{43.589}{90} = 48.432\%$$

5.2.2 Input-side parameters

(a) The input power

$$P_{in} = P_{out}$$
$$= 1 \text{ kW}$$

(b) The average input current is

$$I_{s(av)} = \frac{P_{in}}{V_S} = \frac{1 \text{ kW}}{100}$$
$$= 10 \text{ A}$$

(c) The rms input current is

$$I_{s(rms)} = I_o$$
$$= 10 \text{ A}$$

5.2.3 Inverter parameters

(a) The peak transistor current

$$I_{peak} = \frac{V_S}{R} = \frac{100}{10} = 10 \text{ A}$$

(b) The rms transistor current is

$$I_{T(rms)} = \frac{I_o}{\sqrt{2}} = \frac{10}{\sqrt{2}} = 7.071 \text{ A}$$

(c) The average transistor current is

$$I_{T(av)} = \frac{I_{s(av)}}{2} = \frac{10}{2} = 5 \text{ A}$$

5.2.4 Output filter

An LC filter as shown in Figure 5.1(a) is often connected to the output side to limit the ripple contents of the load resistor, R. The capacitor C provided a low impedance path for the ripple currents to flow through it, and the inductor L provided a high impedance for the ripple currents. The fundamental frequency of output voltage ripples is the same as the output frequency f, and the lower order for $n = 3$ is $3f$. Equation (5.2) gives the rms value of the third harmonic component for $n = 3$ as

$$\begin{aligned} V_{o3} &= \frac{4V_S}{\sqrt{2} \times 3 \times \pi} \\ &= \frac{4 \times 100}{\sqrt{2} \times 3 \times \pi} = 30.011 \text{ V} \end{aligned} \tag{5.5}$$

There are no explicit equations to find the values of L and C. We can find approximate values and then adjust the values to obtain the desired values. Neglecting the effect of capacitor C, the third harmonic voltage will appear across the inductor L and the load resistance R. Assuming the drop across the inductor is larger than that of the resistor R by a factor of 3 such that

$$n\omega L = 3R$$

which gives

$$\begin{aligned} L &= \frac{3R}{n\omega} \\ &= \frac{3 \times 10}{3 \times 366} = 26.526 \text{ mH} \end{aligned} \tag{5.6}$$

Assuming that the impedance of the capacitor is smaller than the load resistor by a factor of 3 such that most of the harmonic current flows through the capacitor. That is

$$\frac{3}{n\omega C} = R$$

which gives

$$\begin{aligned} C &= \frac{3}{n\omega R} = \frac{3}{3\omega R} \\ &= \frac{3}{3 \times 366 \times 10} = 265.258 \text{ µF} \end{aligned} \tag{5.7}$$

The LTspice lots of the output voltage and the current through the load resistor R are shown in Figure 5.2(a) for filter values: $C = 265$ pF, $L = 26.5$ nH, and in Figure 5.2(b) for filter values: $C = 265$ µF, $L = 26.5$ mH.

We can notice that the output voltage is a square wave, whereas the load current is sinusoidal form. The values of L and C can be adjusted to get a waveform closer to a pure sine wave.

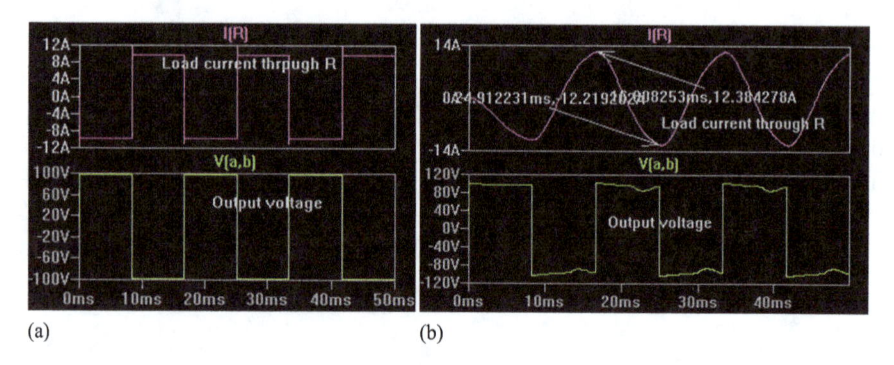

(a) (b)

Figure 5.2 Output voltage and load current: (a) For filter values: C = 265 pF, L = 26.5 nH and (b) for filter values: C = 265 μF, L = 26.5 mH

Fourier analysis: The LTspice command (.four 60 Hz 29 2 $I(R)$) for Fourier Analysis gives

THD: 46.590607% (48.340413%) for $C = 265$ pF and $L = 26.5$ nH
THD: 18.322202% (24.747854%) for $C = 265$ μF and $L = 26.5$ mH

which indicates a significant reduction of the THD with an output filter. LTspice gives the Fourier components as follows:

Fourier components of *I*(*r*)					
DC component: −0.100793					
Harmonic number	**Frequency (Hz)**	**Fourier component**	**Normalized component**	**Phase (degree)**	**Normalized phase (degree)**
1	6.000e+01	1.294e+00	1.000e+00	142.81	0.00
2	1.200e+02	1.142e−01	8.824e−02	−134.01	−276.82
3	1.800e+02	1.653e−01	1.278e−01	109.22	−33.59
4	2.400e+02	5.950e−02	4.599e−02	−134.23	−277.04
5	3.000e+02	3.208e−02	2.480e−02	129.59	−13.22
6	3.600e+02	4.485e−02	3.467e−02	−120.69	−263.50
7	4.200e+02	1.840e−02	1.422e−02	137.93	−4.88
8	4.800e+02	3.899e−02	3.014e−02	−111.89	−254.70
9	5.400e+02	4.544e−02	3.512e−02	93.21	−49.60
10	6.000e+02	8.917e−03	6.893e−03	117.41	−25.40
11	6.600e+02	4.624e−02	3.574e−02	−100.65	−243.46
12	7.200e+02	1.902e−02	1.471e−02	−98.19	−241.00
13	7.800e+02	2.144e−02	1.657e−02	−88.73	−231.54
14	8.400e+02	1.515e−02	1.171e−02	−83.68	−226.49
15	9.000e+02	1.678e−02	1.297e−02	−71.92	−214.74
16	9.600e+02	1.309e−02	1.011e−02	−64.62	−207.43

(Continues)

(*Continued*)

Fourier components of *I*(*r*)					
DC component: −0.100793					
Harmonic number	**Frequency (Hz)**	**Fourier component**	**Normalized component**	**Phase (degree)**	**Normalized phase (degree)**
17	1.020e+03	1.483e−02	1.147e−02	−56.38	−199.19
18	1.080e+03	1.278e−02	9.880e−03	−47.49	−190.31
19	1.140e+03	1.343e−02	1.038e−02	−42.53	−185.34
20	1.200e+03	1.276e−02	9.862e−03	−34.79	−177.60
21	1.260e+03	1.136e−02	8.780e−03	−30.36	−173.17
22	1.320e+03	1.096e−02	8.471e−03	−22.27	−165.08
23	1.380e+03	1.062e−02	8.210e−03	−17.22	−160.03
24	1.440e+03	9.984e−03	7.717e−03	−9.73	−152.54
25	1.500e+03	9.651e−03	7.460e−03	−4.37	−147.19
26	1.560e+03	9.176e−03	7.093e−03	6.89	−135.92
27	1.620e+03	8.731e−03	6.749e−03	8.22	−134.60
28	1.680e+03	9.772e−03	7.553e−03	21.02	−121.79
29	1.740e+03	7.349e−03	5.681e−03	21.35	−121.46
THD: 18.322202%(24.747854%)					

5.2.5 Single-phase inverter circuit model

$R = 10\ \Omega$, $C = 265\ \mu F$ and $L = 26.5$ mH, $f = 60$ Hz

$$\omega_o = 2\pi f_o = 2\pi \times 60 = 377 \text{ rad/s}$$

The nth harmonic impedance of the load with the LC filter is given by

$$Z(n) = jn\omega_o L + \frac{R \| \frac{1}{jn\omega_o C}}{1 + jn\omega_o RC} = \frac{R + jn\omega_o L - (n\omega_o)RLC}{1 + jn\omega_o RC} \tag{5.8}$$

Dividing the voltage expression in Equation (5.2) by the impedance *Z*(*n*) gives the Fourier components of the load currents as

$$i_o(t) = \sum_{n=1,3\ldots}^{\infty} \frac{4V_S}{n\pi Z(n)} \sin(n\omega t) \tag{5.2a}$$

which gives the peak magnitude of the nth harmonic currents as

$$I_m(n) = \frac{4V_S}{n\pi Z(n)} \tag{5.9}$$

which for $n = 1$ and $n = 3$ gives

$$I_m(1) = \frac{4 \times 100}{1 \times \pi \times Z(1)} = 18.01\angle - 44.916^\circ$$

$$I_m(3) = \frac{4 \times 100}{3 \times \pi \times Z(3)} = 1.573\angle - 87.873^\circ$$

Dividing the peak value by $\sqrt{2}$ gives the rms values for $n = 1$ and $n = 3$ as

$$I_{rms}(1) = \frac{I_m(1)}{\sqrt{2}} = \frac{18.01\angle -44.916^\circ}{\sqrt{2}} = 12.739\angle -44.916^\circ$$

$$I_{rms}(3) = \frac{I_m(3)}{\sqrt{2}} = \frac{1.573\angle -87.873^\circ}{\sqrt{2}} = 1.112\angle -87.873^\circ$$

Applying (3.3) for a single-phase full-wave rectifier, the total rms value of the load current can be determined by adding the square of the individual rms values and then taking the square root

$$I_o = \sqrt{\sum_{n=1,3}^{30} (I_{rms}(n))^2} \tag{5.10}$$

$$= 9.273 \text{ A}$$

The average output current $I_{o(av)}$ of a full-wave rectifier with a peak sinusoidal input current I_m is

$$I_{o(av)} = \frac{2I_m}{\pi} \tag{5.11}$$

Since the output current of the inverter contains sinusoidal harmonic components at different frequencies, we can add the corresponding DC components contributing to the output current of the inverter. Applying (5.11) gives the average input current due by n number of harmonic components, $n = 30$

$$I_{s(av)} = \sum_{n=1}^{30} \frac{2 \times |I_m(n)|}{\pi} \tag{5.12}$$

$$= 4.133 \text{ A}$$

5.3 Single-phase PWM inverter

The output voltage of the inverter in Figure 5.1(a) has a positive voltage of V_S for half of the period and a negative voltage of $-V_S$ for another half of the period. The rms output voltage has a fixed value at the switching frequency. For applications requiring a variable rms output voltage, a PWM inverter produces multiple positive pulses during the first half-cycle and the same number of negative pulses during the other half-cycle. By varying the width of the output pulses, we can vary the out rms value of the output voltage. The number of pulses per half-cycle determines the frequency of the lower-order harmonic. Higher the frequency of the lower-order harmonics, lower the values of the filter element, L and C. PWM gating signals are generated by comparing an isosceles triangular wave with a pulse of 50% duty cycle at the invert output frequency. We can use op-amp to generate PWM waveforms, as shown in Figure 5.3(a). An isosceles triangular carrier wave v_{cr1} is compared with a pulse reference wave v_{ref} with 50% on and 50% off. These signal voltages are compared through an op-amp comparator to produce PWM waveforms as shown in Figure 5.3(b) for the first 50% of the time. The {amp} is the amplitude of the reference signal and is defined as a variable so that we can vary the duty cycle of the pulses. For an output frequency of $f = 60$ Hz, the frequency of the carrier waveform is $10f$ (($2 \times 5 \times 60$) = 600 Hz) for five pulses per half-cycle. Figure 5.3(c) shows the waveforms during the second half of the reference wave. The reference signal for the second half of the waveform is obtained by the pulse waveform delayed by 8.33 ms and the pulse description is changed to the statement as follows:

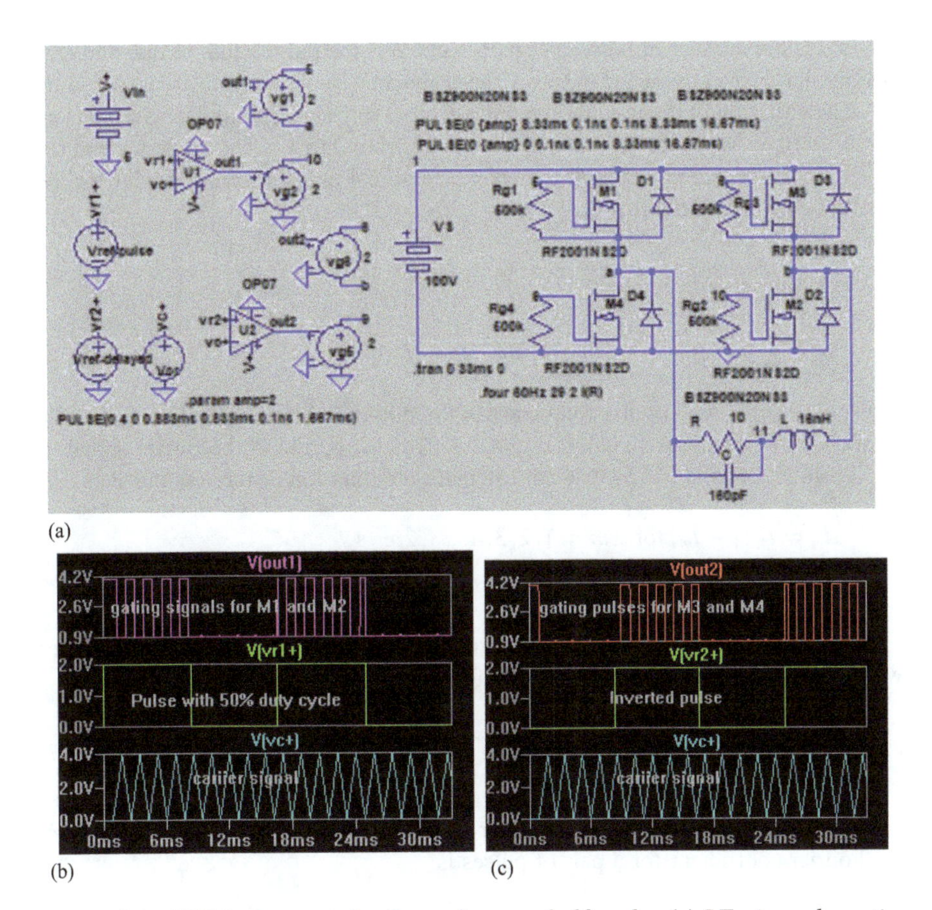

Figure 5.3 PWM generator for five pulses per half-cycle: (a) LTspice schematic, (b) PWM pulses for the first half-cycle, and (c) PWM pulses for the second half-cycle

PULSE(0 {amp} 8.33 ms 0.1 ns 0.1 ns 8.33 ms 16.67 ms)

The ratio of the reference signal A_{ref} to the peak carrier signal A_{cr} is defined as the modulation index M

$$M = \frac{A_{ref}}{A_{cr}}$$

As the reference signal is varied, keeping the carrier signal fixed, the modulation index M and the width of the pulse δ are also varied. If δ is the width of a pulse, the rms value of the output voltage for p pulses per cycle is given by

$$V_o = \sqrt{\frac{2p}{2\pi}\int_{(\pi/p-\delta)/2}^{(\pi/p+\delta)/2} V_S^2 d\theta} = V_S\sqrt{\frac{p\delta}{\pi}} \tag{5.13}$$

where p is the number of pulses per half-cycle, δ is the pulse width in rad, and ω is the angular frequency of the output voltage rad/s.

The variation of the modulation index from 0 to 1 varies the pulse width δ from 0 to $T/2p$ (0 to π/p) and the rms output voltage from 0 to V_S. The angle α_m and the time t_m of intersection between the reference signal and the carrier signal for the mth pulse can be found from [7]

$$t_m = \frac{\alpha_m}{\omega} = (m - M)\frac{T_s}{2} \quad \text{for} \quad m = 1, 3, \ldots, 2p \tag{5.14}$$

$$t_m = \frac{\alpha_m}{\omega} = (m - 1 + M)\frac{T_s}{2} \quad \text{for} \quad m = 2, 4, \ldots, 2p \tag{5.15}$$

where $\omega = 2\pi f$ is the angular frequency of the output voltage, rad/s; $f_s = 2pf$ is the frequency of the carrier signal, Hz; and $T_s = \frac{1}{f_s}$ is the period of the carrier signal, s.

From (5.14) and (5.15), we can calculate the duration of the mth pulse as

$$\begin{aligned} d_m &= t_{m+1} - t_m \ \text{for} \quad m = 1, 2, 3, 4, \ldots, 2p, \ \text{s} \\ \delta_m &= d_m\omega = (t_{m+1} - t_m)\omega \ \text{for} \quad m = 1, 2, 3, 4, \ldots, 2p, \ \text{rad} \end{aligned} \tag{5.16}$$

Due to the symmetry of the output voltage, A_n component of the Fourier series will be zero, and we obtain the general form of the output voltage as [7]

$$v_o(t) = \sum_{n=1,3,5,\ldots}^{\infty} B_n \sin(n\omega t) \tag{5.17}$$

If the positive pulse of the mth pair starts at $\omega t = \alpha_m$ and ends $\omega t = \alpha_m + \delta$, the Fourier coefficient for a pair of pulses is

$$b_n = \frac{2V_S}{\pi}\int_{\alpha_m}^{\alpha_m+\delta} \sin(n\omega t)d(\omega t) = \frac{4V_S}{n\pi}\sin\left(\frac{n\delta}{2}\right)\left[\sin n\left(\alpha_m + \frac{\delta}{2}\right)\right] \tag{5.18}$$

We can find the co-efficient B_n by adding the effects of all pulses as given by

$$B_n = \sum_{m=1}^{2p} \frac{4V_S}{n\pi} \sin\left(\frac{n\delta}{2}\right)\left[\sin n\left(\alpha_m + \frac{\delta}{2}\right)\right] \tag{5.19}$$

The performance parameters of inverters can be divided into three types.

- Output-side parameters,
- Input-side parameters, and
- Inverter parameters.

Let us consider a DC input voltage of $V_S = 100$ V. The output frequency is $f = 60$ Hz, and a load resistance of $R = 10\ \Omega$, $M = 0.5$, and $p = 5$.

The period of the output voltage, $T = 1/f = 1/60 = 16.667$ ms.

The angular frequency of the output voltage, $\omega = 2\pi f = 2 \times \pi \times 60 = 367$ rad/s.

The switching frequency is $f_s = 2pf = 2 \times 5 \times 60 = 6$ kHz.

The switching period is $T_s = 1/f_s = 1/(6 \times 10^3) = 1.667$ ms.

From (5.16) and (5.17), we can calculate these values of the angles α and the pulse widths δ as $\alpha_1 = 9°,\ \alpha_2 = 45°,\ \alpha_3 = 81°,\ \alpha_4 = 117°,\ \alpha_5 = 153°$, and $\delta = \delta_1 = \delta_2 = \delta_3 = \delta_4 = \delta_5 = 18°$.

Substituting these values in (5.19), we obtain the peak value of the fundamental component for $n = 1$ as

$$B_1 == \sum_{m=1}^{2p} \frac{4V_S}{\pi} \sin\left(\frac{\delta}{2}\right)\left[\sin n\left(\alpha_m + \frac{\delta}{2}\right)\right]$$

$$= \sum_{m=1}^{2p} \frac{4V_S}{\pi} \sin\left(\frac{18°}{2}\right)\left[\sin n\left(\alpha_m + \frac{18°}{2}\right)\right] = 0.64456V_S \tag{5.20}$$

which after dividing by $\sqrt{2}$ gives the rms fundamental output voltage as

$$V_{o1} = \frac{B_1}{\sqrt{2}} = \frac{0.64456V_S}{\sqrt{2}} = 0.45577V_S \tag{5.21}$$

5.3.1 Output-side parameters

(a) Equation (5.13) gives the rms output voltage is

$$V_o = V_S\sqrt{\frac{p\delta}{\pi}} = 100\sqrt{\frac{5 \times 18°}{180}} = 70.711 \text{ V}$$

(b) The rms output current is

$$I_o = \frac{V_o}{R} = \frac{70.711}{10} = 7.071 \text{ A}$$

(c) The output power is

$$P_{out} = \frac{V_o^2}{R} = \frac{70.711^2}{10} = 500 \text{ W}$$

(d) Equation (5.21) gives the rms fundamental output voltage

$$V_{o1} = 0.45577 V_S = 0.45577 \times 100 = 45.577 \text{ V}$$

(e) The rms fundamental output current is

$$I_{o1} = \frac{V_{o1}}{R} = \frac{45.577}{10} = 4.558 \text{ A}$$

(f) The fundamental output power is

$$P_{o1} = \frac{V_{o1}^2}{R} = \frac{45.577^2}{10} = 207.726 \text{ W}$$

(g) The rms ripple content of the output voltage

$$V_{ripple} = \sqrt{V_o^2 - V_{o1}^2}$$
$$= \sqrt{70.711^2 - 45.577^2} = 54.062 \text{ V}$$

(h) THD of the output voltage

$$\text{THD} = \frac{V_{ripple}}{V_{o1}} = \frac{54.062}{45.577} = 118.618\%$$

5.3.2 Input-side parameters

(a) The input power,

$$P_{in} = P_{out}$$
$$= 500 \text{ W}$$

(b) The average input current is

$$I_{s(av)} = \frac{P_{in}}{V_S} = \frac{500 \text{ W}}{100}$$
$$= 5 \text{ A}$$

(c) The rms input current is

$$I_{s(rms)} = I_o$$
$$= 7.071 \text{ A}$$

5.3.3 Inverter parameters

(a) The peak transistor current,

$$I_{peak} = \frac{V_S}{R} = \frac{100}{10} = 10 \text{ A}$$

(b) The rms transistor current is

$$I_{T(rms)} = \frac{I_o}{\sqrt{2}} = \frac{7.071}{\sqrt{2}} = 5 \text{ A}$$

(c) The average transistor current is

$$I_{T(av)} = \frac{I_{s(av)}}{2} = \frac{5}{2} = 2.5 \text{ A}$$

5.3.4 Output filter

An LC filter as shown in Figure 5.1(a) is often connected to the output side to limit the ripple contents of the load resistor, R. The capacitor C provided a low impedance path for the ripple currents to flow through it, and the inductor L provided a high impedance for the ripple currents. The fundamental frequency of output voltage ripples is the same as the output frequency f and the lower order for $1 = 3$ is $3f$. Equation (5.19) gives the rms value of the ninth harmonic component for $n = 9$ as

$$B_3 = \sum_{m=1}^{2p} \frac{4V_S}{3\pi} \sin\left(\frac{3\delta}{2}\right)\left[\sin 3\left(\alpha_m + \frac{\delta}{2}\right)\right] = 0.23817V_S \tag{5.22}$$

which after dividing by $\sqrt{2}$ gives the rms value of the third harmonic component

$$V_{o3} = \frac{B_3}{\sqrt{2}} = \frac{0.23817V_S}{\sqrt{2}} = \frac{0.23817 \times 100}{\sqrt{2}} = 16.841 \text{ V}$$

There are no explicit equations to find the values of L and C. We can find approximate values and then adjust the values to obtain the desired values. Neglecting the effect of capacitor C, the third harmonic voltage will appear across the inductor L and the load resistance R. Assuming the drop across the inductor is larger than that of the resistor R by a factor of 3 such that.

$$n\omega L = 3R$$

which gives

$$L = \frac{3R}{n\omega} = \frac{3 \times 10}{3 \times 366} = 26.526 \text{ mH} \tag{5.23}$$

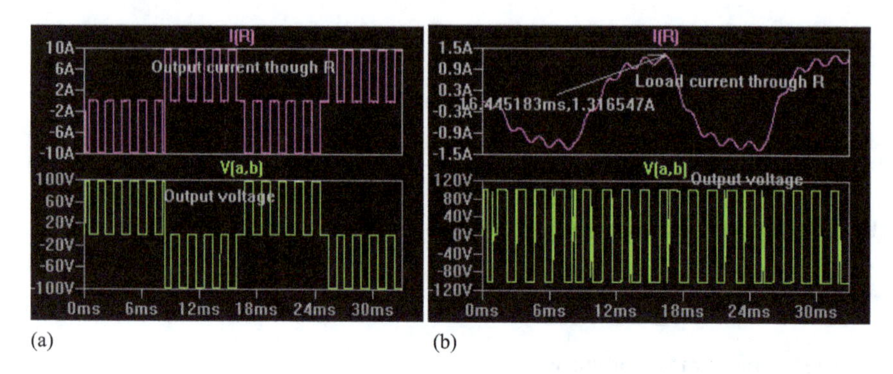

(a) (b)

Figure 5.4 Output voltage and load current: (a) for filter values: C = 265 pF, L = 26.5 nH and (b) For filter values: C = 265 μF, L = 26.5 mH

Assuming that the impedance of the capacitor is smaller than the load resistor by a factor of 3 such that most of the harmonic current flows through the capacitor. That is

$$\frac{3}{n\omega C} = R$$

which gives

$$\begin{aligned} C &= \frac{3}{n\omega R} = \frac{3}{3\omega R} \\ &= \frac{3}{3 \times 366 \times 10} = 265.258\ \mu\text{F} \end{aligned} \tag{5.24}$$

The LTspice lots of the output voltage and the current through the load resistor R are shown in Figure 5.4(a) for filter values: $C = 265$ pF, $L = 26.5$ nH, and in Figure 5.4(b) for filter values: $C = 265$ μF, $L = 26.5$ mH. We can notice that the output voltage is a square wave, whereas the load current is sinusoidal form. The values of L and C can be adjusted to get a waveform closer to a pure sine wave.

Fourier analysis: The LTspice command (.four 60 Hz 29 2 $I(R)$) for Fourier Analysis gives

THD: 80.422993% (118.694925%) for $C = 265$ pF and $L = 26.5$ nH
THD: 18.293770% (24.744076%) for $C = 265$ μF and $L = 26.5$ mH

which indicates a significant reduction of the THD with an output filter. LTspice gives the Fourier components as follows:

Fourier components of *I*(*r*)					
DC component: −0.100793					
Harmonic number	**Frequency (Hz)**	**Fourier component**	**Normalized component**	**Phase (degrees)**	**Normalized phase (degrees)**
1	6.000e+01	1.294e+00	1.000e+00	142.81	0.00
2	1.200e+02	1.142e−01	8.824e−02	−134.01	−276.82
3	1.800e+02	1.653e−01	1.278e−01	109.22	−33.59
4	2.400e+02	5.950e−02	4.599e−02	−134.23	−277.04
5	3.000e+02	3.208e−02	2.480e−02	129.59	−13.22
6	3.600e+02	4.485e−02	3.467e−02	−120.69	−263.50
7	4.200e+02	1.840e−02	1.422e−02	137.93	−4.88
8	4.800e+02	3.899e−02	3.014e−02	−111.89	−254.70
9	5.400e+02	4.544e−02	3.512e−02	93.21	−49.60
10	6.000e+02	8.917e−03	6.893e−03	117.41	−25.40
11	6.600e+02	4.624e−02	3.574e−02	−100.65	−243.46
12	7.200e+02	1.902e−02	1.471e−02	−98.19	−241.00
13	7.800e+02	2.144e−02	1.657e−02	−88.73	−231.54
14	8.400e+02	1.515e−02	1.171e−02	−83.68	−226.49
15	9.000e+02	1.678e−02	1.297e−02	−71.92	−214.74
16	9.600e+02	1.309e−02	1.011e−02	−64.62	−207.43
17	1.020e+03	1.483e−02	1.147e−02	−56.38	−199.19
18	1.080e+03	1.278e−02	9.880e−03	−47.49	−190.31
19	1.140e+03	1.343e−02	1.038e−02	−42.53	−185.34
20	1.200e+03	1.276e−02	9.862e−03	−34.79	−177.60
21	1.260e+03	1.136e−02	8.780e−03	−30.36	−173.17
22	1.320e+03	1.096e−02	8.471e−03	−22.27	−165.08
23	1.380e+03	1.062e−02	8.210e−03	−17.22	−160.03
24	1.440e+03	9.984e−03	7.717e−03	−9.73	−152.54
25	1.500e+03	9.651e−03	7.460e−03	−4.37	−147.19
26	1.560e+03	9.176e−03	7.093e−03	6.89	−135.92
27	1.620e+03	8.731e−03	6.749e−03	8.22	−134.60
28	1.680e+03	9.772e−03	7.553e−03	21.02	−121.79
29	1.740e+03	7.349e−03	5.681e−03	21.35	−121.46
THD: 18.322202%(24.747854%)					

5.3.5 Single-phase PWM inverter circuit model

$R = 10\ \Omega$, $C = 265\ \mu\text{F}$ and $L = 26.5$ mH, $f = 60$ Hz

$$\omega_o = 2\pi f_o = 2\pi \times 60 = 377\ \text{rad/s}$$

The *n*th harmonic impedance of the load with the LC filter is given by

$$Z(n) = jn\omega_o L + \frac{R \| \frac{1}{jn\omega_o C}}{1 + jn\omega_o RC} = \frac{R + jn\omega_o L - (n\omega_o)RLC}{1 + jn\omega_o RC} \tag{5.25}$$

Dividing the voltage expression in (5.19) by impedance $Z(n)$ gives the Fourier components of the load currents as

$$i_o(t) = \sum_{m=1}^{2p} \frac{4V_S}{n\pi Z(n)} \sin\left(\frac{n\delta}{2}\right)\left[\sin n\left(\alpha_m + \frac{\delta}{2}\right)\right]\sin(n\omega t) \tag{5.26}$$

which gives the peak magnitude of the nth harmonic currents

$$I_m(n) = \frac{4V_S}{n\pi Z(n)} \sin\left(\frac{n\delta}{2}\right)\left[\sin n\left(\alpha_m + \frac{\delta}{2}\right)\right] \tag{5.27}$$

which for $n = 1$ and $n = 3$ gives

$$I_m(1) = 9.12\angle - 44.916^\circ$$

$$I_m(3) = 0.883\angle - 87.873^\circ$$

Dividing by $\sqrt{2}$ gives the rms values for $n = 1$ and $n = 3$ as

$$I_{rms}(1) = \frac{9.12\angle - 44.916^\circ}{\sqrt{2}} = 6.449\angle - 44.916^\circ$$

$$I_{rms}(3) = \frac{0.883\angle - 87.873^\circ}{\sqrt{2}} = 0.624\angle - 87.873^\circ$$

Applying (3.3) for a single-phase full-wave rectifier, the total rms value of the load current can be determined by adding the square of the individual rms values and then taking the square root

$$I_o = \sqrt{\sum_{n=1,3}^{30} (|I_{rms}(n)|)^2} = 6.449 \text{ A} \tag{5.28}$$

The average output current $I_{o(av)}$ of a full-wave rectifier with a peak sinusoidal input current I_m is

$$I_{o(av)} = \frac{2I_m}{\pi} \tag{5.29}$$

Since the output current of the inverter contains sinusoidal harmonic components at different frequencies, we can add the corresponding DC components contributing to the output current of the inverter. Applying (5.29) gives the average input current due by n number of harmonic components, $n = 30$.

$$I_{s(av)} = \sum_{n=1}^{30} \frac{2 \times |I_m(n)|}{\pi} = 2.74 \text{ A} \tag{5.30}$$

5.4 Single-phase sinusoidal PWM (SPWM) inverter

A PWM inverter as shown in Figure 5.3(a) produces multiple positive pulses during the first half-cycle and the same number of negative pulses during the other half-cycle. PWM gating signals are generated by comparing an isosceles triangular carrier wave with a pulse of 50% duty cycle with a DC pulse signal of 50% duty cycle. The widths of the pulses are equal. In a SPWM inverter, the reference signal is a sine wave, and the widths of the pulses follow the sine wave. That is, the widths of the pulses in the middle close to 90° are the widest and become narrower toward 0° and 180°. As the amplitude of the reference sine wave varies, the widths of the pulses follow the sinusoidal function. The amplitude {amp} of the reference signal is defined as a variable so that we can vary the widths of the pulses. We can use op-amp to generate SPWM waveforms, as shown in Figure 9.5(a). By varying the width of the output pulses, we can vary the rms value of the output voltage. The number of pulses per half-cycle determines the frequency of the lower-order harmonic. Higher the frequency of the lower-order harmonics, lower the values of the filter elements, L and C.

SPWM gating signals are generated by comparing an isosceles triangular wave with a pulse of 50% duty cycle at the switching frequency. An isosceles triangular carrier wave v_{cr1} is compared with a sinusoidal reference wave v_{ref}. These signal voltages are compared through an op-amp comparator to produce SPWM waveforms as shown in Figure 5.5(b) for the first 50% of the time. The {amp} is the amplitude of the reference signal and is defined as a variable so that we can vary the duty cycle of the pulses. For an output frequency of $f = 60$ Hz, the frequency of the carrier waveform is $10f$ $(2 \times 5 \times 60 = 600$ Hz) for five pulses per half-cycle. Figure 5.5(c) shows the waveforms during the second half of the reference wave. The reference signal for the second half of the waveform is obtained by the pulse waveform delayed by 8.33ms and the pulse description is changed to the statement as follows. The reference signal for the second half of the waveform is obtained by the pulse waveform delayed by 8.33ms and the pulse description is changed to the statement as follows.

PULSE(0 {amp} 8.33 ms 0.1 ns 0.1 ns 8.33 ms 16.67 ms)

The ratio of the reference signal A_{ref} to the peak carrier signal A_{cr} is defined as the modulation index M

$$M = \frac{A_{ref}}{A_{cr}}$$

As the peak value of the reference sinewave signal is varied keeping the carrier signal fixed, the modulation index M and the width of the mth pulse δ_m are also varied. If δ_m is the width of a pulse, the rms value of the output voltage for p pulses per cycle, the rms value of the output voltage is given by

$$V_o = V_S \sqrt{\sum_{m=1}^{2p-1} \frac{\omega \delta_m}{\pi}} \tag{5.31}$$

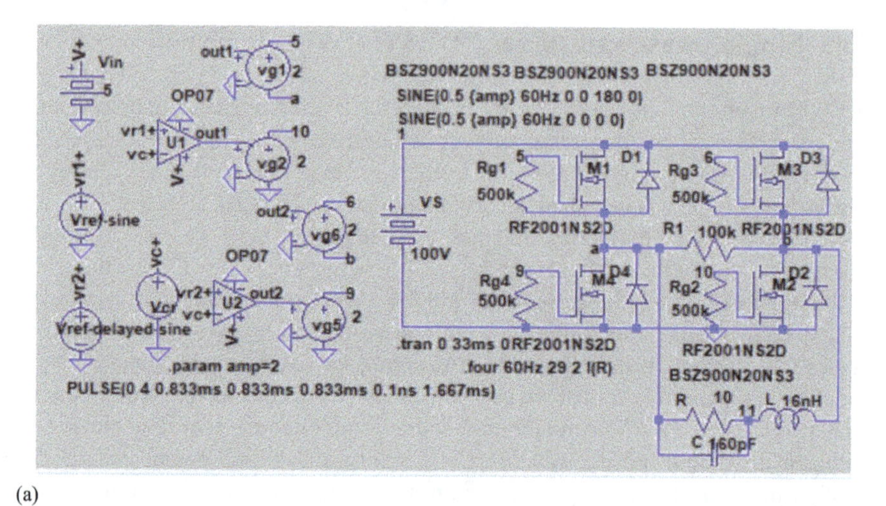

(a)

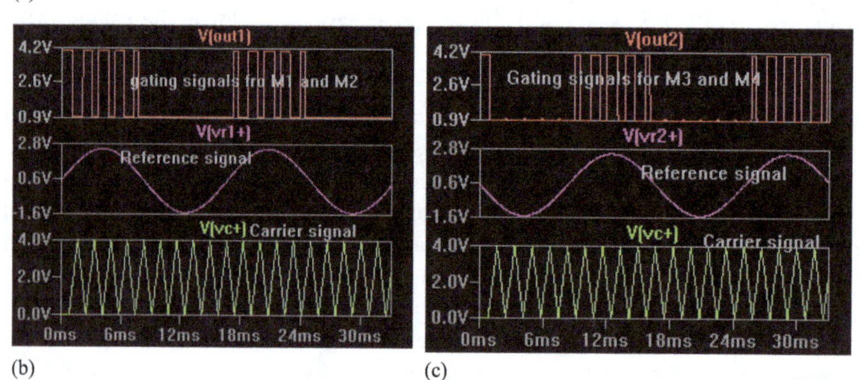

(b) (c)

Figure 5.5 SPWM generator for 5-pulses per half-cycle: (a) LTspice schematic, (b) SPWM pulses for the first half cycle, and (c) SPWM pulses for the second half-cycle

where p is the number of pulses per half-cycle,

δ is the pulse width in rad, and

ω is the angular frequency of the output voltage rad/s.

The mth time t_m and the angle α_m of intersection of the reference signal and the carrier signal can be determined from [7]

$$t_m = \frac{\alpha_m}{\omega} = t_x + m\frac{T_s}{2} \quad \text{for} \quad m = 1, 2, 3, \ldots, p \tag{5.32}$$

where t_x can be solved from

$$1 - \frac{2t}{T_s} = M \sin\left(\omega\left(t_x + \frac{mT_s}{2} - \frac{T_s}{2}\right)\right) \quad \text{for} \quad m = 1, 3, \ldots, p \tag{5.33}$$

$$\frac{2t}{T_s} = M \sin\left(\omega\left(t_x + \frac{mT_s}{2} - \frac{T_s}{2}\right)\right) \quad \text{for} \quad m = 2, 4, \ldots, p \tag{5.34}$$

where $\omega = 2\pi f$ is the angular frequency of the output voltage, rad/s; $f_s = 2pf$ is the frequency of the carrier signal, Hz; And $T_s = \frac{1}{f_s}$ is the period of the carrier signal, s.

Note: The reference sine wave in (5.33) and (5.34) is delayed by $T_s/2$ to match the intersection with a carrier signal. Otherwise, the generated pulses may not be symmetrical.

By using the iterative method of solving non-linear equations (5.33) and (5.34) in the Mathcad software tool, the duration of the mth pulse can be calculated as

$$\begin{aligned} d_m &= t_{m+1} - t_m \quad \text{for} \quad m = 1, 2, 3, 4, \ldots, 2p, \ \text{s} \\ \delta_m &= d_m\omega = (t_{m+1} - t_m)\omega \quad \text{for} \quad m = 1, 2, 3, 4, \ldots, 2p, \ \text{rad} \end{aligned} \tag{5.35}$$

Due to the symmetry of the output voltages, the A_n component of the Fourier series will be zero, and we get the general form of the output voltage as

$$v_o(t) = \sum_{n=1,3,5\ldots}^{\infty} B_n \sin(n\omega t) \tag{5.36}$$

If the positive pulse of the mth pair starts at $\omega t = \alpha_m$ and ends at $\omega t = \alpha_m + \delta_m$, the Fourier coefficient for a pair of pulses is [7]

$$b_n = \frac{2V_S}{\pi} \int_{\alpha_m}^{\alpha_m+\delta_m} \sin(n\omega t)d(\omega t) = \frac{4V_S}{n\pi} \sin\left(\frac{n\delta_m}{2}\right)\left[\sin n\left(\alpha_m + \frac{\delta_m}{2}\right)\right] \tag{5.37}$$

We can find the coefficient B_n by adding the effects of all pulses as given by

$$B_n = \sum_{m=1}^{2p-1} \frac{4V_S}{n\pi} \sin\left(\frac{n\delta_m}{2}\right)\left[\sin n\left(\alpha_m + \frac{\delta_m}{2}\right)\right] \tag{5.38}$$

The performance parameters of inverters can be divided into three types.

- Output-side parameters,
- Input-side parameters, and
- Inverter parameters.

Let us consider a DC input voltage of $V_S = 100$ V. The output frequency is $f = 60$ Hz with a load resistance of $R = 10\ \Omega$, $M = 0.5$, and $p = 5$.

The period of the output voltage, $T = 1/f = 1/60 = 16.667$ ms.

The angular frequency of the output voltage, $\omega = 2\pi f = 2 \times \pi \times 60 = 367$ rad/s.

The switching frequency is $f_s = 2pf = 2 \times 5 \times 60 = 6$ kHz.

The switching period is $T_s = 1/f_s = 1/(6 \times 10^3) = 1.667$ ms.

Using the Mathcad software tool and (5.32)–(5.35), we can calculate these values of the angles α_n and the pulse widths δ_n as $\alpha_1 = 33.582°$, $\alpha_2 = 65.378°$,

$\alpha_3 = 99.108°$, $\alpha_4 = 136.058°$, $\alpha_5 = 176.736°$ and $\delta_1 = 5.682°$, $\delta_2 = 14.565°$, $\delta_3 = 17.784°$, $\delta_4 = 14.565°$, $\delta_5 = 5.682°$.

Substituting these values in (5.38), we obtain the peak value of the fundamental component for $n = 1$ as

$$B_1 == \sum_{m=1}^{2p} \frac{4V_S}{\pi} \sin\left(\frac{\delta_m}{2}\right)\left[\sin n\left(\alpha_m + \frac{\delta_m}{2}\right)\right] = 0.47553V_S \tag{5.39}$$

The rms fundamental output voltage is

$$V_{o1} = \frac{B_1}{\sqrt{2}} = \frac{0.47553V_S}{\sqrt{2}} = 0.33625V_S \tag{5.40}$$

5.4.1 Output-side parameters

(a) Equation (5.31) gives the rms output voltage is

$$V_o = V_S\sqrt{\sum_{m=1}^{2p-1} \frac{\omega\delta_m}{\pi}} = 56.9 \text{ V}$$

(b) The rms output current is

$$I_o = \frac{V_o}{R} = \frac{56.9}{10} = 5.69 \text{ A}$$

(c) The output power is

$$P_{out} = \frac{V_o^2}{R} = \frac{56.9^2}{10} = 323.758 \text{ W}$$

(d) Equation (5.40) gives the rms fundamental output voltage

$$V_{o1} = 0.33625V_S = 0.33625 \times 100 = 33.625 \text{ V}$$

(e) The rms fundamental output current is

$$I_{o1} = \frac{V_{o1}}{R} = \frac{33.625}{10} = 3.3625 \text{ A}$$

(f) The fundamental output power is

$$P_{o1} = \frac{V_{o1}^2}{R} = \frac{33.625^2}{10} = 113.064 \text{ W}$$

(g) The rms ripple content of the output voltage

$$V_{ripple} = \sqrt{V_o^2 - V_{o1}^2}$$
$$= \sqrt{56.9^2 - 33.625^2} = 45.901 \text{ V}$$

(h) THD of the output voltage

$$\text{THD} = \frac{V_{ripple}}{V_{o1}} = \frac{45.901}{33.625} = 136.51\%$$

5.4.2 Input-side parameters

(a) The input power

$$P_{in} = P_{out} = 323.758 \text{ W}$$

(b) The average input current is

$$I_{s(av)} = \frac{P_{in}}{V_S} = \frac{323.758 \text{ W}}{100} = 3.238 \text{ A}$$

(c) The rms input current is

$$I_{s(rms)} = I_o = 5.69 \text{ A}$$

5.4.3 Inverter parameters

(a) The peak transistor current, $I_{peak} = \frac{V_S}{R} = \frac{100}{10} = 10$ A
(b) The rms transistor current is $I_{T(rms)} = \frac{I_o}{\sqrt{2}} = \frac{5.69}{\sqrt{2}} = 4.023$ A
(c) The average transistor current is $I_{T(av)} = \frac{I_{s(av)}}{2} = \frac{3.238}{2} = 1.619$ A

5.4.4 Output filter

An LC filter as shown in Figure 5.1(a) is often connected to the output side to limit the ripple contents of the load resistor, R. The capacitor C provided a low impedance path for the ripple currents to flow through it, and the inductor L provided a high impedance for the ripple currents. The fundamental frequency of output voltage ripples is the same as the output frequency f, and the lower order for $n = 9$ is $9f$. Equation (5.38) gives the rms value of the ninth harmonic component for $n = 9$ as

$$B_9 = \sum_{m=1}^{2p-1} \frac{4V_S}{9\pi} \sin\left(\frac{9\delta_m}{2}\right)\left[\sin 9\left(\alpha_m + \frac{\delta_m}{2}\right)\right] \tag{5.41}$$

which gives the rms value of the ninth harmonic component

$$V_{o9} = \frac{B_9}{\sqrt{2}} = \frac{0.34323 V_S}{\sqrt{2}} = \frac{0.34323 \times 100}{\sqrt{2}} = 24.267 \text{ V}$$

There are no explicit equations to find the values of L and C. We can find approximate values and then adjust the values to obtain the desired values.

Neglecting the effect of capacitor C, this ninth harmonic voltage will appear across the inductor L and the load resistance R. Assuming the drop across the inductor is larger than that of the resistor R by a factor of 3 such that

$$n\omega L = 3R$$

which gives

$$\begin{aligned} L &= \frac{3R}{n\omega} \\ &= \frac{3 \times 10}{9 \times 366} = 8.842 \text{ mH} \end{aligned} \tag{5.42}$$

Assuming that the impedance of the capacitor is smaller than the load resistor by a factor of 3 such that most of the harmonic current flows through the capacitor. That is

$$\frac{3}{n\omega C} = R$$

which gives

$$\begin{aligned} C &= \frac{3}{n\omega R} = \frac{3}{9\omega R} \\ &= \frac{3}{9 \times 366 \times 10} = 88.4 \text{ µF} \end{aligned} \tag{5.43}$$

The LTspice lots of the output voltage and the current through the load resistor R are shown in Figure 5.6(a) for filter values C = 88.4 pF, L = 8.8 nH, and in Figure 5.6(b) for filter values C = 88.4 µF, L = 8.8 mH. We can notice that the output voltage is a square wave, whereas the load current is sinusoidal form. The values of L and C can be adjusted to get a waveform closer to a pure sine wave.

Fourier analysis: The LTspice command (.four 60 Hz 29 2 *I*(*R*)) for Fourier Analysis gives

THD: 94.194074% (100.378612%) for C = 88.4 pF and L = 8.8 nH

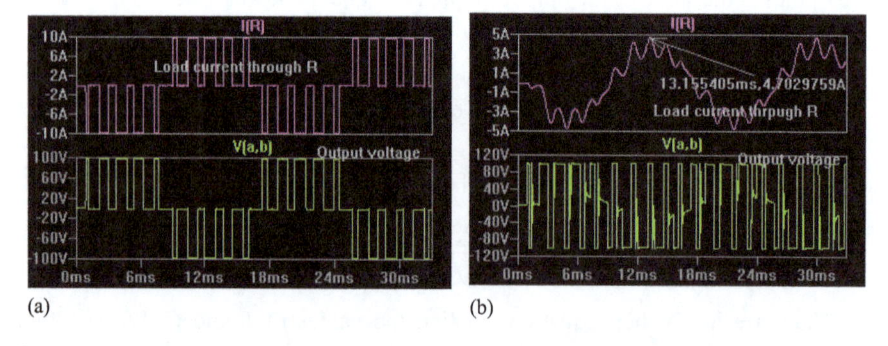

Figure 5.6 Output voltage and load current: (a) for filter values: C = 88.4 pF, L = 8.8 nH and (b) for filter values: C = 88.4 µF, L = 8.8 mH

THD: 24.631648% (23.609495%) for $C = 88.4\ \mu F$ and $L = 8.8$ mH

which indicates a significant reduction of the THD with an output filter. LTspice gives the Fourier components as follows:

Fourier components of *I*(*r*)

DC component: −0.0127733

Harmonic number	Frequency (Hz)	Fourier component	Normalized component	Phase (degrees)	Normalized phase (degrees)
1	6.000e+01	3.688e+00	1.000e+00	159.85	0.00
2	1.200e+02	8.550e−02	2.318e−02	−53.15	−213.00
3	1.800e+02	9.451e−02	2.563e−02	−29.38	−189.23
4	2.400e+02	1.272e−01	3.447e−02	−83.81	−243.66
5	3.000e+02	1.447e−01	3.924e−02	−127.41	−287.26
6	3.600e+02	1.013e−01	2.747e−02	−116.11	−275.96
7	4.200e+02	1.901e−01	5.155e−02	121.02	−38.83
8	4.800e+02	5.922e−02	1.606e−02	−130.17	−290.02
9	5.400e+02	6.242e−01	1.693e−01	36.01	−123.84
10	6.000e+02	2.591e−02	7.024e−03	−130.02	−289.87
11	6.600e+02	5.401e−01	1.464e−01	−148.87	−308.72
12	7.200e+02	1.203e−02	3.262e−03	−67.36	−227.21
13	7.800e+02	1.268e−01	3.438e−02	−100.22	−260.07
14	8.400e+02	2.500e−02	6.778e−03	−38.84	−198.69
15	9.000e+02	7.359e−02	1.995e−02	−99.10	−258.95
16	9.600e+02	2.827e−02	7.666e−03	−41.83	−201.68
17	1.020e+03	6.355e−02	1.723e−02	−156.03	−315.88
18	1.080e+03	2.529e−02	6.858e−03	−45.01	−204.86
19	1.140e+03	4.582e−02	1.242e−02	35.53	−124.32
20	1.200e+03	1.855e−02	5.030e−03	−37.38	−197.23
21	1.260e+03	3.866e−02	1.048e−02	−93.48	−253.33
22	1.320e+03	1.509e−02	4.091e−03	−17.15	−177.00
23	1.380e+03	7.226e−02	1.959e−02	−12.84	−172.69
24	1.440e+03	1.478e−02	4.007e−03	−1.19	−161.04
25	1.500e+03	4.456e−02	1.208e−02	−7.00	−166.85
26	1.560e+03	1.392e−02	3.774e−03	11.45	−148.40
27	1.620e+03	1.204e−02	3.264e−03	3.19	−156.66
28	1.680e+03	1.189e−02	3.224e−03	23.76	−136.09
29	1.740e+03	1.414e−02	3.834e−03	70.61	−89.24

THD: 24.631648% (23.609495%)

5.4.5 Single-phase sinusoidal PWM inverter circuit model

$R = 10\ \Omega$, $C = 88.4\ \mu F$, and $L = 8.9$ mH, $f = 60$ Hz

$$\omega_o = 2\pi f_o = 2\pi \times 60 = 377\ \text{rad/s}$$

The nth harmonic impedance of the load with the LC filter is given by

$$Z(n) = jn\omega_o L + \frac{R \| \frac{1}{jn\omega_o C}}{1 + jn\omega_o RC} = \frac{R + jn\omega_o L - (n\omega_o)RLC}{1 + jn\omega_o RC} \tag{5.44}$$

Dividing the voltage expression in (5.38) by the load impedance $Z(n)$ gives the Fourier components of the load currents as

$$i_o(t) = \sum_{m=1}^{2p} \frac{4V_S}{n\pi Z(n)} \sin\left(\frac{n\delta}{2}\right)\left[\sin n\left(\alpha_m + \frac{\delta}{2}\right)\right] \sin(n\omega t) \tag{5.45}$$

which gives the peak magnitude of the nth harmonic currents

$$I_m(n) = \frac{4V_S}{n\pi Z(n)} \sin\left(\frac{n\delta}{2}\right)\left[\sin n\left(\alpha_m + \frac{\delta}{2}\right)\right] \tag{5.46}$$

which for $n = 1$, $n = 9$, and $n = 11$ gives

$$I_m(1) = 5.279\angle -2.263^\circ$$
$$I_m(9) = 1.261\angle 92.107^\circ$$
$$I_m(11) = 0.998\angle -88.846^\circ$$

Dividing by $\sqrt{2}$ gives the rms values for $n = 1$, $n = 9$, and $n = 1$ as

$$I_{rms}(1) = \frac{5.279\angle - 2.263^\circ}{\sqrt{2}} = 3.733\angle - 2.263^\circ$$

$$I_{rms}(9) = \frac{1.261\angle 92.107^\circ}{\sqrt{2}} = 0.892\angle 92.107^\circ$$

$$I_{rms}(11) = \frac{0.998\angle - 88.846^\circ}{\sqrt{2}} = 0.706\angle - 88.846^\circ$$

The total rms value of the load current can be determined by adding the square of the individual rms values and then taking the square root

$$I_o = \sqrt{\sum_{n=1,3}^{30} (|I_{rms}(n)|)^2} \tag{5.47}$$
$$= 4.085 \text{ A}$$

Applying (3.3) for a single-phase full-eave rectifier, the average output current $I_{o(av)}$ of a full-wave rectifier with a peak sinusoidal input current of I_m is

$$I_{o(av)} = \frac{2I_m}{\pi}. \tag{5.48}$$

Since the output current of the inverter contains sinusoidal harmonic components at different frequencies, we can add the corresponding DC components contributing to the output current of the inverter. Applying (5.48) gives the average input current due by n number of harmonic components, $n = 30$

$$I_{s(av)} = \sum_{n=1}^{30} \frac{2 \times |I_m(n)|}{\pi} \tag{5.49}$$
$$= 2.74 \text{ A}$$

5.5 Three-phase voltage-source inverter

The LTspice schematic of a three-phase inverter is shown in Figure 5.7(a). The addition of a small inductor $L = 15.9$ nH and a small capacitor $C = 159$ pF makes

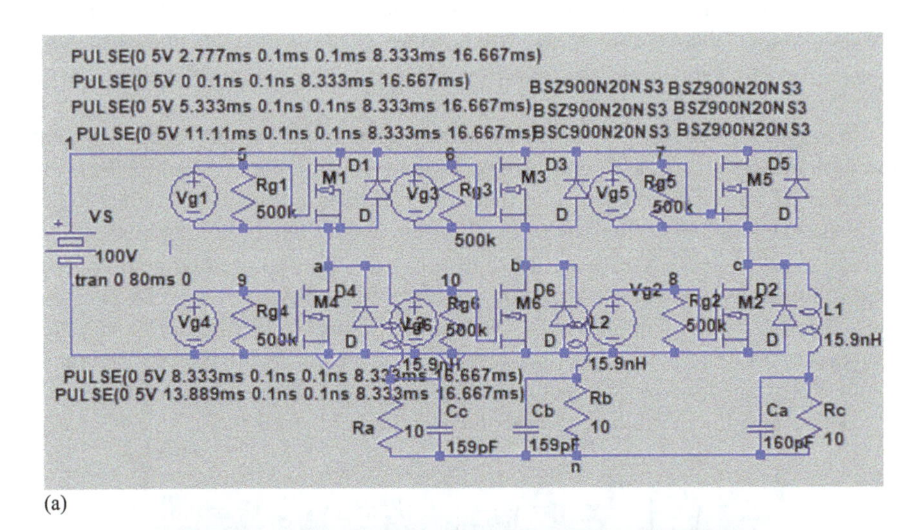

(a)

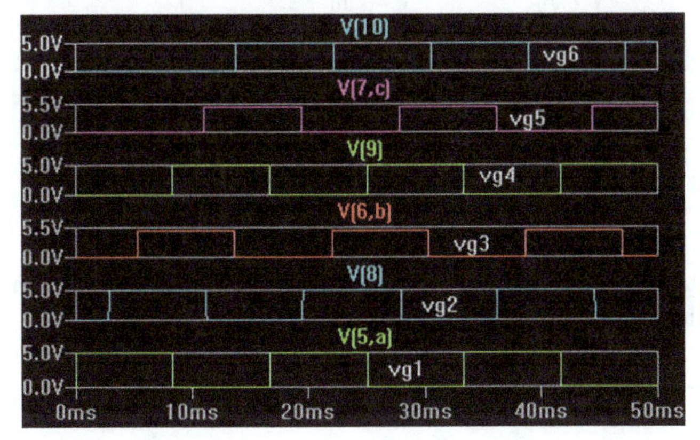

(b)

Figure 5.7 Three-phase inverter: (a) schematic and (b) gating waveforms the circuit in a generalized form, and these elements are often connected to serve as a filter. We have observed in Section 5.2 for a single-phase inverter consisting of two pairs of four switches. For a pair of switches, switching on one switch from the top left arm and one switch from the bottom right arm gives a positive output voltage v_{ab} between terminals a and b, whereas switching on one switch from the top right arm and one switch from the bottom left arm gives a negative output voltage. Adding one more pair of switches to a single-phase inverter makes six pairs of switches and allows three output voltages between terminals a and b, b and c, and c and a. That is voltages, v_{ab}, v_{bc}, and v_{ca}. There are three switches in the top row.

For the continuity of the current flow, each top switch needs to conduct $360°/6 = 60°$. Similarly, there are three switches on the bottom row. For the continuity of the current flow, each bottom switch also needs to conduct $360°/6 = 60°$. To produce an output voltage between the output terminals, one switch from the top and one switch from the bottom must be conducted. Since there are six switches, there would be six switch pairs – 12, 23, 34, 45, 56, and 61 thereby completing one cycle and then repeating the cycle. First gating pulse starts with zero delay and then each pulse by delayed by 60°.

Six gating pulses of 5 V with a 50% duty cycle are shown in Figure 5.7(b) at the desired period ($T = 1/f$) of the output voltage. As shown in Figure 5.7(a), the switches are numbered according to the order in which the switches conduct – 135 top switches and 246 bottom switches. The line-to-line output voltages are shown in Figure 5.8(a) with a positive voltage of $+V_S$ for 120° followed by zero for 60°, then followed by a negative voltage of $-V_S$ for 120° followed by zero for 60°, and then followed by 60° completing the cycle and repeating. The line-to-line voltages are shown in Figure 5.8(a). The line-to-neutral or the phase voltages are shown in Figure 5.8(b), and the waveforms of the output phase voltages are of stair-case type [8].

The rms value of the line-to-line output voltage is given by

$$V_L = V_{ab} = \sqrt{\frac{2}{T}\int_0^{2\pi/3} v_o^2 dt} = \sqrt{\frac{2}{2\pi}\int_0^{2\pi/3} V_S^2 d\theta} = \sqrt{\frac{2}{3}}\,V_S = 0.8165\,V_S \tag{5.50}$$

which, for a balanced three-phase source, gives the rms value of line-to-neutral voltage as

$$V_p = \frac{V_L}{\sqrt{3}} = \frac{1}{\sqrt{3}}\sqrt{\frac{2}{3}}\,V_S = \frac{\sqrt{2}V_S}{3} = 0.4714\,V_S \tag{5.51}$$

The line-to-line output voltage v_{ab} can be expanded to express the instantaneous output voltage in a Fourier series as [7]

$$v_{ab}(t) = \sum_{n=1,3\ldots}^{\infty} \frac{4V_S}{n\pi}\sin\left(\frac{n\pi}{2}\right)\sin\left(\frac{n\pi}{3}\right)\sin n\left(\omega t + \frac{\pi}{6}\right) \tag{5.52}$$

where angular frequency $\omega = 2\pi f$.

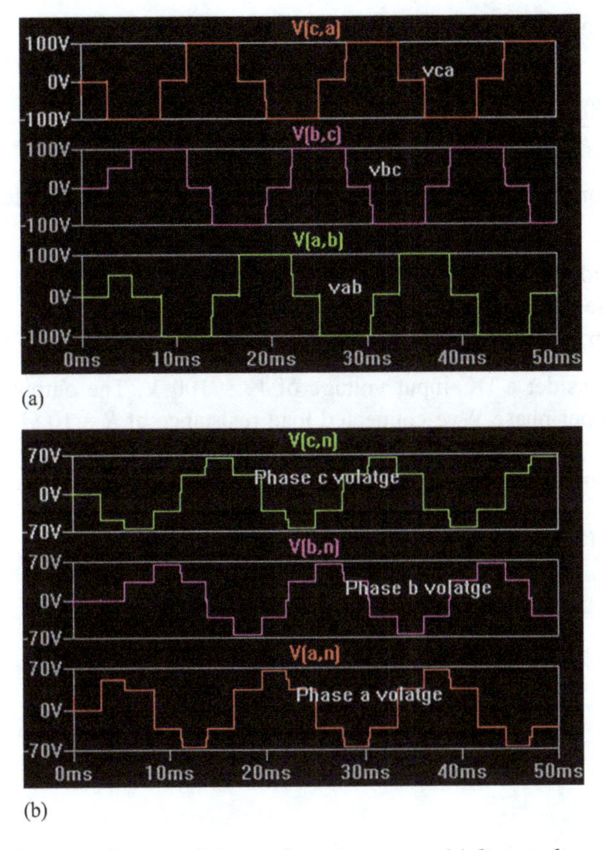

Figure 5.8 Output voltages of three-phase inverter: (a) line-to-line voltages and (b) phase voltages

Dividing the line-to-line output voltage $v_{ab}(t)$ by $\sqrt{3}$ and phase shifting by $\pi/6$ gives the instantaneous phase voltage in a Fourier series as

$$v_{an}(t) = \sum_{n=1,3\ldots}^{\infty} \frac{4V_S}{n\pi\sqrt{3}} \sin\left(\frac{n\pi}{2}\right) \sin\left(\frac{n\pi}{3}\right) \sin n(\omega t) \tag{5.53}$$

Equation (5.52) gives the peak value as the fundamental component of the line-to-line voltage for $n = 1$ as

$$V_{ab(peak)} = \frac{4V_S}{\pi} \sin\left(\frac{\pi}{2}\right) \sin\left(\frac{\pi}{3}\right) = \frac{4V_S}{\pi} \sin\left(\frac{\pi}{3}\right) = 1.103\ V_S \tag{5.54}$$

Dividing the peak value by $\sqrt{2}$, we get the rms fundamental value of the line-to-line voltage as

$$V_{L1} = \frac{V_{ab(peak)}}{\sqrt{2}} = \frac{1.103\ V_S}{\sqrt{2}} = 0.7797\ V_S \tag{5.55}$$

which dividing by $\sqrt{3}$, we get the rms fundamental component of the phase voltage as

$$V_{p1} = \frac{V_{L1}}{\sqrt{3}} = \frac{0.7797\ V_S}{\sqrt{3}} = 0.45\ V_S \tag{5.56}$$

The performance parameters of three-phase inverters can be divided into three types

- Output-side parameters,
- Input-side parameters, and
- Inverter parameters.

Let us consider a DC input voltage of $V_s = 100$ V. The output frequency is 60 Hz with a per-phase Wye-connected load resistance of $R = 10\ \Omega$.

5.5.1 Output-side parameters

(a) Equation (5.50) gives the rms line voltage is

$$V_L = \sqrt{\frac{2}{3}} V_S$$
$$= \sqrt{\frac{2}{3}} \times 100 = 81.65\ \text{V}$$

(b) Equation (5.51) gives the rms phase voltage is

$$V_p = \frac{\sqrt{2}}{3} V_S$$
$$= \frac{\sqrt{2}}{3} \times 100 = 47.14\ \text{V}$$

(c) Equation (5.55) gives the rms line fundamental current

$$V_{L1} = \frac{1.103 V_S}{\sqrt{2}}$$
$$= \frac{1.103 \times 100}{\sqrt{2}} = 77.97\ \text{V}$$

(d) Equation (5.56) gives the rms phase fundamental current

$$V_{p1} = \frac{1.103 V_S}{\sqrt{2} \times \sqrt{3}}$$
$$= \frac{1.103 \times 100}{\sqrt{2} \times \sqrt{3}} = 45.016\ \text{V}$$

(e) The rms output phase current is

$$I_p = \frac{V_p}{R}$$
$$= \frac{47.14}{10} = 4.714 \text{ A}$$

(f) The rms fundamental output phase current is

$$I_{p1} = \frac{V_{p1}}{R}$$
$$= \frac{45.016}{10} = 4.5016 \text{ A}$$

(g) The output power for three phases is

$$P_{out} = 3 \times \frac{V_p^2}{R}$$
$$= 3 \times \frac{47.14^2}{10} = 666.667 \text{ W}$$

(h) The fundamental output power for three-phase is

$$P_{out1} = 3 \times \frac{V_{p1}^2}{R}$$
$$= 3 \times \frac{45.016^2}{10} = 666.667 \text{ W}$$

(i) The rms ripple content of the line voltage

$$V_{L(ripple)} = \sqrt{V_L^2 - V_{L1}^2}$$
$$= \sqrt{81.65^2 - 77.97^2} = 24.236 \text{ V}$$

(j) The rms ripple content of the phase voltage

$$V_{p(ripple)} = \sqrt{V_p^2 - V_{p1}^2}$$
$$= \sqrt{47.14^2 - 45.016^2} = 13.993 \text{ V}$$

(k) The ripple factor of the line-to-line output voltage

$$RF_L = \frac{V_{L(ripple)}}{V_{L1}} = \frac{24.236}{77.97} = 31.084\%$$

(l) The ripple factor of the phase output voltage

$$RF_p = \frac{V_{p(ripple)}}{V_{p1}} = \frac{13.993}{45.016} = 31.084\%$$

5.5.2 Input-side parameters

(a) The input power,

$$P_{in} = P_{out}$$
$$= 666.667 \text{ W}$$

(b) The average input current is

$$I_{s(av)} = \frac{P_{in}}{V_S} = \frac{666.667}{100} = 6.667 \text{ A}$$

(c) Since each phase current I_p is supplied with two switches, the rms current of each switch is $I_p/\sqrt{2}$. Thus, the rms input current drawn from the DC source through three switches is

$$I_{s(rms)} = \frac{\sqrt{3}}{\sqrt{2}} I_p$$
$$= \sqrt{\frac{3}{2}} \times 4.714 = 5.774 \text{ A}$$

5.5.3 Inverter parameters

(a) Since the peak phase voltage is $2V_S/3$, the peak transistor current

$$I_{peak} = \frac{2V_S}{3R}$$
$$= \frac{2 \times 100}{3 \times 10} = 6.667 \text{ A}$$

(b) The rms transistor current is

$$I_{T(rms)} = \frac{I_p}{\sqrt{2}}$$
$$= \frac{4.714}{\sqrt{2}} = 3.333 \text{ A}$$

(c) The average transistor current is

$$I_{T(av)} = \frac{I_{s(av)}}{3}$$
$$= \frac{6.667}{3} = 2.222 \text{ A}$$

5.5.4 Output filter

An LC filter as shown in Figure 5.9 is often connected to the output side to limit the ripple contents of the load resistor, R. The capacitor C provided a low impedance

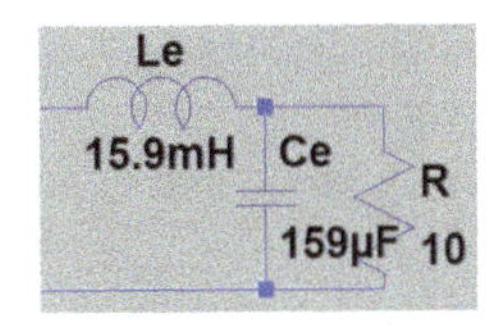

Figure 5.9 An LC filter

path for the ripple currents to flow through it, and the inductor L provided a high impedance for the ripple currents. The fundamental frequency of output voltage ripples is the same as the output frequency f, and the lower order for $n = 5$ is $5f$. Due to the symmetry of the waveform, even harmonic components and the third harmonic and the multiple of 3d harmonics are not present. Equation (5.52) gives the rms value of the 5d harmonic component for $n = 5$ as

$$V_{L5} = \frac{4V_S}{5\pi\sqrt{2}} \sin\left(\frac{5\pi}{2}\right) \sin\left(\frac{5\pi}{3}\right) = 0.156V_S \tag{5.57}$$

There are no explicit equations to find the values of L and C. We can find approximate values and then adjust the values to obtain the desired values. Neglecting the effect of capacitor C, this fifth harmonic voltage will appear across the inductor L and the load resistance R. Assuming the drop across the inductor is larger than that of the resistor R by a factor of 3 such that

$$n\omega L = 3R$$

which gives

$$\begin{aligned} L &= \frac{3R}{n\omega} \\ &= \frac{3 \times 10}{5 \times 366} = 16 \text{ mH} \end{aligned} \tag{5.58}$$

Assuming that the impedance of the capacitor is smaller than the load resistor by a factor of 3 such that most of the harmonic current flows through the capacitor. That is

$$\frac{3}{n\omega C} = R$$

which gives

$$\begin{aligned} C &= \frac{3}{n\omega R} = \frac{3}{5\omega R} \\ &= \frac{3}{5 \times 366 \times 10} = 160\ \mu\text{F} \end{aligned} \tag{5.59}$$

The LTspice plots of the output voltage and the current through the load resistor R are shown in Figure 5.10(a) for filter values C = 160 pF, L = 16 nH $C = 160$ pF, $L = 16$ nH, and in Figure 5.10(b) for filter values C = 160 µF, L = 16 mH. We can notice that the output voltage is a square wave, whereas the load current is sinusoidal form. The values of L and C can be adjusted to get a waveform closer to a pure sine wave.

Fourier analysis: The LTspice command (.four 60 Hz 29 2 $I(R)$) for Fourier Analysis gives

$$\text{THD: } 29.892153\%(31.347041\%) \text{ for } C_a = C_b = C_c = C = 160 \text{ pF and } L_a = L_b = L_c = L = 16 \text{ nH}$$

$$\text{THD: } 2.478307\%(2.478377\%) \text{ for } C_a = C_b = C_c = C = 160\ \mu\text{F and } L_a = L_b = L_c = L = 16 \text{ mH}$$

which indicates a significant reduction of the THD with an output filter. LTspice gives the Fourier components as follows:

Fourier components of *I*(*ra*)

DC component: 0.0272655

Harmonic number	Frequency (Hz)	Fourier component	Normalized component	Phase (degrees)	Normalized phase (degrees)
1	6.000e+01	6.998e+00	1.000e+00	−42.82	0.00
2	1.200e+02	2.447e−02	3.497e−03	128.58	171.40
3	1.800e+02	3.867e−02	5.526e−03	138.68	181.50
4	2.400e+02	3.811e−03	5.446e−04	−168.27	−125.45
5	3.000e+02	1.551e−01	2.217e−02	−158.78	−115.95
6	3.600e+02	4.952e−03	7.077e−04	−83.26	−40.44
7	4.200e+02	3.687e−02	5.269e−03	−170.31	−127.49
8	4.800e+02	1.077e−02	1.540e−03	−176.25	−133.43
9	5.400e+02	8.305e−03	1.187e−03	96.40	139.22
10	6.000e+02	6.520e−03	9.318e−04	−171.34	−128.52
11	6.600e+02	4.546e−03	6.497e−04	−119.00	−76.18
12	7.200e+02	2.697e−03	3.854e−04	3.82	46.64
13	7.800e+02	6.027e−03	8.613e−04	−71.85	−29.03
14	8.400e+02	3.682e−03	5.262e−04	39.34	82.16
15	9.000e+02	5.470e−04	7.817e−05	97.96	140.78
16	9.600e+02	1.834e−03	2.622e−04	20.32	63.14
17	1.020e+03	6.447e−03	9.213e−04	−138.49	−95.67
18	1.080e+03	8.434e−04	1.205e−04	−86.72	−43.90
19	1.140e+03	7.851e−04	1.122e−04	143.12	185.94
20	1.200e+03	9.054e−04	1.294e−04	−154.84	−112.01
21	1.260e+03	1.100e−03	1.572e−04	−168.99	−126.17
22	1.320e+03	1.629e−03	2.328e−04	117.05	159.88
23	1.380e+03	2.608e−03	3.726e−04	−100.85	−58.03
24	1.440e+03	1.270e−04	1.814e−05	134.59	177.41
25	1.500e+03	5.767e−04	8.241e−05	−112.05	−69.23
26	1.560e+03	1.364e−03	1.949e−04	29.27	72.09
27	1.620e+03	6.554e−04	9.367e−05	−68.08	−25.26
28	1.680e+03	2.280e−04	3.258e−05	6.52	49.34
29	1.740e+03	1.164e−03	1.663e−04	−146.88	−104.06

THD: 2.388146% (2.391763%)

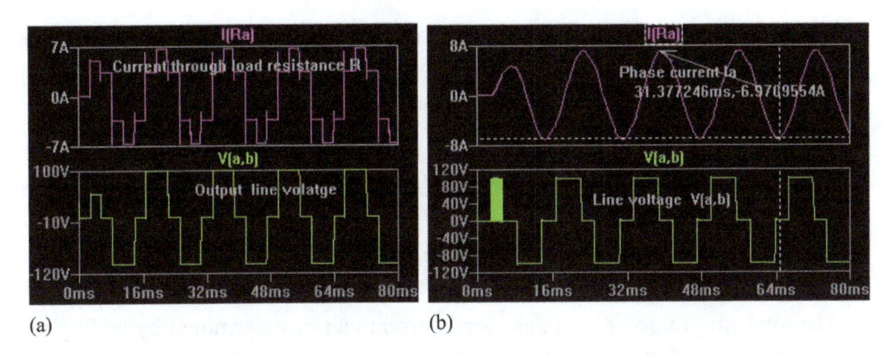

(a) (b)

Figure 5.10 Output line voltage and load current: (a) for filter values: C = 160 pF, L = 16 nH and (b) for filter values: C = 160 μF, L = 16 mH

5.5.5 Three-phase inverter circuit model

$R = 10\ \Omega$, $C = 160\ \mu F$ and $L = 16$ mH, $f = 60$ Hz

$$\omega_o = 2\pi f_o = 2\pi \times 60 = 377\ \text{rad/s}$$

The nth harmonic impedance of the load with the LC filter is given by

$$Z(n) = jn\omega_o L + \frac{R \| \frac{1}{jn\omega_o C}}{1 + jn\omega_o RC} = \frac{R + jn\omega_o L - (n\omega_o)RLC}{1 + jn\omega_o RC} \tag{5.60}$$

Dividing the voltage expression in (5.53) by the impedance $Z(n)$ gives the Fourier components of the phase load currents as

$$i_a(t) = \sum_{n=1,3\ldots}^{\infty} \frac{4V_S}{n\pi\sqrt{3}Z(n)} \sin\left(\frac{n\pi}{2}\right) \sin\left(\frac{n\pi}{3}\right) \tag{5.61}$$

which gives the peak magnitude of the nth harmonic currents of the phase current

$$I_m(n) = \frac{4V_S}{n\pi\sqrt{3}Z(n)} \sin\left(\frac{n\pi}{2}\right) \sin\left(\frac{n\pi}{3}\right) \tag{5.62}$$

which for $n = 1$, $n = 5$, and $n = 7$ gives

$$I_m(1) = \frac{4 \times 100}{1 \times \pi \times \sqrt{3} \times Z(1)} = 8.481\angle - 12.378^\circ$$

$$I_m(5) = \frac{4 \times 100}{3 \times \pi \times \sqrt{3} \times Z(5)} = 0.468\angle 92.088^\circ$$

$$I_m(7) = \frac{4 \times 100}{3 \times \pi \times \sqrt{3} \times Z(7)} = 0.227\angle 90.761^\circ$$

Dividing the peak value by $\sqrt{2}$ gives the rms values for $n = 1$, $n = 5$, and $n = 7$ as

$$I_{rms}(1) = \frac{I_m(1)}{\sqrt{2}} = \frac{8.481\angle -12.378^\circ}{\sqrt{2}} = 5.997\angle -12.378^\circ$$

$$I_{rms}(5) = \frac{I_m(3)}{\sqrt{2}} = \frac{0.468\angle 92.088^\circ}{\sqrt{2}} = 0.331\angle 92.088^\circ$$

$$I_{rms}(7) = \frac{I_m(3)}{\sqrt{2}} = \frac{0.227\angle 90.761^\circ}{\sqrt{2}} = 0.161\angle 90.761^\circ$$

The total rms value of the phase load current can be determined by adding the square of the individual rms values and then taking the square root

$$I_o = \sqrt{\sum_{n=1,3}^{30} (I_{rms}(n))^2} \tag{5.63}$$
$$= 6.009 \text{ A}$$

Applying (3.3) for a single-phase full-eave rectifier, the average output current $I_{o(av)}$ of a full-wave rectifier with a peak sinusoidal input current of I_m is

$$I_{o(av)} = \frac{2I_m}{\pi} \tag{5.64}$$

Since the output current of the inverter contains sinusoidal harmonic components at different frequencies, we can add the corresponding DC components contributing to the output current of the inverter. Applying (5.64) for three phases gives the average input current due by n number of harmonic components, $n = 30$

$$I_{s(av)} = \sum_{n=1}^{30} \frac{3 \times 2 \times |I_m(n)|}{\pi} \tag{5.65}$$
$$= 4.247 \text{ A}$$

5.6 Three-phase PWM inverter

The three-phase inverter in Figure 5.7(a) has a positive voltage of V_S for 1/3 of the period and a negative voltage of $-V_S$ for another 1/3 of the period. A three-phase PWM inverter produces multiple positive pulses during 1/3 of the period, the same number of negative pulses during another 1/3 of the period, and zero voltage for another 1/3 of the period. By varying the width of the output pulses, we can vary the rms value of the output voltage. The number of pulses per half-cycle determines the frequency of the lower-order harmonic. Higher the frequency of the lower-order harmonics, lower the values of the filter element, L and C. An isosceles triangular carrier wave v_{cr1} at a switching frequency is compared with a pulse reference wave v_{ref} of 50% on and 50% off at the output frequency. Although the comparator generates pulses during the complete period T of the output frequency, the positive pulses of $+V_S$ appear on the output of the inverter for only an interval of $T/3$, and the negative pulses of $-V_S$ appear on the output of the inverter for another interval of $T/3$.

If p_c is the number of gating pulses generated by the comparator during the first half-cycle of the output voltage, the number of pulses p_o appearing on the output of the inverter is

$$p_o = \frac{120^\circ}{180^\circ} \times p_r = \frac{2}{3} p_r \tag{5.66}$$

which gives

$$p_r = \frac{180^\circ}{120^\circ} \times p_o = \frac{3}{2} p_o \tag{5.67}$$

For example, for $p_o = 4$, $p_r = 6$ and for $p_o = 6$, $p_r = 9$

For an output frequency of $f = 60$ H, and 9 pulses per half cycle, the frequency of the carrier waveform is $9f$ (540 Hz). Figure 5.11(a) shows the generation of the

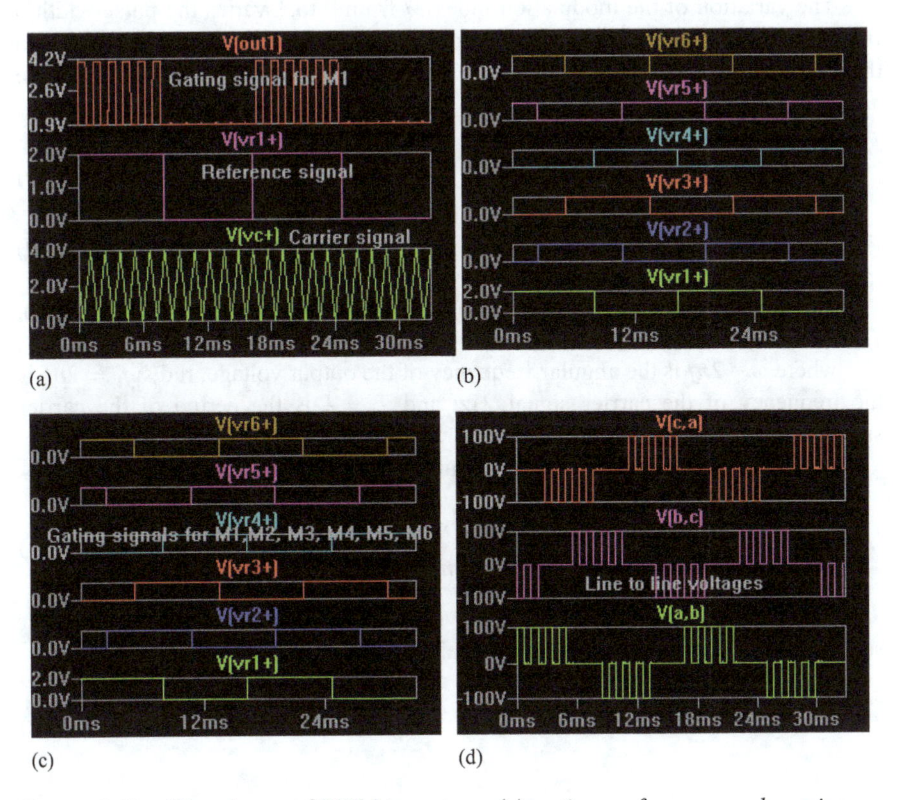

Figure 5.11 Waveforms of PWM inverters: (a) gating, reference, and carrier signals; (b) reference signals delayed by 60 degrees; (c) six gating signals; and (d) line-to-line output voltages [8]

gating signals during the first half cycle of the reference wave. The reference signals for generating gating signals for other switches, as shown in Figure 5.11(b), are obtained by the pulse waveforms delayed by 60° (2.778 ms), and the pulse description is changed to the statement as follows:

PULSE (0 {amp} 2.778 ms 0.1 ns 0.1 ns 8.33 ms 16.67 ms)

Figure 5.11(c) shows the six gating signals, and Figure 5.11(d) shows the line-to-line output voltages.

The ratio of the reference signal A_{ref} to the peak carrier signal A_{cr} is defined as the modulation index $M = A_{ref}/A_{cr}$. As the reference signal is varied, keeping the carrier signal fixed, the modulation index M and the width of the pulse δ are also varied. If δ is the width of a pulse, the rms value of the output voltage for p pulses per cycle and the rms value of the output voltage per phase is given by

$$V_o = \sqrt{\frac{2p_o}{2\pi}\int_0^{\delta} V_S^2 d\theta} = V_S\sqrt{\frac{p_o\delta}{\pi}} \tag{5.68}$$

where p_o is the number of pulses per half-cycle of the output voltage; δ is the pulse width in rad; and ω is the angular frequency of the output voltage rad/s.

The variation of the modulation index M from 0 to 1 varies the pulse width δ from 0 to $T/2p$ (0 to π/p) and the rms output voltage from 0 to V_S. The angle α_m and the time t_m of intersection between the reference signal and the carrier signal for the mth pulse can be found from [7]

$$t_m = \frac{\alpha_m}{\omega} = \frac{T}{12} + (m - M)\frac{T_s}{2} \quad \text{for} \quad m = 1, 3, \ldots, 2p_o \tag{5.69}$$

$$t_m = \frac{\alpha_m}{\omega} = \frac{T}{12} + (m - 1 + M)\frac{T_s}{2} \quad \text{for} \quad m = 2, 4, \ldots, 2p_o \tag{5.70}$$

Note: Since the pulses for the output voltage are delayed by 30 from the zero, $T/12$ is added to (5.69) and (5.70).

where $\omega = 2\pi f$ is the angular frequency of the output voltage, rad/s; $f_s = 2pf$ is the frequency of the carrier signal, Hz; and $T_s = \frac{1}{f_s}$ is the period of the carrier signal, s.

From (5.69) and (5.70), we can calculate the duration of the mth pulse as

$$\begin{aligned} d_m &= t_{m+1} - t_m \quad \text{for} \quad m = 1, 2, 3, 4, \ldots, 2p_o, \ \text{s} \\ \delta_m &= d_m\omega = (t_{m+1} - t_m\ )\omega \quad \text{for} \quad m = 1, 2, 3, 4, \ldots, 2p_o, \ \text{rad} \end{aligned} \tag{5.71}$$

Due to the symmetry of the output voltage, the A_n component of the Fourier series will be zero, and we get the general form of the output voltage as

$$v_o(t) = \sum_{n=1,3,5\ldots}^{\infty} B_n \sin(n\omega t) \tag{5.72}$$

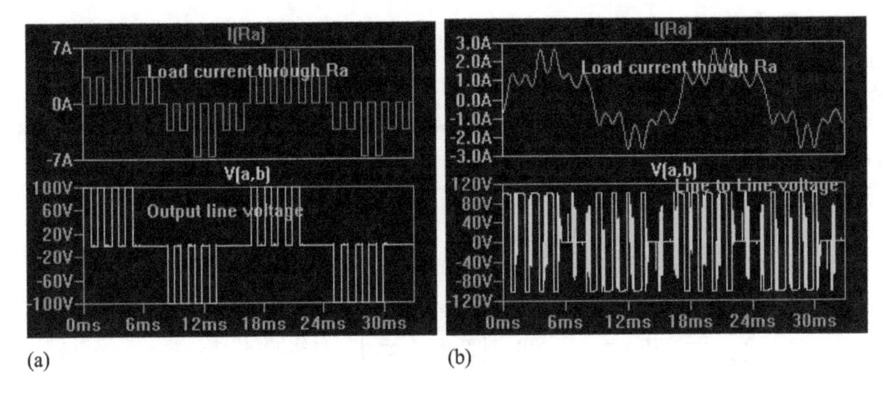

(a) (b)

Figure 5.12 Output voltage and load current: (a) for filter values: C = 72.3 pF, L = 7.2 nH and (b) for filter values: C = 72.3 μF, L = 7.2 mH

If the positive pulse of the mth pair starts at $\omega t = \alpha_m$ and ends $\omega t = \alpha_m + \delta_m$, the Fourier coefficient for a pair of pulses is

$$b_n = \frac{2V_S}{\pi}\int_{\alpha_m}^{\alpha_m+\delta_m} \sin(n\omega t)d(\omega t) = \frac{4V_S}{n\pi}\sin\left(\frac{n\delta_m}{2}\right)\left[\sin n\left(\alpha_m + \frac{\delta_m}{2}\right)\right] \tag{5.73}$$

We can find the coefficient B_n by adding the effects of all pulses as given by

$$B_n = \sum_{m=1}^{2p_o} \frac{4V_S}{n\pi}\sin\left(\frac{n\delta_m}{2}\right)\left[\sin n\left(\alpha_m + \frac{\delta_m}{2}\right)\right] \tag{5.74}$$

Dividing the line-to-line output voltage $v_{ab}(t)$ by $\sqrt{3}$ and phase shifting by $\pi/6$ gives the instantaneous phase voltage in a Fourier series as

$$v_{an}(t) = \sum_{n=1,3\ldots}^{\infty} \frac{4V_S}{n\pi\sqrt{3}}\sin\left(\frac{n\delta_m}{2}\right)\left[\sin n\left(\alpha_m + \frac{\delta_m}{2}\right)\right]\sin n(\omega t) \tag{5.75}$$

The performance parameters of inverters can be divided into three types:

- Output-side parameters,
- Input-side parameters, and
- Inverter parameters.

Let us consider a DC input voltage of $V_S = 100$ V. The output frequency is $f = 60$ Hz with a per-phase load resistance of $R = 10\ \Omega$, $M = 0.5$, $p_o = 4$, and $p_r = 6$.

The period of the output voltage, $T = 1/f = 1/60 = 16.667$ ms.

The angular frequency of the output voltage, $\omega = 2\pi f = 2 \times \pi \times 60 = 367$ rad/s.

The switching frequency is $f_s = 2p_r f = 2 \times 6 \times 60 = 720$ Hz.

The switching period is $T_s = 1/f_s = 1/(720) = 1.389$ ms.

From (5.69) to (5.71), we can calculate these values of the angles α and the pulse widths δ as $\alpha_1 = 37.5°$, $\alpha_2 = 67.5°$, $\alpha_3 = 97.5°$, $\alpha_4 = 127.5°$, and $\delta = \delta_1 = \delta_2 = \delta_3 = \delta_4 = 15°$.

Substituting these values in (5.74), we get the peak value of the fundamental component for $n = 1$ as

$$B_1 = \sum_{m=1}^{2p_o} \frac{4V_S}{\pi} \sin\left(\frac{\delta_m}{2}\right)\left[\sin n\left(\alpha_m + \frac{\delta_m}{2}\right)\right]$$

$$= \sum_{m=1}^{2\times 4} \frac{4V_S}{\pi} \sin\left(\frac{15°}{2}\right)\left[\sin n\left(\alpha_m + \frac{15°}{2}\right)\right] = 0.55609V_S \tag{5.76}$$

Dividing by $\sqrt{2}$ gives the rms fundamental output voltage as

$$V_{o1} = \frac{B_1}{\sqrt{2}} = \frac{0.55609V_S}{\sqrt{2}} = 0.39321V_S \tag{5.77}$$

5.6.1 *Output-side parameters*

(a) Equation (5.68) gives the rms output voltage

$$V_o = V_S\sqrt{\frac{p\delta}{\pi}} = 100\sqrt{\frac{5 \times 15°}{180}} = 57.735 \text{ V}$$

(b) The rms output current is

$$I_o = \frac{V_o}{R} = \frac{57.735}{10} = 5.7735 \text{ A}$$

(c) The output power is

$$P_{out} = \frac{V_o^2}{R} = \frac{57.735^2}{10} = 333.333 \text{ W}$$

(d) Equation (5.77) gives the rms fundamental output voltage

$$V_{o1} = 0.39321V_S = 0.39321 \times 100 = 39.321 \text{ V}.$$

(e) The rms fundamental output current is

$$I_{o1} = \frac{V_{o1}}{R} = \frac{45.577}{10} = 4.558 \text{ A}$$

(f) The fundamental output power is

$$P_{o1} = \frac{V_{o1}^2}{R} = \frac{39.321^2}{10} = 154.616 \text{ W}$$

(g) The rms ripple content of the output voltage

$$V_{ripple} = \sqrt{V_o^2 - V_{o1}^2}$$
$$= \sqrt{57.735^2 - 39.321^2} = 42.275 \text{ V}$$

(h) THD of the output voltage

$$\text{THD} = \frac{V_{ripple}}{V_{o1}} = \frac{42.275}{39.321} = 107.512\%$$

5.6.2 Input-side parameters

(d) The input power

$$P_{in} = P_{out}$$
$$= 333.333 \text{ W}$$

(e) The average input current is

$$I_{s(av)} = \frac{P_{in}}{V_S} = \frac{333.333 \text{ W}}{100}$$
$$= 3.33333 \text{ A}$$

(f) The rms input current is

$$I_{s(rms)} = I_o$$
$$= 5.7735 \text{ A}$$

5.6.3 Inverter parameters

(a) The peak transistor current,

$$I_{peak} = \frac{V_S}{R} = \frac{100}{10} = 10 \text{ A}$$

(b) The rms transistor current is

$$I_{T(rms)} = \frac{I_o}{\sqrt{3}} = \frac{7.071}{\sqrt{3}} = 3.333 \text{ A}$$

(c) The average transistor current is

$$I_{T(av)} = \frac{I_{s(av)}}{3} = \frac{3.33333}{3} = 1.111 \text{ A}$$

5.6.4 Output filter

An LC filter as shown in Figure 5.9 is often connected to the output side to limit the ripple contents of the load resistor, R. The capacitor C provided a low impedance path for the ripple currents to flow through it, and the inductor L provided a high impedance for the ripple currents. The fundamental frequency of output voltage ripples is the same as the output frequency f, and the lower order for $1 = 3$ is $3f$. Equation (5.74) gives the rms value of the 11th harmonic component for $n = 11$ as

$$B_{11} = \sum_{m=1}^{2p_o} \frac{4V_S}{11 \times \pi} \sin\left(\frac{3\delta_m}{2}\right)\left[\sin 3\left(\alpha_m + \frac{\delta_m}{2}\right)\right] = 0.38399V_S \tag{5.78}$$

which gives the rms value of the 11th harmonic component

$$V_{o11} = \frac{B_{11}}{\sqrt{2}} = \frac{0.38399V_S}{\sqrt{2}} = \frac{0.38399 \times 100}{\sqrt{2}} = 27.152 \text{ V}$$

There are no explicit equations to find the values of L and C. We can find approximate values and then adjust the values to obtain the desired values. Neglecting the effect of capacitor C, the 11th harmonic voltage will appear across the inductor L and the load resistance R. Assuming the drop across the inductor is larger than that of the resistor R by a factor of 3 such that

$$n\omega L = 3R$$

which gives

$$\begin{aligned} L &= \frac{3R}{n\omega} \\ &= \frac{3 \times 10}{11 \times 366} = 7.2 \text{ mH} \end{aligned} \tag{5.79}$$

Assuming that the impedance of the capacitor is smaller than the load resistor by a factor of 3 such that most of the harmonic current flows through the capacitor. That is

$$\frac{3}{n\omega C} = R$$

which gives

$$\begin{aligned} C &= \frac{3}{n\omega R} = \frac{3}{3\omega R} \\ &= \frac{3}{11 \times 366 \times 10} = 72.3\ \mu\text{F} \end{aligned} \tag{5.80}$$

The LTspice lots of the output voltage and the current through the load resistor R are shown in Figure 5.12(a) for filter values $C = 72.3$ pF, $L = 7.2$ nH, and in Figure 5.12(b) for filter values $C = 72.3$ μF, $L = 7.2$ mH. We can notice that the output voltage is a square wave, whereas the load current is sinusoidal form. The values of L and C can be adjusted to get a waveform closer to a pure sine wave.

Fourier analysis: The LTspice command (.four 60 Hz 29 2 *I*(*R*)) for Fourier Analysis gives

THD: 98.160903% (105.899331%) for $C = 72.3$ pF and $L = 7.2$ nH
THD: 35.965743% (34.509790%) for $C = 72.3$ μF and $L = 7.2$ mH

which indicates a significant reduction of the THD with an output filter. LTspice gives the Fourier components as follows:

Fourier components of *I*(*ra*)

DC component: 0.0101671

Harmonic number	Frequency (Hz)	Fourier component	Normalized component	Phase (degrees)	Normalized phase (degrees)
1	6.000e+01	1.861e+00	1.000e+00	−6.63	0.00
2	1.200e+02	2.059e−02	1.106e−02	97.26	103.89
3	1.800e+02	2.037e−02	1.095e−02	101.05	107.67
4	2.400e+02	2.037e−02	1.095e−02	104.86	111.48
5	3.000e+02	3.905e−01	2.099e−01	−40.24	−33.61
6	3.600e+02	2.010e−02	1.080e−02	112.66	119.29
7	4.200e+02	2.584e−01	1.389e−01	−60.87	−54.24
8	4.800e+02	1.929e−02	1.037e−02	120.55	127.18
9	5.400e+02	1.929e−02	1.037e−02	124.36	130.99
10	6.000e+02	1.923e−02	1.033e−02	128.44	135.06
11	6.600e+02	3.579e−01	1.923e−01	−61.83	−55.21
12	7.200e+02	1.777e−02	9.550e−03	136.99	143.61
13	7.800e+02	2.970e−01	1.596e−01	127.65	134.27
14	8.400e+02	1.792e−02	9.632e−03	144.25	150.87
15	9.000e+02	1.747e−02	9.391e−03	148.35	154.97
16	9.600e+02	1.707e−02	9.175e−03	152.34	158.96
17	1.020e+03	5.327e−02	2.863e−02	123.06	129.69
18	1.080e+03	1.631e−02	8.768e−03	160.24	166.87
19	1.140e+03	3.114e−02	1.674e−02	129.88	136.51
20	1.200e+03	1.546e−02	8.307e−03	167.84	174.46
21	1.260e+03	1.500e−02	8.060e−03	172.05	178.68
22	1.320e+03	1.449e−02	7.787e−03	176.25	182.88
23	1.380e+03	3.139e−02	1.687e−02	114.23	120.86
24	1.440e+03	1.345e−02	7.231e−03	−176.33	−169.70
25	1.500e+03	3.959e−02	2.128e−02	−135.27	−128.64
26	1.560e+03	1.254e−02	6.740e−03	−167.66	−161.04
27	1.620e+03	1.198e−02	6.438e−03	−163.46	−156.84
28	1.680e+03	1.147e−02	6.163e−03	−159.12	−152.50
29	1.740e+03	1.778e−02	9.554e−03	−165.54	−158.92

THD: 35.965743%(34.509790%)

5.6.5 Three-phase PWM inverter circuit model

$R = 10\ \Omega$, $C = 72.3$ μF and $L = 7.2$ mH, $f = 60$ Hz

$$\omega_o = 2\pi f_o = 2\pi \times 60 = 377 \text{ rad/s}$$

The nth harmonic impedance of the load with the LC-filter is given by

$$Z(n) = jn\omega_o L + \frac{R \| \frac{1}{jn\omega_o C}}{1 + jn\omega_o RC} = \frac{R + jn\omega_o L - (n\omega_o)RLC}{1 + jn\omega_o RC} \tag{5.81}$$

Dividing the voltage expression in (5.75) by the impedance $Z(n)$ gives the Fourier components of the phase load currents as

$$i_a(t) = \sum_{n=1,3\ldots}^{\infty} \frac{4V_S}{n\pi\sqrt{3}Z(n)} \sin\left(\frac{n\delta_m}{2}\right)\left[\sin n\left(\alpha_m + \frac{\delta_m}{2}\right)\right] \tag{5.82}$$

which gives the peak magnitude of the nth harmonic currents of the phase current

$$I_m(n) = \sum_{n=1,3\ldots}^{\infty} \frac{4V_S}{n\pi\sqrt{3}Z(n)} \sin\left(\frac{n\delta_m}{2}\right)\left[\sin n\left(\alpha_m + \frac{\delta_m}{2}\right)\right] \tag{5.83}$$

which for $n = 1$, $n = 5$, $n = 7$, $n = 11$, and $n = 13$ gives

$$I_m(1) = 4.277\angle - 12.378^\circ$$
$$I_m(5) = 0.295\angle 92.088^\circ$$
$$I_m(7) = 0.187\angle 90.761^\circ$$
$$I_m(11) = 0.342\angle - 89.804^\circ$$
$$I_m(13) = 0.243\angle 90.119^\circ$$

Dividing the peak value by $\sqrt{2}$ gives the rms values for $n = 1$, $n = 5$, $n = 7$, $n = 11$, and $n = 13$ as

$$I_{rms}(1) = \frac{I_m(1)}{\sqrt{2}} = \frac{4.277\angle - 12.378^\circ}{\sqrt{2}} = 3.024\angle - 12.378^\circ$$

$$I_{rms}(5) = \frac{I_m(5)}{\sqrt{2}} = \frac{0.295\angle 92.088^\circ}{\sqrt{2}} = 0.209\angle 92.088^\circ$$

$$I_{rms}(7) = \frac{I_m(7)}{\sqrt{2}} = \frac{0.187\angle 90.761^\circ}{\sqrt{2}} = 0.132\angle 90.761^\circ$$

$$I_{rms}(11) = \frac{I_m(11)}{\sqrt{2}} = \frac{0.342\angle - 89.804^\circ}{\sqrt{2}} = 0.242\angle - 89.804^\circ$$

$$I_{rms}(13) = \frac{I_m(13)}{\sqrt{2}} = \frac{0.243\angle 90.119^\circ}{\sqrt{2}} = 0.172\angle 90.119^\circ$$

The total rms value of the phase load current can be determined by adding the square of the individual rms values and then taking the square root

$$I_o = \sqrt{\sum_{n=1,3}^{30} (I_{rms}(n))^2} \tag{5.84}$$
$$= 3.049 \text{ A}$$

Applying (3.3) for a single-phase full-wave rectifier, the average output current $I_{o(av)}$ of a full-wave rectifier with a peak sinusoidal input current of I_m is

$$I_{o(av)} = \frac{2I_m}{\pi} \tag{5.85}$$

Since the output current of the inverter contains sinusoidal harmonic components at different frequencies, we can add the corresponding DC components contributing to the output current of the inverter. Applying (5.85) for three phases gives the average input current due by n number of harmonic components, $n = 30$

$$\begin{aligned} I_{s(av)} &= \sum_{n=1}^{30} \frac{3 \times 2 \times |I_m(n)|}{\pi} \\ &= 3.217 \text{ A} \end{aligned} \tag{5.86}$$

5.7 Three-phase sinusoidal PWM inverter

The three-phase inverter in Figure 5.7(a) has a positive voltage of V_S for 1/3 of the period and a negative voltage V_S of $-V_S$ for another 1/3 of the period. A PWM three-inverter produces multiple positive pulses during 1/3 of the period, the same number of negative pulses during another 1/3 of the period, and zero voltage for another 1/3 of the period. By varying the width of the output pulses, we can vary the rms value of the output voltage. The number of pulses per half-cycle determines the frequency of the lower-order harmonic. Higher the frequency of the lower-order harmonics, lower the values of the filter elements, L and C. An isosceles triangular carrier wave v_{cr1} at a switching frequency is compared with a pulse reference sine wave v_{ref} of 4 V peak at the output frequency. Although the comparator generates pulses during the complete period T of the output frequency, the positive pulses of $+V_S$ appear on the output of the inverter for only an interval of $T/3$ and the negative pulses of $-V_S$ appear on the output of the inverter for another interval of $T/3$.

If p_c is the number of gating pulses generated by the comparator during the first half-cycle of the output voltage, the number of pulses p_o appearing on the output of the inverter is

$$p_o = \frac{120^\circ}{180^\circ} \times p_r = \frac{2}{3} p_r \tag{5.87}$$

which gives

$$p_r = \frac{180^\circ}{120^\circ} \times p_o = \frac{3}{2} p_o \tag{5.88}$$

For examples, for $p_o = 4$, $p_r = 6$ and for $p_o = 6$, $p_r = 9$.

For an output frequency of $f = 60$ Hz, and nine pulses per half cycle, the frequency of the carrier waveform is $9f$ (540 Hz). Figure 5.13(a) shows the generation of the gating signals during the first half cycle of the reference wave. The

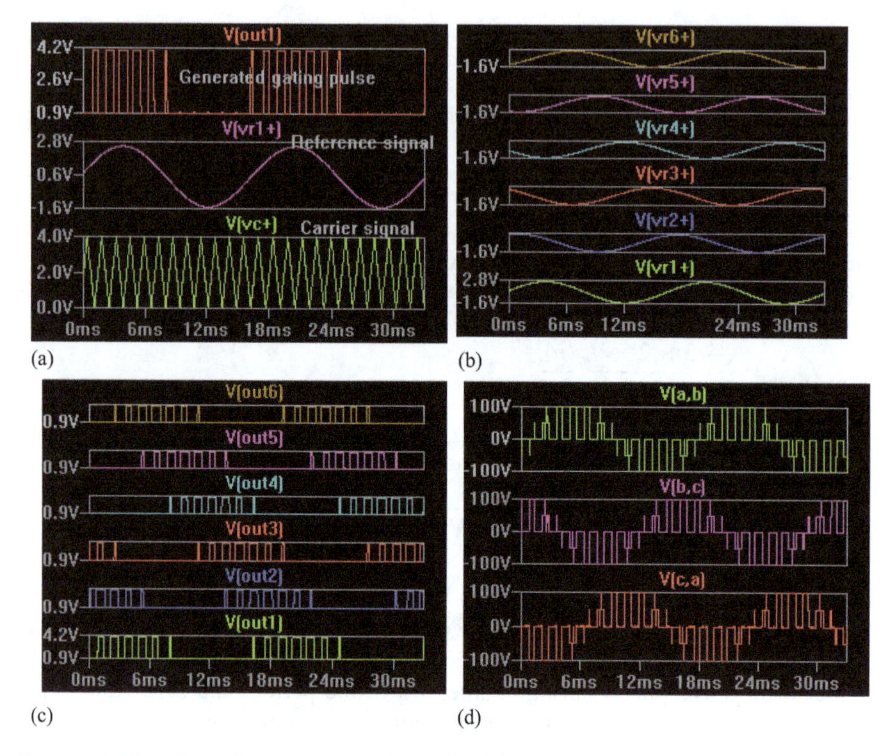

Figure 5.13 Waveforms of SPWM inverters: (a) gating, reference, and carrier signals, (b) reference signals delayed by 60°, (c) six gating signals, and (d) line-to-line output voltages

reference sinewave signals for generating gating signals for other switches as shown in Figure 5.13(b) are obtained by the pulse waveforms delayed by 60° (2.778 ms) and the pulse description is changed to the statement as follows.

PULSE(0 {amp} 2.778 ms 0.1 ns 0.1 ns 8.33 ms 16.67 ms)

Figure 5.13(c) shows the six gating signals, and Figure 5.13(d) shows the line-to-line output voltages.

The ratio of the reference signal A_{ref} to the peak carrier signal A_{cr} is defined as the modulation index M, $M = A_{ref}/A_{cr}$. As the reference signal is varied keeping the carrier signal fixed, the modulation index M and the width of the kth pulse δ are also varied. If δ_k is the width of the kth pulse, the rms value of the output voltage for p pulses per cycle, the rms value of the output voltage per phase is given by

$$V_o = \sqrt{\frac{2p_o}{2\pi}\int_0^{\delta_k} V_S^2 d\theta} = V_S\sqrt{\sum_{m=1}^{2p-1}\frac{\omega\delta_m}{\pi}} \tag{5.89}$$

where p_o is the number of pulses per half-cycle of the output voltage, δ_m is the pulse width of the mth pulse in rad, and ω is the angular frequency of the output voltage rad/s.

The variation of the modulation index M from 0 to 1 varies the pulse width δ from 0 to $T/2p$ (0 to π/p) and the rms output voltage from 0 to V_S. The mth time t_m and the angle α_m of the intersection of the reference signal and the carrier signal can be determined from [7]

$$t_m = \frac{\alpha_m}{\omega} = t_x + m\frac{T_s}{2} + \frac{T}{12} \quad \text{for} \quad m = 1, 2, 3, \ldots, p \tag{5.90}$$

where t_x can be solved from

$$1 - \frac{2t}{T_s} = M \sin\left(\omega\left(t_x + \frac{mT_s}{2} - \frac{T_s}{2}\right)\right) \quad \text{for} \quad m = 1, 3, \ldots, p \tag{5.91}$$

$$\frac{2t}{T_s} = M \sin\left(\omega\left(t_x + \frac{mT_s}{2} - \frac{T_s}{2}\right)\right) \quad \text{for} \quad m = 2, 4, \ldots, p \tag{5.92}$$

where $\omega = 2\pi f$ is the angular frequency of the output voltage, rad/s; $f_s = 2pf$ is the frequency of the carrier signal, Hz; and $T_s = 1/f_s$ is the period of the carrier signal, s.

Notes:

The reference sine wave in (5.91) and (5.92) is delayed by $T_s/2$ to match the intersection with carrier signal. Otherwise, the generated pulses may not be symmetrical.

Since the pulses for the output voltage is delayed by 30 from the zero, $T/12$ is added to (5.90).

By using the iterative method of solving non-linear equations (5.90), (5.91), and (5.92) in the Mathcad software tool, the duration of the mth pulse can be calculated as

$$\begin{aligned} d_m &= t_{m+1} - t_m \quad \text{for} \quad m = 1, 2, 3, 4, \ldots, 2p, \ \text{s} \\ \delta_m &= d_m\omega = (t_{m+1} - t_m)\omega \quad \text{for} \quad m = 1, 2, 3, 4, \ldots, 2p, \ \text{rad} \end{aligned} \tag{5.93}$$

Due to the symmetry of the output voltage, the A_n component of the Fourier series will be zero and we get the general form of the output voltage as

$$v_o(t) = \sum_{n=1,3,5,\ldots}^{\infty} B_n \sin(n\omega t) \tag{5.94}$$

If the positive pulse of the mth pair starts at $\omega t = \alpha_m$ and ends $\omega t = \alpha_m + \delta_m$, the Fourier co-efficient for a pair of pulses is

$$b_n = \frac{2V_S}{\pi}\int_{\alpha_m}^{\alpha_m+\delta_m} \sin(n\omega t)d(\omega t) = \frac{4V_S}{n\pi} \sin\left(\frac{n\delta_m}{2}\right)\left[\sin n\left(\alpha_m + \frac{\delta_m}{2}\right)\right] \tag{5.95}$$

We can find the co-efficient B_n by adding the effects of all pulses as given by

$$B_n = \sum_{m=1}^{2p-1} \frac{4V_S}{n\pi} \sin\left(\frac{n\delta_m}{2}\right)\left[\sin n\left(\alpha_m + \frac{\delta_m}{2}\right)\right] \tag{5.96}$$

The performance parameters of inverters can be divided into three types.

- Output-side parameters,
- Input-side parameters, and
- Inverter parameters.

Let us consider a DC input voltage of $V_S = 100$ V. The output frequency is $f = 60$ Hz and a load resistance of $R = 10\ \Omega$, $M = 0.5$, $p_o = 4$, and $p_r = 6$.

The period of the output voltage, $T = 1/f = 1/60 = 16.667$ ms.

The angular frequency of the output voltage, $\omega = 2\pi f = 2 \times \pi \times 60 = 367$ rad/s.

The switching frequency is $f_s = 2p_r f = 2 \times 6 \times 60 = 720 Hz$.

The switching period is $T_s = 1/f_s = 1/(720) = 1.389$ ms.

From (5.90) to (5.93), we can calculate these values of the angles α and the pulse widths δ as $\alpha_1 = 58.277°,\ \alpha_2 = 85.163°,\ \alpha_3 = 113.044°,\ \alpha_4 = 142.565°$ and $\delta_1 = 3.943°,\ \delta_2 = 10.65°,\ \delta_3 = 14.391°,\ \delta_4 = 14.391°$.

Substituting these values in (5.96), we get the peak value of the fundamental component for $n = 1$ as

$$\begin{aligned} B_1 &= \sum_{m=1}^{2p_o} \frac{4V_S}{\pi} \sin\left(\frac{\delta_m}{2}\right)\left[\sin n\left(\alpha_m + \frac{\delta_m}{2}\right)\right] \\ &= 0.37428 V_S \end{aligned} \tag{5.97}$$

which dividing by $\sqrt{2}$ gives the rms fundamental output voltage as

$$\begin{aligned} V_{o1} &= \frac{B_1}{\sqrt{2}} \\ &= \frac{0.37428 V_S}{\sqrt{2}} = 0.26466 V_S \end{aligned} \tag{5.98}$$

5.7.1 *Output-side parameters*

(a) Equation (5.89) gives the rms output voltage is

$$\begin{aligned} V_o &= V_S \sqrt{\sum_{m=1}^{2p-1} \frac{\omega \delta_m}{\pi}} \\ &= 49.089 \text{ V} \end{aligned}$$

(b) The rms output current is

$$I_o = \frac{V_o}{R} = \frac{49.089}{10} = 4.9089 \text{ A}$$

(c) The output power is

$$P_{out} = \frac{V_o^2}{R} = \frac{49.089^2}{10} = 240.971 \text{ W}$$

(d) Equation (5.98) gives the rms fundamental output voltage

$$V_{o1} = 0.26466V_S = 0.26466 \times 100 = 26.466 \text{ V}$$

(e) The rms fundamental output current is

$$I_{o1} = \frac{V_{o1}}{R} = \frac{26.466}{10} = 2.6466 \text{ A}$$

(f) The fundamental output power is

$$P_{o1} = \frac{V_{o1}^2}{R} = \frac{26.466^2}{10} = 70.044 \text{ W}$$

(g) The rms ripple content of the output voltage

$$\begin{aligned} V_{ripple} &= \sqrt{V_o^2 - V_{o1}^2} \\ &= \sqrt{49.089^2 - 26.466^2} = 41.343 \text{ V} \end{aligned}$$

(h) The THD of the output voltage

$$\text{THD} = \frac{V_{ripple}}{V_{o1}} = \frac{41.343}{26.466} = 156.214\%$$

5.7.2 Input-side parameters

(a) The input power

$$\begin{aligned} P_{in} &= P_{out} \\ &= 240.971 \text{ W} \end{aligned}$$

(b) The average input current is

$$\begin{aligned} I_{s(av)} &= \frac{P_{in}}{V_S} = \frac{240.971 \text{ W}}{100} \\ &= 2.40971 \text{ A} \end{aligned}$$

(c) The rms input current is

$$\begin{aligned} I_{s(rms)} &= I_o \\ &= 4.9089 \text{ A} \end{aligned}$$

5.7.3 Inverter parameters

(a) The peak transistor current

$$I_{peak} = \frac{V_S}{R} = \frac{100}{10} = 10 \text{ A}$$

(b) The rms transistor current is

$$I_{T(rms)} = \frac{I_o}{\sqrt{3}} = \frac{4.9089}{\sqrt{3}} = 2.834 \text{ A}$$

(c) The average transistor current is

$$I_{T(av)} = \frac{I_{s(av)}}{3} = \frac{2.40971}{3} = 0.803 \text{ A}$$

5.7.4 Output filter

An LC-filter as shown in Figure 5.1(a) is often connected to the output side to limit the ripple contents of the load resistor, R. The capacitor C provided a low impedance path for the ripple currents to flow through it and the inductor L provides a high impedance for the ripple currents. The fundamental frequency of output voltage ripples is the same as the output frequency f and the lower order for 1 = 11 is $11f$. Equation (5.96) gives the rms value of the 11th harmonic component for $n = 11$ as

$$\begin{aligned} B_{11} &= \sum_{m=1}^{2p_o} \frac{4V_S}{11 \times \pi} \sin\left(\frac{11\delta_m}{2}\right)\left[\sin 11\left(\alpha_m + \frac{\delta_m}{2}\right)\right] \\ &= 0.28753 V_S \end{aligned} \tag{5.99}$$

which gives the rms value of the 11th harmonic component

$$\begin{aligned} V_{o11} &= \frac{B_{11}}{\sqrt{2}} = \frac{0.28753 V_S}{\sqrt{2}} \\ &= \frac{0.28753 \times 100}{\sqrt{2}} = 20.332 \text{ V} \end{aligned}$$

There are no explicit equations to find the values of L and C. We can find approximate values and then adjust the values to obtain the desired values. Neglecting the effect of capacitor C, the 11th harmonic voltage will appear across the inductor L and the load resistance R. Assuming the drop across the inductor is larger than that of the resistor R by a factor of 3 such that.

$$n\omega L = 3R$$

which gives

$$\begin{aligned} L &= \frac{3R}{n\omega} \\ &= \frac{3 \times 10}{11 \times 366} = 7.2 \text{ mH} \end{aligned} \tag{5.100}$$

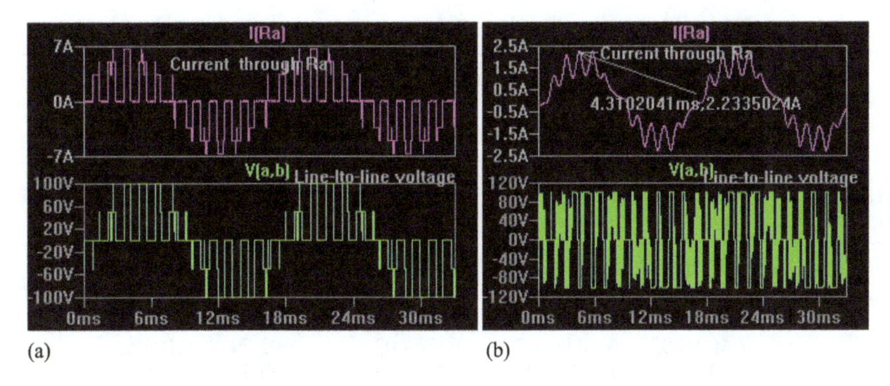

(a) (b)

Figure 5.14 Output voltage and load current

Assuming that the impedance of the capacitor is smaller than the load resistor by a factor of 3 such that most of the harmonic current flows through the capacitor. That is

$$\frac{3}{n\omega C} = R$$

which gives

$$\begin{aligned} C &= \frac{3}{n\omega R} = \frac{3}{3\omega R} \\ &= \frac{3}{11 \times 366 \times 10} = 72.3\ \mu\text{F} \end{aligned} \tag{5.101}$$

The LTspice lots of the output voltage and the current through the load resistor R are shown in Figure 5.14(a) for filter values of C = 72.3 pF, L = 7.2 nH, and in Figure 5.14(b) for filter values of C = 72.3 µF, L = 7.2 mH. We can notice that the output voltage is a square wave, whereas the load current is sinusoidal form. The values of L and C can be adjusted to get a wave form closer to a pure sine wave.

For filter values: C = 72.3 pF, L = 7.2 nH(b). For filter values: C = 72.3 µF, L = 7.2 mH

Fourier analysis: The LTspice command (.four 60 Hz 29 2 $I(R)$) for Fourier Analysis gives

THD: 98.887769% (108.045099%) for C = 72.3 pF and L = 7.2 nH
THD: 26.158940% (23.860521%) for C = 72.3 µF and L = 7.2 mH

which indicates a significant reduction of the THD with an output filter. LTspice gives the Fourier components as follows:

Fourier components of *I(ra)*					
DC component: 0.00405413					
Harmonic number	**Frequency (Hz)**	**Fourier component**	**Normalized component**	**Phase (degrees)**	**Normalized phase (degrees)**
1	6.000e+01	1.735e+00	1.000e+00	−10.76	0.00
2	1.200e+02	7.497e−03	4.320e−03	88.14	98.90
3	1.800e+02	7.438e−03	4.286e−03	99.82	110.57
4	2.400e+02	6.304e−03	3.633e−03	92.79	103.55
5	3.000e+02	1.132e−01	6.526e−02	163.38	174.13
6	3.600e+02	5.374e−03	3.097e−03	118.81	129.57
7	4.200e+02	5.080e−02	2.928e−02	−134.45	−123.70
8	4.800e+02	5.855e−03	3.374e−03	129.42	140.17
9	5.400e+02	4.908e−03	2.829e−03	130.07	140.83
10	6.000e+02	6.098e−03	3.514e−03	134.23	144.98
11	6.600e+02	3.393e−01	1.955e−01	115.61	126.37
12	7.200e+02	7.049e−03	4.062e−03	133.39	144.15
13	7.800e+02	2.697e−01	1.554e−01	−60.05	−49.30
14	8.400e+02	5.795e−03	3.340e−03	140.58	151.33
15	9.000e+02	6.707e−03	3.865e−03	142.37	153.13
16	9.600e+02	5.803e−03	3.344e−03	152.33	163.08
17	1.020e+03	8.654e−03	4.987e−03	114.11	124.87
18	1.080e+03	5.429e−03	3.128e−03	156.97	167.72
19	1.140e+03	2.911e−02	1.678e−02	153.28	164.04
20	1.200e+03	4.623e−03	2.664e−03	163.92	174.68
21	1.260e+03	4.285e−03	2.469e−03	−178.55	−167.79
22	1.320e+03	4.102e−03	2.364e−03	172.71	183.47
23	1.380e+03	2.267e−02	1.307e−02	69.93	80.68
24	1.440e+03	3.978e−03	2.293e−03	−169.61	−158.86
25	1.500e+03	2.756e−02	1.588e−02	−126.34	−115.58
26	1.560e+03	4.691e−03	2.703e−03	−160.30	−149.54
27	1.620e+03	4.040e−03	2.328e−03	−168.82	−158.06
28	1.680e+03	4.473e−03	2.578e−03	−160.02	−149.27
29	1.740e+03	4.213e−03	2.428e−03	26.29	37.05
THD: 26.158940% (23.860521%)					

5.7.5 *Three-phase sinusoidal PWM inverter circuit model*

$R = 10\ \Omega$, $C = 72.3\ \mu\text{F}$, and $L = 7.2$ mH, $f = 60$ Hz

$$\omega_o = 2\pi f_o = 2\pi \times 60 = 377\ \text{rad/s}$$

The nth harmonic impedance of the load with the LC-filter is given by

$$Z(n) = jn\omega_o L + \frac{R \| \frac{1}{jn\omega_o C}}{1 + jn\omega_o RC} = \frac{R + jn\omega_o L - (n\omega_o)RLC}{1 + jn\omega_o RC} \tag{5.102}$$

Dividing the voltage expression in (5.96) by the impedance $Z(n)$ and by $\sqrt{3}$ for converting to phase current gives the Fourier components of the phase load currents as

$$i_a(t) = \sum_{n=1,3\ldots}^{\infty} \frac{4V_S}{n\pi\sqrt{3}Z(n)} \sin\left(\frac{n\delta_m}{2}\right)\left[\sin n\left(\alpha_m + \frac{\delta_m}{2}\right)\right] \tag{5.103}$$

which gives the peak magnitude of the nth harmonic currents of the phase current

$$I_m(n) = \sum_{n=1,3\ldots}^{\infty} \frac{4V_S}{n\pi\sqrt{3}Z(n)} \sin\left(\frac{n\delta_m}{2}\right)\left[\sin n\left(\alpha_m + \frac{\delta_m}{2}\right)\right] \tag{5.104}$$

which for $n = 1$, $n = 5$, $n = 7$, $n = 11$, and $n = 13$ gives

$$I_m(1) = 2.879\angle - 12.378^\circ$$
$$I_m(5) = 0.038\angle 87.912^\circ$$
$$I_m(7) = 0.038\angle 90.761^\circ$$
$$I_m(11) = 0.256\angle 90.196^\circ$$
$$I_m(13) = 0.198\angle - 89.881^\circ$$

Dividing the peak value by $\sqrt{2}$ gives the rms values for $n = 1$, $n = 5$, $n = 7$, $n = 11$, and $n = 13$ as

$$I_{rms}(1) = \frac{I_m(1)}{\sqrt{2}} = \frac{2.879\angle - 12.378^\circ}{\sqrt{2}} = 2.035\angle - 12.378^\circ$$

$$I_{rms}(5) = \frac{I_m(5)}{\sqrt{2}} = \frac{0.038\angle 87.912^\circ}{\sqrt{2}} = 0.027\angle 87.912^\circ$$

$$I_{rms}(7) = \frac{I_m(7)}{\sqrt{2}} = \frac{0.038\angle 90.761^\circ}{\sqrt{2}} = 0.027\angle 90.761^\circ$$

$$I_{rms}(11) = \frac{I_m(11)}{\sqrt{2}} = \frac{0.256\angle 90.196^\circ}{\sqrt{2}} = 0.181\angle 90.196^\circ$$

$$I_{rms}(13) = \frac{I_m(13)}{\sqrt{2}} = \frac{0.198\angle - 89.881^\circ}{\sqrt{2}} = 0.14\angle - 89.881^\circ$$

Applying (3.3) for a single-phase full-wave rectifier, the total rms value of the phase load current can be determined by adding the square of the individual rms values and then taking the square root

$$I_o = \sqrt{\sum_{n=1,3}^{30} (I_{rms}(n))^2} \tag{5.105}$$
$$= 2.377 \text{ A}$$

The average output current $I_{o(av)}$ of a full-wave rectifier with a peak sinusoidal input current of I_m is

$$I_{o(av)} = \frac{2I_m}{\pi} \tag{5.106}$$

Since the output current of the inverter contains sinusoidal harmonic components at different frequencies, we can add the corresponding DC components contributing to the output current of the inverter. Applying (5.106) for three phases gives the average input current due by n number of harmonic components, $n = 30$

$$\begin{aligned} I_{s(av)} &= \sum_{n=1}^{30} \frac{3 \times 2 \times |I_m(n)|}{\pi} \\ &= 3.349 \text{ A} \end{aligned} \tag{5.107}$$

5.8 Summary

The DC–AC converters normally use controlled switches. A single-phase inverter consists of four switches forming a bridge. One top switch from left arm of the bridge and one bottom switch from the other arm are turned on to connect Dc source producing a positive voltage of V_S for the first half the period. One bottom switch from left arm of the bridge and one top switch from the other arm are turned on to connect the DC source producing a negative voltage V_S of $-V_S$ for another half the period. The rms output voltage has a fixed value at the switching frequency. A PWM inverter produces multiple positive pulses during the first half-cycle and the same number of negative pulses during the other half cycle. By varying the width of the output pulses, we can vary the out rms value of the output voltage. The number of pulses per half-cycle determines the frequency of the lower-order harmonic.

The PWM gating signals are generated by comparing a triangular carrier signal with a pulse signal of 50% duty cycle at the output frequency. The number of PWM gate pulses depend on the frequency of the triangular carrier frequency. The SPWM gating signals are generated by comparing a triangular carrier signal with a sine wave at the output frequency of the inverter output voltage. A three-phase inverter has three switches in the top row and three switches in the bottom row. To produce an output voltage between the output terminals, one switch from the top and one switch from the bottom must be conducted. Since there are six switches for a three-phase inverter, there would be six switch pairs – 12, 23, 34, 45, 56, and 61 thereby completing one cycle and then then repeating the cycle. First gating pulse starts with zero delay and then each pulse by delayed by 60°. The higher the number of pulses, the higher the frequency of the ripple components thereby reducing the sizes of filter requirements. However, to minimize the switching losses of the switching devices, the number of pulses are normally limited to 10.

References

[1] Strollo, A.G.M., A new IGBT circuit model for SPICE simulation, *Power Electronics Specialists Conference*, June 1997, Vol. 1, pp. 133–138.

[2] Sheng, K., Finney, S.J., and Williams, B.W., A new analytical IGBT model with improved electrical characteristics, *IEEE Transactions on Power Electronics*, 14(1), 1999, 98–107.

[3] Sheng, K., Williams, B.W., and Finney, S.J., A review of IGBT models, *IEEE Transactions on Power Electronics*, 15(6), 2000, 1250–1266.

[4] Infineon-IKW30N65EL5-DataSheet-v02_02-EN. https://www.infineon.com/dgdl/Infineon-IGW30N65L5-DataSheet-v02_02-EN.pdf?fileId=5546d4624b0b249c014b11cd55583ac9

[5] *IGBT Spice Model*, https://www.globalspec.com/industrial-directory/igbt_spice_model

[6] *Discrete IGBT P-Spice Models*, https://www.fujielectric.com/products/semiconductor/model/igbt/technical/design_tool.html

[7] Rashid, M.H. *Power Electronics: Circuits, Devices, and Applications*, 4th ed. Englewood Cliffs, NJ: Prentice-Hall, 2014, Chapter 5.

[8] M. H. Rashid, *SPICE and LTspice for Power Electronics and Electric Power*. Boca Raton, FL: CRC Press, 2024.

Chapter 6
Resonant pulse inverters

6.1 Introduction

The input to a resonant-pulse inverter is a fixed direct current (DC) voltage or a DC current source, and the output is a voltage or a current of resonant pulse. Power semiconductor devices perform the switching action, and the desired output is obtained by varying their turn-on and turn-off times. The commonly used devices are bipolar transistors (BJTs), metal–oxide–semiconductor field-effect transistor (MOSFETs), or insulated gate bipolar transistors (IGBTs). Thus, the voltage and the current on the input side are of DC types, and the voltage and current on the output side are of alternating current (AC) types. We will use power MOSFETs, for example, as switches to obtain the voltage and the current waveforms of the inverters of the following types:

half-bridge resonant pulse inverter,
half-bridge bidirectional resonant pulse inverter,
full-bridge bidirectional resonant pulse inverter,
parallel resonant pulse inverter, and
class-e resonant pulse inverter.

6.2 Resonant circuits

We have observed in Chapter 2 that when a DC voltage is switched on an LC circuit, a sine-wave current through the circuit oscillates at a resonant frequency $\omega_o = 1/\sqrt{LC}$ rad/s. The current *i(t)* is given by

$$i(t) = V_S\sqrt{\frac{C}{L}}\sin(\omega_n t) \tag{6.1}$$

where V_S is the DC input voltage; C is the capacitance, F; and L is the inductance.

A resonant inverter forms an RLC circuit with the addition of a small resistance R as a load resistance to an LC circuit. The value of R is such that it forms an under-damped RLC circuit so that the load current is closer to a sine-wave of Equation (3.39) as derived in Chapter 2 as follows:

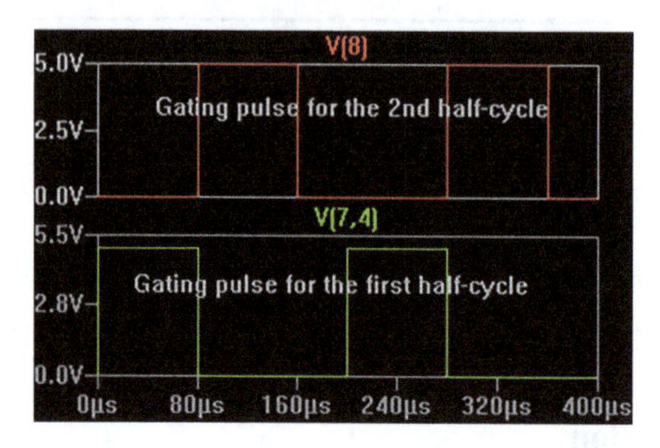

Figure 6.1 Gating pulses for resonant pulse inverters

$$i(t) = \frac{V_S}{L} e^{-\alpha t} \sin \omega_n t \tag{6.2}$$

where $\alpha = \frac{R}{2L}$ is the damping factor; $\omega_n = \frac{1}{\sqrt{LC}}$ is the undamped resonant frequency, rad/s.

One or more switches of resonant inverters are turned on during the first cycle of the output waveform to initiate resonant oscillations in an underdamped RLC circuit. The switches are kept turned-on and maintained the on-state condition to complete the oscillations, producing a positive load current. One or more switches are turned-on to allow the resonant current to flow in the opposite direction. Figure 6.1 shows the gating pulses. The output waveform depends on the circuit parameters and the input source. The on-time and switching frequency of power devices must match the resonant frequency of the circuit.

6.3 Half-bridge resonant pulse inverter

The LTspice schematic of a half-bridge resonant pulse inverter is shown in Figure 6.2(a). The plots of the gating pulses, the output voltage, and the load current are shown in Figure 6.2(b). The performance parameters of the resonant inverter can be divided into three types [1]:

- Inverter parameters,
- Output-side parameters, and
- Input-side parameters.

Let us consider a DC input voltage of $V_S = 100$ V, $L = L_1 = L_2 = 50$ µH, $C_1 = C_2 = 4$ µF, and $R = 10\ \Omega$. The inverter is operating at an output frequency of $f_o = 7$ kHz, a switching period of $T_o = 1/f_o = 143$ µs, and a duty cycle of $k = 50\%$. Thus, the on-time of the switch is $t_1 = kT = 0.5 \times 143$ µs $= 71$ µs, and the off-time of the switch is $t_2 = (1 - k)T = (1 - 0.5) \times 143$ µs $= 71$ µs. Allowing a dead time of $t_d = 6$ µs, the delay time of the gating switch M_2 is $t_{delay} = t_2 - t_d = 71$ µs $- 6$ µs $= 65$ µs.

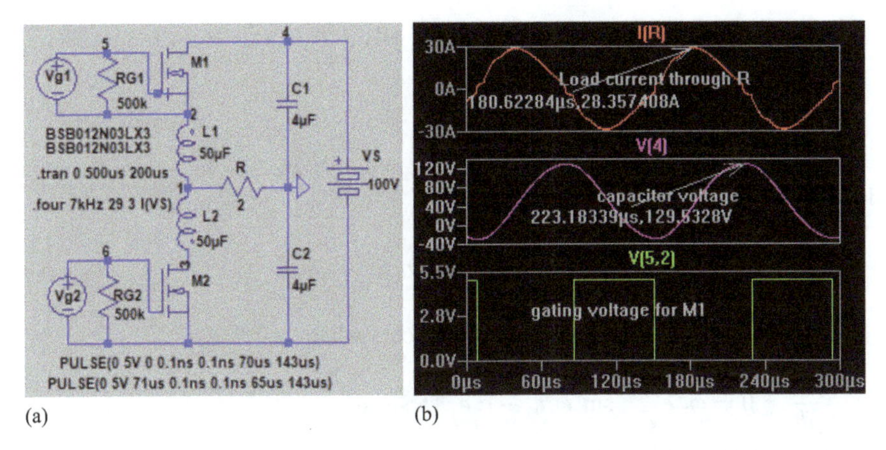

(a) (b)

Figure 6.2 Half-bridge resonant-pulse inverter: (a) schematic and (b) waveforms

The LTspice gating pulse: PULSE (0 5V 0 0.1 ns 0.1 ns 65 µs 143 µs)

Depending on whether the switch is turned on or off, there are two modes of operation. During mode 1, the switch M_1 is turned on, and during mode 2, the switch M_1 is turned off.

Mode 1: When the switch M_1 is turned on, the DC supply voltage V_S is connected to the load consisting of *R*, *L*, and *C*. Capacitor C_1 forms a parallel circuit with capacitor C_2 through the DC supply source V_S and the effective capacitor $C = C_1 + C_2$. The resonant current *i*(*t*) of an *RLC* circuit during mode 1 can be described by

$$L\frac{di_1}{dt} + Ri_1 + \frac{1}{C}\int i_1 \ dt + v_{C1}(t = 0) = V_S \tag{6.3}$$

Solving for $i_1(t)$ with an initial capacitor initial voltage $v_C(t = 0) = V_C$ gives the resonant current of an *RLC* circuit as [1]

$$\begin{aligned} i_1(t) &= A_1 e^{-tR/2L} \sin \omega_r t \\ &= \frac{V_S + V_c}{\omega_r L} e^{-at} \sin \omega_r t \end{aligned} \tag{6.4}$$

where ω_r is the damped resonant frequency

$$\omega_r = \sqrt{\frac{1}{LC} - \frac{R^2}{4L^2}} \tag{6.5}$$

And the damping factor *r*

$$\alpha = \frac{R}{2L} \tag{6.6}$$

The circuit must be underdamped such that,

$$\frac{R^2}{4L^2} \gg \frac{1}{LC} \quad \text{or} \quad R \gg \frac{4L}{C} \tag{6.7}$$

Let us define a parameter ξ, called *damping ratio*, as the ratio of actual damping,

$$\xi = \frac{\textit{damping factor}}{\textit{natural resonant frequency}} = \frac{\alpha}{\omega_n} \textit{ or } \alpha = \xi\omega_n \tag{6.8}$$

The peak of the current $i_1(t)$ in (6.4) occurs when

$$\frac{di_1}{dt} = 0 = \omega_r\, e^{-\alpha t_m} \sin\, \omega_r t_m - \alpha\, e^{-\alpha t_m} \sin\, \omega_r t_m \tag{6.9}$$

which gives the time t_m when the peak occurs as

$$t_m = \frac{1}{\omega_r} \tan^{-1}\frac{\omega_r}{\alpha} \tag{6.10}$$

The instantaneous capacitor voltage during mode 1 can be found from

$$\begin{aligned} v_{C1}(t) &= \frac{1}{C}\int_0^t i_1(t)dt - V_c \\ &= -\frac{V_S + V_c}{\omega_r}\; e^{-\alpha t}(\alpha \sin\, \omega_r t + \omega_r \cos\, \omega_r t) + V_S \end{aligned} \tag{6.11}$$

When V_c is the initial capacitor voltage at the beginning of mode 1.
At the end of mode 2 at $t = t_{1m}$, the resonant current falls to zero

$$i_1(t = t_{1m}) = 0 \tag{6.12}$$

And the capacitor voltage becomes

$$\begin{aligned} v_{C1}(t = t_{1m}) &= V_{c1} \\ &= (V_S + V_c)\, e^{-\alpha\pi/\omega_r} + V_S \end{aligned} \tag{6.13}$$

Mode 2: When the switch M_1 is turned off, the DC supply voltage V_S is disconnected and the switch M_2 is turned on. The capacitor with its initial voltage at the end of mode 1 forces the current flow, and the capacitor voltage is reversed. The resonant current $i_2(t)$ during mode 2 can be described by

$$L\frac{di_2}{dt} + Ri_2 + \frac{1}{C}\int i_2\; dt + v_{C2}(t = 0) = 0 \tag{6.14}$$

Solving for $i_2(t)$ with an initial capacitor initial voltage $v_{C2}(t = 0) = -V_{c1}$ gives the resonant current of an *RLC* circuit as [1]

$$i_2(t) = \frac{V_{c1}}{\omega_r L} e^{-\alpha t} \sin\, \omega_r t \tag{6.15}$$

which gives the instantaneous capacitor voltage during mode 2 as

$$v_{C2}(t) = \frac{1}{C}\int_0^t i_2(t)dt - V_{c1} = \frac{-V_{c1}}{\omega_r}\; e^{-at}(\alpha \sin \omega_r t + \omega_r \cos \omega_r t) + V_S \tag{6.16}$$

At the end of mode 2 at $t = t_{2m}$, the resonant current falls to zero

$$i_2(t = t_{2m}) = 0 \tag{6.17}$$

And the capacitor voltage becomes

$$\begin{aligned} v_{C2}(t = t_{2m}) &= V_{c2} \\ &= V_c\; e^{-\alpha\pi/\omega_r} \end{aligned} \tag{6.18}$$

The cycle then repeats and $v_{C2}(t = t_{2m}) = v_C(t = 0)$.
Equating (6.18)–(6.13) and solving for V_c and V_{c1}, we obtain

$$V_c = V_s \frac{1 + e^{-z}}{e^z - e^{-z}} = V_s \frac{e^z + 1}{e^{2z} - 1} = \frac{V_s}{e^z - 1} \tag{6.19}$$

$$V_{c1} = V_s \frac{1 + e^z}{e^z - e^{-z}} = V_s \frac{e^z(1 + e^z)}{e^{2z} - 1} = \frac{V_s\; e^z}{e^z - 1} \tag{6.20}$$

where

$$z = \frac{\alpha\pi}{\omega} \tag{6.21}$$

$$V_s + V_c = V_{c1} \tag{6.22}$$

With a higher switching frequency, there may not be enough time for the resonant oscillation to complete the cycle and the peak current is lower. The resonant current may also be discontinuous. The load current $i_1(t)$ for mode 1 must be zero before switch M_1 must be turned off, and M_2 is turned on. The dead zone t_d is the time between the completion of the resonant cycle after the switch M_1 is turned off and the turning on the switch M_2. t_d must be greater than the turn-off time of the switching device.

$$\frac{\pi}{\omega_o} - \frac{\pi}{\omega_r} = t_d > t_{off} \quad (t_{off} = \text{turn-off time of the device}) \tag{6.23}$$

Note: Turn-off time of the device is normally specified in the device datasheet.
Let us consider the following circuit parameters:

$$R = 2,\; L = 50\ \mu\text{H},\; C = 8\ \mu\text{F},\; f_o = 7\ \text{kHz},\; V_s = 100\text{V},\; \omega_o = 2\pi f_o$$
$$= 43{,}982\ \text{rad/s}$$

Equation (6.5) gives

$$\omega_r = \sqrt{\frac{1}{LC} - \frac{R^2}{4L^2}} = 4.5830 \text{ rad/s}$$
$$f_r = \frac{\omega_r}{2 \times \pi} = 7.293 \text{ kHz} \quad T_r = \frac{1}{f_r} = 142.857 \text{ µs}$$

Equation (6.6) gives

$$\alpha = \frac{R}{2L} = 20{,}000$$

Equation (6.8) gives

$$\xi = \frac{\alpha}{\omega_n} = 0.436$$

6.3.1 Timing of gating signals

The operation of the resonant inverter involves three frequencies: the frequency of the switching devices, f_s, the resonant frequency of the RLC circuit, f_r, and the output frequency of the inverter, f_o. Thus, there are three corresponding periods: switching period, $T_s = 1/f_s$, the period of the resonant circuit, $T_r = 1/f_r$, and the period of the output waveform, $T_o = 1/f_o$. Switching on M_1 starts the resonant oscillation and M_1 must remain turned on until the positive resonant current through R falls to zero. There should be a small dead time after M_1 is turned off, until M_2 is turned on to initial the reverse resonant through R. M_1 conducts only during the first period of the output, and M_2 conducts only during the second part of the output period. To obtain the desired output, the switching devices must be gated with appropriate gating pulses of duration. The gate pulses must ensure that the device is turned fully off = turn-off time of the device turned on and off at the desired time as follows [2]:

Resonant period, $T_r = 1/f_r = 143$ µs
Period of output waveform, $T_o = 1/f_o = 143$ µs

$$\text{The online of } M_1,\ t_1 = \frac{T_r}{2} = 70 \text{ µs}$$

$$\text{The online of } M_2,\ t_2 = \frac{T_r}{2} \text{ delayed by } \frac{T_o}{2} = 71 \text{ µs} = 70 \text{ µs}$$

The LTspice descriptions of the gating pulses for M_1 and M_2 are

PULSE (0 5V 0 0.1 ns 0.1 ns 70 µs 143 µs)
PULSE (0 5V 71 µs 0.1 ns 0.1 ns 65 µs 143 µs)

6.3.2 Converter parameters

(a) Equation (6.10) gives the time when the load current peaks,
$t_m = \frac{1}{\omega_r}\tan^{-1}\frac{\omega_r}{\alpha} = 25.3\ \mu s$

(b) Equation (6.4) gives the peak transistor current at $t = t_m$,
$I_p = \frac{V_S+V_c}{\omega_r L}e^{-\alpha t}\sin\omega_r t_m = 32.321$ A

(c) The average transistor current

$$I_{av} = \frac{1}{T_o}\left(\int_0^{\frac{T_r}{2}} \frac{V_S + V_c}{\omega_r L} e^{-\alpha t}\sin\omega_r t\ \mathrm{dt}\right) = 9.41\ \mathrm{A}$$

(d) The rms transistor current

$$I_{rms} = \sqrt{\frac{1}{T_o}\int_0^{\frac{T_r}{2}}\left(\frac{V_S + V_c}{\omega_r L}e^{-\alpha \cdot t}\sin(\omega_r t)\right)^2 dt} = 15.338\ \mathrm{A}$$

(e) Equation (6.20) gives the peak switch voltage $V_{peak} = V_{c1} = \frac{V_s e^z}{e^z - 1} = 134.017$ V

(f) Equation (6.23) gives the dead time between the gating of the switches $t_d = \frac{\pi}{\omega_o} - \frac{\pi}{\omega_r} = 6\ \mu s$

6.3.3 Output-side parameters

(a) The rms output voltage, $V_{rms} = V_S = 100$ V
(b) The rms load current is contributed by two transistors, $I_o = \sqrt{2} \times I_{rms} = \sqrt{2} \times 15.338 = 21.691$ A

(c) Equation (6.19) gives the minimum value of the capacitor voltage
$V_{\min} = V_c = \frac{V_s}{e^z - 1} = 34.017$ V

(d) Equation (6.20) gives the maximum value of the capacitor voltage
$V_{\max} = V_{c1} = \frac{V_s e^z}{e^z - 1} = 134.017$ V

(e) Assuming triangular waveform, the rms value of the capacitor voltage

$$V_{rms} = \sqrt{\frac{{V_c}^2 + V_c V_{c1} + {V_{c1}}^2}{3}}$$
$$= \sqrt{\frac{34.017^2 + 34.017 \times 134.017 + 134.017^2}{3}} = 76.936\ \mathrm{V}$$

(f) Power delivered to the load $P_{out} = {I_o}^2 R = 21.691^2 \times 2 = 940.992$ W

6.3.4 Input-side parameters

(a) The average input current is

$$\begin{aligned} I_s &= I_{av} \\ &= 8.82\ \text{A} \end{aligned}$$

(b) The rms input current is

$$\begin{aligned} I_i &= I_r \\ &= 11.402\ \text{A} \end{aligned}$$

(c) The input power,

$$\begin{aligned} P_{in} &= P_{out} \\ &= 940.992\ \text{W} \end{aligned}$$

(d) The average input current is

$$\begin{aligned} I_s &= \frac{P_{in}}{V_S} = \frac{940.992}{100} \\ &= 9.41\ \text{A} \end{aligned}$$

(e) The rms input current is

$$\begin{aligned} I_i &= I_{rms} \\ &= 15.338\ \text{A} \end{aligned}$$

(f) The ripple content of the input current

$$\begin{aligned} I_{ripple} &= \sqrt{I_i^2 - I_s^2} \\ &= \sqrt{15.338^2 - 9.41^2} = 12.112\ \text{A} \end{aligned}$$

(g) The ripple factor of the input current

$$\begin{aligned} RF_i &= \frac{I_{ripple}}{I_s} \\ &= \frac{12.112}{9.41} \times 100 = 128.716\% \end{aligned}$$

6.3.5 Resonant pulse inverter circuit model

The inductor L ($= L_1 = L_2$) as shown in Figure 6.2(a) provides a high impedance for the ripple currents and limits the ripple current of the load resistor, R. Capacitor C provides a low impedance path for the ripple currents and limits the ripple voltage of the load resistor, R. The nth harmonic impedance for the RLC-load is given by

$$Z(n) = R + jn\omega_o L + \frac{1}{jn\omega_o C} = R + j\left(n\omega_o L - \frac{1}{n\omega_o C}\right) \tag{6.24}$$

At the resonant frequency

$$n\omega_o L = \frac{1}{n\omega_o C} \tag{6.25}$$

which gives the nth harmonic resonant frequency

$$\omega_{on} = \frac{1}{n\sqrt{LC}} \tag{6.26}$$

The DC supply is connected to the load only when M_1 is turned on during the first half-cycle of the output frequency. The a_n component of the Fourier series is given by

$$a_n = \frac{1}{\pi}\int_{-\frac{\pi}{2}}^{\frac{\pi}{2}} V_s \cos n(\theta) d\theta$$

$$= \frac{1}{\pi \times n\omega_o} \sin n(\theta)\Bigg|_{-\frac{\pi}{2}}^{\frac{\pi}{2}} = \frac{1}{\pi \times n}(\sin n\left(\frac{\pi}{2}\right) + \sin n\left(\frac{\pi}{2}\right) = \frac{2V_s}{\pi \times n} \sin\left(\frac{n\pi}{2}\right) \tag{6.27}$$

Thus, we can express the instantaneous output voltage in a Fourier series as

$$v_o(t) = \sum_{n=1}^{\infty} \frac{2V_s}{\pi \times n} \sin\left(\frac{n\pi}{2}\right) \sin(n\omega_o t) \tag{6.28}$$

where angular output frequency $\omega_o = 2\pi f_o$.

Dividing the voltage expression in (5.28) by the load impedance $Z(n)$ in (6.24) gives the Fourier components of the load currents as

$$i_o(t) = v_o(t) = \sum_{n=1}^{\infty} \frac{2 \times V_s}{\pi \times n \times |Z(n)|} \sin\left(\frac{n\pi}{2}\right) \sin(n\omega_o t - \theta_n) \tag{6.29}$$

where θ_n is the impedance angle of the nth harmonic component, rad.

Equation (6.29) gives the peak magnitude of the nth harmonic currents

$$I_m(n) = \frac{2 \times V_s}{\pi \times n \times |Z(n)|} \sin\left(\frac{n\pi}{2}\right) \tag{6.30}$$

which for $n = 1$, $n = 3$, and $n = 5$ gives

$$I_m(1) = \frac{2 \times 100}{1 \times \pi \times Z(1)} = 30.304\angle 17.821^\circ$$

$$I_m(3) = \frac{2 \times 100}{3 \times \pi \times Z(3)} = 3.541\angle 109.493^\circ$$

$$I_m(5) = \frac{2 \times 100}{1 \times \pi \times Z(5)} = 1.199\angle - 79.142^\circ$$

Dividing the peak values by $\sqrt{2}$ gives the rms values for $n = 1$, $n = 3$, and $n = 5$ as

$$I_{rms}(1) = \frac{I_m(1)}{\sqrt{2}} = \frac{30.304\angle 17.821^\circ}{\sqrt{2}} = 21.428\angle 17.821^\circ$$

$$I_{rms}(3) = \frac{I_m(3)}{\sqrt{2}} = \frac{3.541\angle 109.493^\circ}{\sqrt{2}} = 2.504\angle 109.493^\circ$$

$$I_{rms}(5) = \frac{I_m(5)}{\sqrt{2}} = \frac{1.199\angle -79.142^\circ}{\sqrt{2}} = 0.848\angle -79.142^\circ$$

The total rms value of the load current can be determined by adding the square of the individual rms values and then taking the square root

$$\begin{aligned} I_o &= \sqrt{\sum_{n=1}^{30} (I_{rms}(n))^2} \\ &= 21.598 \text{ A} \end{aligned} \tag{6.31}$$

Equation (6.30) related the average output current $I_{o(av)}$ of a full-wave rectifier with a peak sinusoidal input current of I_m as

$$I_{o(av)} = \frac{2I_m}{\pi} \tag{6.32}$$

Since the output current of the inverter contains sinusoidal harmonic components at different frequencies, we can add the corresponding DC components contributing to the output current of the inverter. Applying (6.32) gives the average input current due by n number of harmonic components, $n = 30$.

$$\begin{aligned} I_{s(av)} &= \sum_{n=1}^{30} \frac{2 \times |I_m(n)|}{\pi} \\ &= 4.851 \text{ A} \end{aligned} \tag{6.33}$$

6.3.6 Resonant pulse inverter voltage gain

$$\omega_o = 2\pi f_o = 2\pi \times 7000 = 377 \text{ rad/s}$$

The load resistance R forms a series circuit with L and C. Using the voltage dividing rule, we can find the output voltage across R in the frequency domain as

$$G(j\omega) = \frac{V_o}{V_S}(j\omega) = \frac{R}{1 + j\omega L - \frac{j}{\omega C}} = \frac{1}{1 + j\omega \frac{L}{R} - \frac{j}{\omega CR}} \tag{6.34}$$

Defining $\omega_o = \frac{1}{\sqrt{LC}}$ as the resonant frequency and $Q_s = \frac{\omega_o L}{R}$ as the quality factor give the gain as

$$G(j\omega) = \frac{V_o}{V_S}(j\omega) = \frac{R}{1 + jQ_s\left(\frac{\omega}{\omega_o} - \frac{\omega_o}{\omega}\right)} \tag{6.35}$$

Defining $u = \frac{\omega}{\omega_o}$ as the normalized frequency, we obtain

$$G(j\omega) = \frac{V_o}{V_S}(j\omega) = \frac{R}{1 + jQ_s\left(u - \frac{1}{u}\right)} \tag{6.36}$$

which gives the gain magnitude as

$$M = \left|\frac{R}{1 + jQ_s\left(u - \frac{1}{u}\right)}\right| = \frac{1}{\sqrt{1 + Q_s^2\left(u - \frac{1}{u}\right)^2}} \tag{6.37}$$

For $R = 2\ \Omega$, $C = 8\ \mu F$ and $L = 50\ \mu m\ f_o = 7$ kHz

$\omega_r = \sqrt{\frac{1}{LC} - \frac{R^2}{4L^2}} = 45{,}830\ \text{rad/s}, \quad \omega_o = 2\pi f_o = 43{,}982\ \text{rad/s}, \quad Q_s = \frac{\omega_r L}{R} = 1.146$
$u = \frac{\omega_o}{\omega_r} = 0.96$

Equation (5.37) gives the gain magnitude $M = \frac{1}{\sqrt{1+Q_s^2\left(u-\frac{1}{u}\right)^2}} = 0.996$

Note: By varying the normalized frequency, u, that is, varying the switching frequency, we can vary the voltage gain M of the inverter.

6.4 Half-bridge bidirectional resonant pulse inverter

A half-bridge resonant pulse inverter with the bidirectional switching is shown in Figure 6.3(a). The addition of the antiparallel diodes with the switching devices allows the resonant oscillation through M_1 during the positive current through R and then continues the oscillation during the negative current through R. The current through the load R flows during the complete resonant cycle and is a better representation of the resonant waveform. This circuit arrangement can be operated to yield the output waveforms of Figure 6.3(b) if we use the same gating pulses. These gating signals do not allow time for the oscillation to continue. The gate pulses must ensure that the device is turned on and off at the desired time as follows:

Resonant period, $T_r = 1/f_r = 143\ \mu s$
Period of the output waveform, $T_o \geq T_r = 143 = 143\ \mu s$
The online of M_1,

$$t_1 = \frac{T_r}{2}$$
$$= 71\ \mu s$$

The online of M_2, $t_2 = \frac{T_r}{2}$ delayed by $\frac{T_o}{2} = 71\ \mu s$
$= 71\ \mu s$

The LTspice descriptions of the gating pulses for M_1 and M_2 are:

PULSE (0 5V 0 0.1 ns 0.1 ns 70 μs 143 μs)
PULSE (0 5V 71 μs 0.1 ns 0.1 ns 65 μs 143 μs)

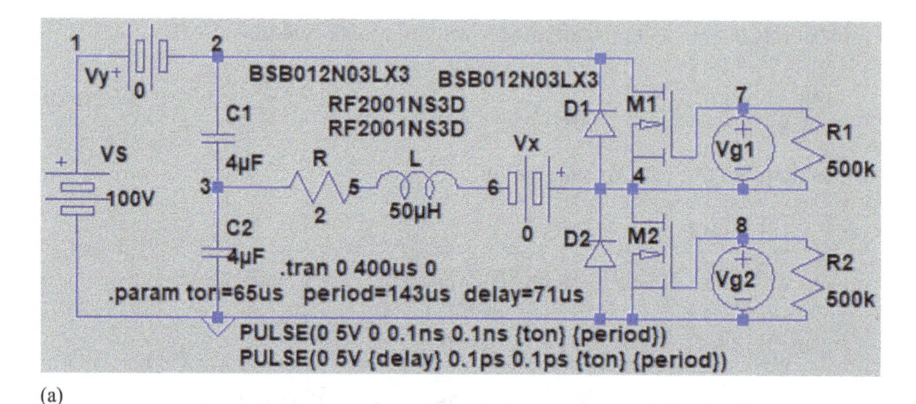

(a)

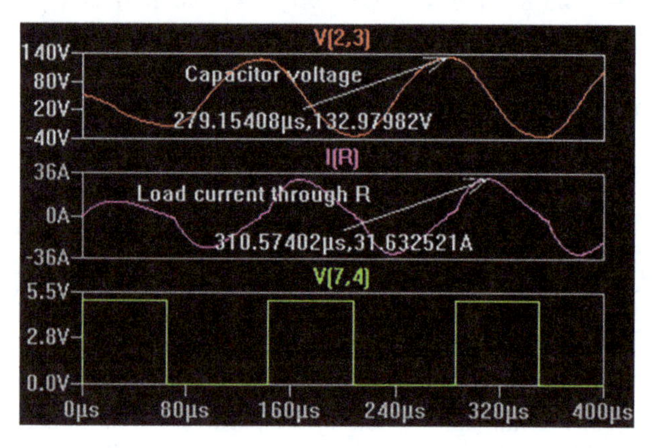

(b)

Figure 6.3 A half-bridge bidirectional resonant pulse inverter for $T_o \geq T_r$: (a) schematic and (b) inverter waveforms [1]

For bidirectional operation, switch M_1 and diode D_1 are conducted during the first resonant cycle. Switching on M_1 starts the resonant oscillation, and M_1 must remain turned on until the positive resonant current through R falls to zero. M_1 is conducted only during the first half of the resonant period. Due to the antiparallel diode D_1, the resonant oscillation continues and the reverse resonant flows through R during the second half of the resonant period. D_1 is conducted only during the second half of the resonant period. Switch M_2 and diode D_2 are conducted during the second half of the resonant cycle. Thus, the waveform of the output current would have two positive resonant pulses and two negative resonant pulses, and the output period should be twice the resonant period. For the bidirectional operation, the gate pulses must ensure that the device is turned on and off at the desired time as follows:

Resonant period, $T_r = 1/f_r = 143$ µs

Period of the output waveform, $T_o \geq 2 \times T_r = 2 \times 143 = 286$ μs

$$\text{The online of } M_1,\ t_1 = \frac{T_r}{2} = 71\ \mu\text{s}$$

$$\text{The online of } M_2,\ t_2 = \frac{T_r}{2} \text{ delayed by } \tfrac{T_o}{2} = 143\ \mu\text{s} = 71\ \mu\text{s}$$

The LTspice descriptions of the gating pulses for M_1 and M_2 are:

PULSE (0 5V 0 0.1 ns 0.1 ns 71 μs 286 μs)
PULSE (0 5V 143 μs 0.1 ns 0.1 ns 710 μs 286 μs)

Depending on whether the switch is turned on or off, there are two modes of operation. During mode 1, the switch is turned on, and during mode 2, the switch is turned off [1–4]. Figure 6.4 shows the schematic and the waveforms of the half-bridge bidirectional resonant pulse inverter for $T_o \geq 2T_r$.

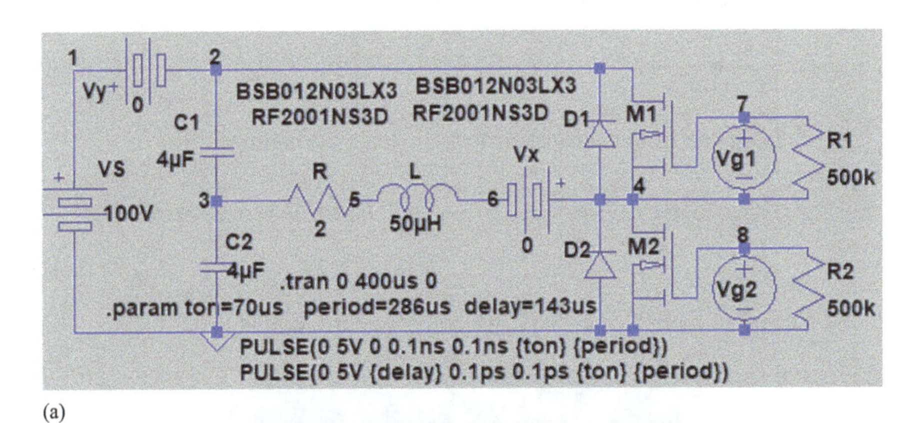

(a)

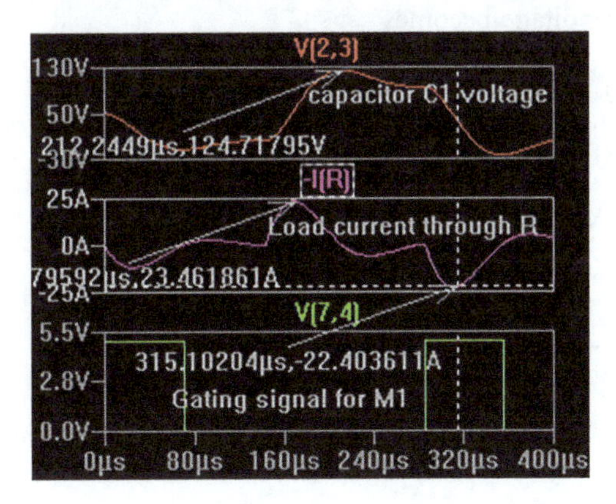

(b)

Figure 6.4 A half-bridge bidirectional resonant pulse inverter for $T_o \geq 2T_r$: (a) schematic and (b) inverter waveforms

Mode 1: When the switch M_1 is turned on, the DC supply voltage V_S is connected to the load consisting of R, L, and C. Capacitor C_1 forms a parallel circuit with capacitor C_2 through the DC supply source V_S and the effective capacitor $C = C_1 + C_2$. We can use the equation in section 6.3 for Mode 1 [1].

$$\begin{aligned} v_{C1}(t = t_{1m}) &= V_{c1} \\ &= (V_S + V_c)\, e^{-\alpha\pi/\omega_r} + V_S \end{aligned} \tag{6.38}$$

Mode 2: When the switch M_1 is turned off, the DC supply voltage V_S is disconnected, and the antiparallel diode D_1 is turned on. The capacitor with its initial voltage at the end of mode 1 forces current, and the capacitor voltage is reversed. The resonant current $i_2(t)$ during mode 2 can be described by [1]

$$L\frac{di_2}{dt} + Ri_2 + \frac{1}{C}\int i_2\, dt + v_{C2}(t = 0) = -V_S \tag{6.39}$$

Solving for $i_2(t)$ with an initial capacitor initial voltage $v_{C2}(t = 0) = -V_{c1}$ gives the resonant current of an RLC circuit as

$$i_2(t) = \frac{V_{c1} - V_S}{\omega_r L} e^{-at} \sin \omega_r t \tag{6.40}$$

which gives the instantaneous capacitor voltage during mode 2 as

$$v_{C2}(t) = \frac{1}{C}\int_0^t i_2(t)dt - V_{c1} = \frac{V_S - V_{c1}}{\omega_r}\, e^{-at}(\alpha \sin \omega_r t + \omega_r \cos \omega_r t) + V_S \tag{6.41}$$

At the end of mode 2 at $t = t_{2m}$, the resonant current falls to zero

$$i_2(t = t_{2m}) = 0 \tag{6.42}$$

The capacitor voltage becomes

$$v_{C2}(t = t_{2m}) = V_{c2} = (V_S - V_{c1})\, e^{-\alpha\pi/\omega_r} \tag{6.43}$$

Solving (6.38) and (6.43) gives

$$V_c = \frac{V_s e^{-2z}}{1 + e^{-2z}} \tag{6.44}$$

$$V_{c1} = V_s \frac{(1 + e^{-z} + e^{-2z})}{1 + e^{-2z}} \tag{6.45}$$

The circuit parameters are $R = 2$, $L = 50\ \mu\text{H}$, $C = 8\ \mu\text{F}$, $f_o = 3.5\ \text{kHz}$, $V_s = 100\ \text{V}$, $\omega_o = 2\pi f_o = 21{,}990\ \text{rad/s}$.

Equation (6.5) gives

$$\omega_r = \sqrt{\frac{1}{LC} - \frac{R^2}{4L^2}} = 4.5830 \text{ rad/s}$$

$$f_r = \frac{\omega_r}{2 \times \pi} = 7.293 \text{ kHz } T_r = \frac{1}{f_r} = 142.857 \text{ µs}$$

Equation (6.6) gives

$$\alpha = \frac{R}{2L} = 20{,}000$$

Equation (6.8) gives

$$\xi = \frac{\alpha}{\omega_n} = 0.436$$

$$z = \frac{\alpha\pi}{\omega} = 1.371$$

$$V_c = \frac{V_s e^{-2z}}{1 + e^{-2z}} = 6.053 \text{ V}$$

$$V_{c1} = V_s \frac{(1 + e^{-z} + e^{-2z})}{1 + e^{-2z}} = 123.846 \text{ V}$$

6.4.1 Converter parameters

(a) Equation (6.10) gives the time when the load current peaks,

$$t_m = \frac{1}{\omega_r} \tan^{-1} \frac{\omega_r}{\alpha} = 25.3 \text{ µs}$$

(b) Equation (6.4) gives the peak transistor current at $t = t_m$,

$$I_p = \frac{V_S + V_c}{\omega_r L} e^{-\alpha t} \sin \omega_r t_m = 32.321 \text{ A}$$

(c) The average transistor current

$$I_{av} = \frac{1}{T_o} \left(\int_0^{\frac{T_r}{2}} \frac{V_{c1}}{\omega_r L} \times e^{-\alpha . t} \sin(\omega_r . t)\, dt \right) = 3.723 \text{ A}$$

(d) The rms transistor current

$$I_{rms} = \sqrt{\frac{1}{T_o} \int_0^{\frac{T_r}{2}} \left(\frac{V_S + V_c}{\omega_r L} e^{-\alpha \cdot t} \sin(\omega_r \cdot t) \right)^2 dt} = 8.582 \text{ A}$$

(e) The average diode current

$$I_d = \frac{1}{T_o}\left(\int_0^{\frac{T_r}{2}} \frac{V_{c1} - V_s}{\omega_r L} \times e^{-a.t} \sin(\omega_r.t)\, dt\right) = 0.915 \text{ A}$$

(f) The rms diode current

$$I_r = \sqrt{\frac{1}{T_o}\int_0^{\frac{T_r}{2}} \left(\frac{V_{c1} - V_S}{\omega_r L} e^{-a \cdot t} \sin(\omega_r \cdot t)\right)^2 dt} = 1.93\text{A}$$

(g) Equation (6.45) gives the peak switch voltage

$$V_{peak} = V_{c1} = \frac{V_s \times (1 + e^{-z} + e^{-2z})}{1 + e^{-2z}} = 123.846 \text{ V}$$

(h) Equation (6.23) gives the dead-time between the gating of the switches

$$t_d = \frac{\pi}{\omega_o} - \frac{\pi}{2\omega_r} = 11.494 \text{ µs}$$

6.4.2 *Output-side parameters*

(a) The rms output voltage, $V_{rms} = V_S = 100$ V
(b) The rms load current is contributed by two transistors and two diodes,

$$I_o = \sqrt{2 \times (I_{rms}^2 + I_r^2)} = \sqrt{2 \times (8.582^2 + 1.93^2)} = 12.44 \text{ A}$$

(c) Equation (6.44) gives the minimum value of the capacitor voltage

$$V_{\min} = V_c = \frac{V_s e^{-2z}}{1 + e^{-2z}} = 6.053 \text{ V}$$

(d) Equation (6.45) gives the maximum value of the capacitor voltage

$$V_{\max} = V_{c1} = V_s \frac{(1 + e^{-z} + e^{-2z})}{1 + e^{-2z}} = 123.846 \text{ V}$$

(e) Assuming a triangular waveform, the rms value of the capacitor voltage

$$V_{rms} = \sqrt{\frac{V_c^2 + V_c V_{c1} + V_{c1}^2}{3}}$$
$$= \sqrt{\frac{6.053^2 + 6.053 \times 123.846 + 123.846^2}{3}} = 63.49 \text{ V}$$

(f) Power delivered to the load $P_{out} = I_o^2 R = 12.44^2 \times 2 = 309.528$ W

6.4.3 Input-side parameters

(a) The input power,

$$P_{in} = P_{out} = 309.528 \text{ W}$$

(b) The average input current is

$$I_s = \frac{P_{in}}{V_S} = \frac{309.528}{100} = 3.095 \text{ A}$$

6.5 Full-bridge bidirectional resonant pulse inverter

A full-bridge resonant pulse inverter with the bidirectional switching is shown in Figure 6.5. Four transistors with four antiparallel diodes form a full bridge and allow the current flow. The antiparallel diodes allow the current through the load R to flow during the complete resonant cycles. Transistors M_1 and M_2 with the diodes allow current through the load R to flow for one complete resonant cycle. Transistors M_3 and M_4 with the diodes allow the current through the load R to flow for another complete resonant cycle. Thus, the load current consists of two positive resultant pulses and two negative resultant pulses during one cycle of the output waveform.

The gate pulses must ensure that the device is turned on and off at the desired time as follows:

Resonant period, $T_r = 1/f_r = 143$ μs

Period of output waveform, $T_o = 286$ μs

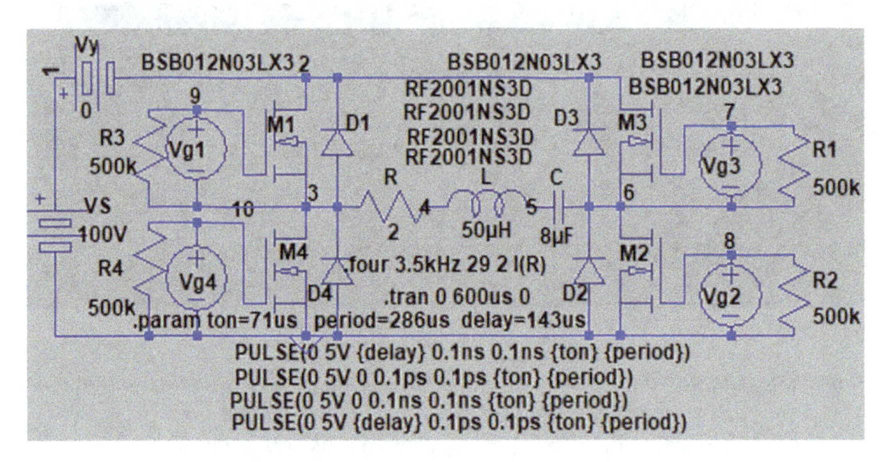

Figure 6.5 A full-bridge bidirectional resonant pulse inverter for $T_o = T_r$

$$\text{The online of } M_1, \ t_1 = \frac{T_r}{2} = 71\ \mu s$$

$$\text{The online of } M_2, \ t_2 = \frac{T_r}{2} \text{ delayed by } \frac{T_o}{2} = 143\ \mu\text{s}. = 71\ \mu\text{s}$$

The LTspice descriptions of the gating pulses for M_1 and M_3 are

PULSE (0 5V 0 0.1 ns 0.1 ns 71 us 286 μs)
PULSE (0 5V 143 μs 0.1 ns 0.1 ns 71 μs 286 μs)

For bidirectional operation, M_1, M_2, D_1, and D_2 conduct during the first-half of the resonant cycle. M_3, M_4, D_3, and D_4 conduct during the second-half of the resonant cycle. Once the resonant circuit is designed, the resonant frequency is fixed by the values of the RLC components. The output frequencyf_o or the period T_o can be varied to vary the output current and the output power. However, the delay of gate pulses for M_3 and M_4 should adjusted $t_{delay} = \frac{T_o}{2}$ as shown in Figure 6.6.

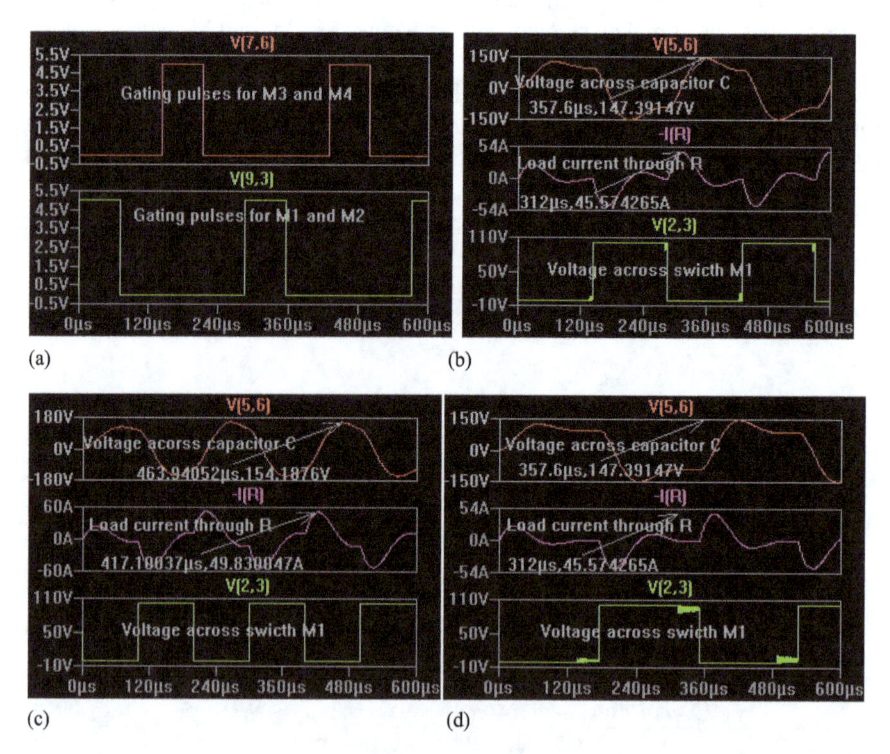

Figure 6.6 Output voltage and current for full-bridge bidirectional resonant pulse inverter: (a) gating pulse for $T_o = 2T_r$*; (b) output for* $T_o = 2T_r$, $T_o = 286\ \mu s$, $t_{delay} = 143\ \mu s$*; (c) output for* $T_o > 2T_r$, $T_o = 350\ \mu s$, $t_{delay} = 175\ \mu s$*; and (d) output for* $T_o < 2T_r$, $T_o = 200\ \mu s$, $t_{delay} = 100\ \mu s$

During the first resonant cycle, there are two modes of operation: During mode 1, switches M_1 and M_2 are turned on, and the DC supply voltage V_S is connected to the load consisting of R, L, and C. During mode 2, diodes D_1 and D_2 are turned on, and the resonant oscillation continues through diodes and feeding back the current to the DC supply voltage V_S. We can use the equations for the half-bridge bidirectional resonant pulse inverter in Section 6.4. During the second resonant cycle, there are two identical modes of operation: During mode 3, switches M_3 and M_4 are turned on, the DC supply voltage V_S is connected to the load consisting of R, L, and C. During mode 4, diodes D_3 and D_4 are turned on, and the resonant oscillation continues through diodes and feeding back the current to the DC supply voltage V_S. The direction of the load current and the polarity of the capacitor voltage for the second resonant cycle are opposite that of the first resonant cycle. The LTspice gives the Fourier components of the load current $I(R)$ at the fundamental frequency of 3.5 kHz as follows:

Fourier components of *I*(*R*)

DC component: 2.11006

Harmonic number	Frequency (Hz)	Fourier component	Normalized component	Phase (degrees)	Normalized phase (degrees)
1	3.500e+03	8.266e+00	1.000e+00	−114.18	0.00
2	7.000e+03	4.875e+00	5.897e−01	25.69	139.87
3	1.050e+04	6.070e+00	7.344e−01	134.01	248.19
4	1.400e+04	2.098e+00	2.538e−01	−23.50	90.68
5	1.750e+04	1.412e+00	1.708e−01	78.31	192.49
6	2.100e+04	1.453e+00	1.757e−01	−53.40	60.78
7	2.450e+04	7.121e−02	8.615e−03	85.67	199.85
8	2.800e+04	9.425e−01	1.140e−01	−92.51	21.67
9	3.150e+04	4.476e−01	5.415e−02	168.71	282.89
10	3.500e+04	4.707e−01	5.694e−02	−139.80	−25.61
11	3.850e+04	5.618e−01	6.797e−02	129.14	243.32
12	4.200e+04	1.431e−01	1.731e−02	131.35	245.53
13	4.550e+04	4.565e−01	5.523e−02	83.42	197.60
14	4.900e+04	2.294e−01	2.775e−02	5.84	120.02
15	5.250e+04	2.669e−01	3.229e−02	28.74	142.92
16	5.600e+04	3.225e−01	3.902e−02	−50.84	63.34
17	5.950e+04	1.356e−01	1.641e−02	−62.48	51.70
18	6.300e+04	3.006e−01	3.636e−02	−101.43	12.75
19	6.650e+04	1.774e−01	2.146e−02	−164.17	−49.99
20	7.000e+04	2.054e−01	2.485e−02	−158.24	−44.06
21	7.350e+04	2.241e−01	2.712e−02	132.86	247.04
22	7.700e+04	1.175e−01	1.421e−02	118.33	232.51
23	8.050e+04	2.116e−01	2.560e−02	77.30	191.48
24	8.400e+04	1.338e−01	1.619e−02	19.11	133.29
25	8.750e+04	1.591e−01	1.924e−02	16.03	130.21
26	9.100e+04	1.744e−01	2.109e−02	−46.82	67.36

(Continues)

(*Continued*)

Fourier components of *I*(*R*)					
DC component: 2.11006					
Harmonic number	**Frequency (Hz)**	**Fourier component**	**Normalized component**	**Phase (degrees)**	**Normalized phase (degrees)**
27	9.450e+04	1.088e−01	1.316e−02	−64.52	49.66
28	9.800e+04	1.722e−01	2.083e−02	−103.51	10.67
29	1.015e+05	1.133e−01	1.371e−02	−154.72	−40.54
Total harmonic distortion: 102.382196% (180.037609%)					

6.6 Parallel resonant pulse inverter

In a series resonant circuit, the load resistance R forms a series circuit with L and C. The current through the series resonant circuit depends on the impedance. In a parallel resonant circuit, the load resistance R is in parallel with both L and C. A parallel circuit is a dual of the series resonant circuit. The parallel resonant circuit is supplied with a current source so that the circuit offers a high impedance to the switching circuit [1]. The transfer function between the output voltage V_o and the input current source I_S is given by [3]

$$\frac{V_o}{I_S}(j\omega) = Z((j\omega) = R\|j\omega L\|\frac{1}{j\omega C} = R\frac{1}{1 + jR/(\omega L) + j\omega CR} \tag{6.46}$$

The undamped resonant frequency

$$\omega_o = \sqrt{\frac{1}{LC}} \tag{6.47}$$

The damped resonant frequency

$$\omega_r = \sqrt{\frac{1}{LC} - \frac{1}{4R^2C^2}} \tag{6.48}$$

The quality factor Q_p of the parallel resonant circuit is

$$Q_p = \omega_o CR = \frac{R}{\omega_o L} = R\sqrt{\frac{C}{L}} = 2\xi \tag{6.49}$$

where ξ is the damping factor and $\xi = \frac{R}{2}\sqrt{\frac{C}{L}}$. Substituting L, C, and R in terms of Q_p and ω_o in (6.46), we obtain

$$\frac{V_o}{I_S}(j\omega) = Z(j\omega) = R\frac{1}{1 + jQ_p(\omega/\omega_o - \omega_o/\omega)} = \frac{1}{1 + jQ_p(u - 1/u)} \tag{6.50}$$

where $u = \omega/\omega_o$ is the frequency ratio.

The magnitude of $Z(j\omega)$ in (6.50) can be found from

$$|Z(j\omega)| = \frac{1}{\sqrt{1 + Q_p^2\left(u - \frac{1}{u}\right)^2}} \tag{6.51}$$

which is identical to the voltage gain $G(j\omega)$ in (6.37).

For $R = 1$ kΩ, $C = 0.4$ µF, $L_s = 4$ mH, and $L = 0.5$ mH, $f = 30$ kHz

$\omega_r = \sqrt{\frac{1}{LC} - \frac{1}{4R^2C^2}} = 5 \times 10^{11}$ rad/s

$\omega = 2\pi f = 1.885 \times 10^5$ rad/s, $T = \frac{1}{30k} = 34$ µs $\quad t_{on} = 17$ µs

Duty cycle, $k = \frac{t_{on}}{T} = \frac{17}{34} = 0.5$ $Q_p = R\sqrt{\frac{C}{L}} = 282.843$ $u = \frac{\omega}{\omega_o} = 3.77 \times 10^{-7}$.

Thus, the magnitude of $Z(j\omega)$ in (6.50) is

$$|Z(j\omega)| = \frac{1}{\sqrt{1 + Q_p^2\left(u - \frac{1}{u}\right)^2}}$$

$$= \frac{1}{\sqrt{1 + 282.843^2 \times \left(3.77 \times 10^{-7} - \frac{1}{3.77 \times 10^{-7}}\right)^2}} = 1.333 \times 10^{-6}$$

The parallel resonant inverter is shown in Figure 6.7. The voltage and current waveforms are shown in Figure 6.8. The inductor L_s serves as a current source and maintains a current approximately constant at $I_S = 994$ mA. The transistors M_1 and M_2 are switched on and off at a period of 34 µs with a 50% duty cycle. The gating signals and the voltages across the transistors are shown in Figure 6.8(a). As the transistors M_1 and M_2 are switched on, a square wave of constant current I_S appears

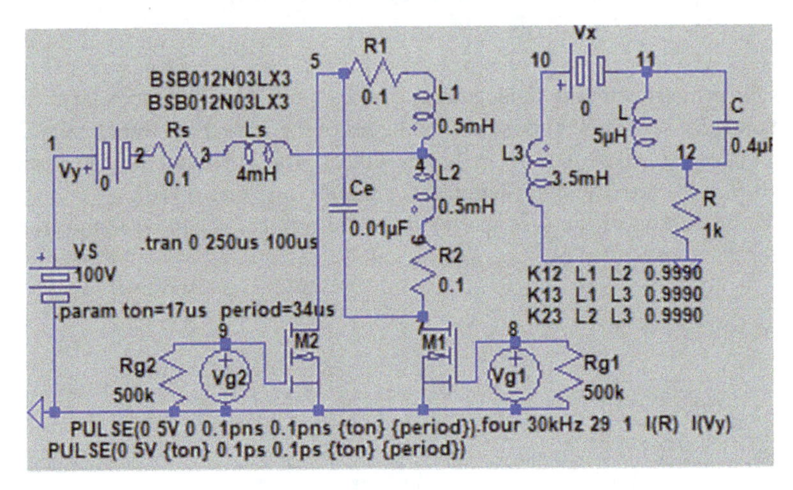

Figure 6.7 Parallel resonant circuit [2,4]

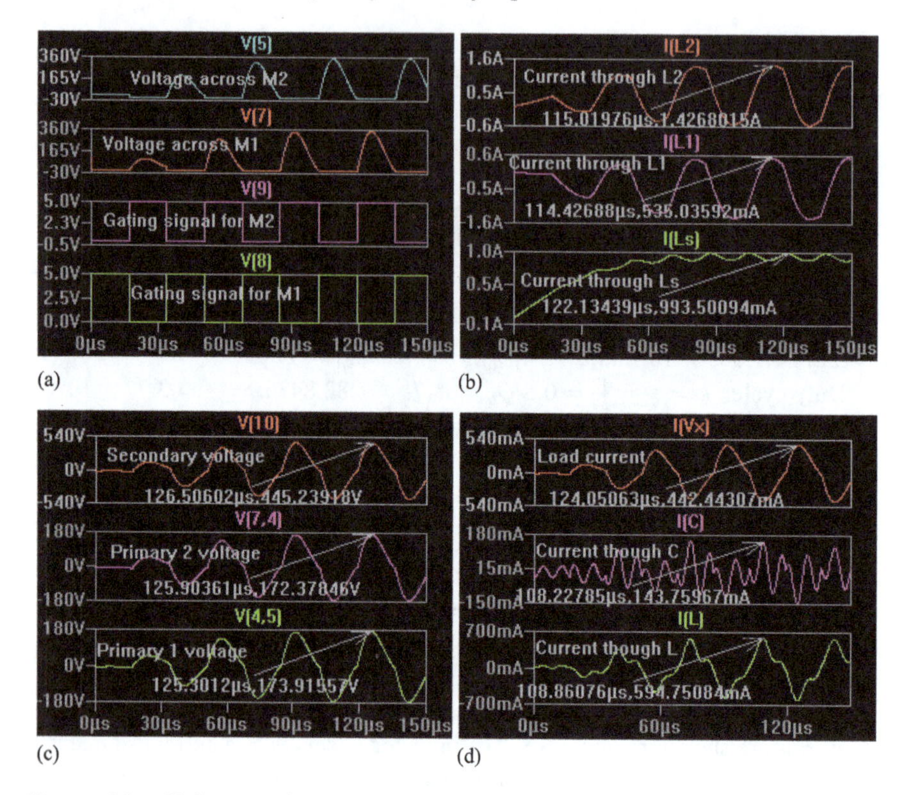

(a) (b) (c) (d)

Figure 6.8 Voltage and current waveforms for parallel resonant inverter: (a) gate pulses and voltages for M1 and M2, (b) current through inductors Ls, L_1 and L_2, (c) primary and secondary voltages, and (d) secondary currents through Vx, L, and C

across the primary winding inductances, $L_1 = L_2 = 0.5$ mH. The secondary winding inductance, $L_3 = 3.5$ mH. The induced voltages of the primary 1, the primary 2, and the secondary sides are shown in Figure 6.8(c). The corresponding load current through R, inductor L, and capacitor C are shown in Figure 6.8(d).

The turns ratio of the secondary side to the primary 1 can be found from the inductor values as given by

$$\begin{aligned} a &= \sqrt{\frac{L_3}{L_1}} \\ &= \sqrt{\frac{3.5 \times 10^{-3}}{0.5 \times 10^{-3}}} = 2.646 \end{aligned} \tag{6.52}$$

The load impedance Z_L is given by

$$Z_L = j\omega L \| \frac{1}{j\omega C} + R = \frac{j\omega L \times \frac{1}{j\omega C}}{j\omega L + \frac{1}{j\omega C}} + R \tag{6.53}$$
$$= 1000 - j15.435$$

Equating volt-amps in the primary side to volt-amps on the secondary side, we obtain

$$V_1 I_1 + V_2 I_2 = V_3 I_3 \tag{6.54}$$

Which for two identical primary sides $V_1 I_1 = V_2 I_2$ becomes $2V_1 I_1 = V_3 I_3$

$$\text{Since } V_2 = aV_1 \text{ and } I_1 = \frac{V_3}{V_1}\frac{I_3}{2} = \frac{aI_3}{2} \tag{6.55}$$

which gives the equivalent load impedance referred to primary impedance as

$$Z_1 = \frac{V_1}{I_1} = \frac{V_3/a}{aI_3/2} = \frac{2}{a^2}\frac{V_3}{I_3} = \frac{2}{a^2} Z_L$$
$$= \frac{2}{2.646^2} \times 10^3 \angle - 0.884^\circ = 285.748 \angle - 0.884^\circ \tag{6.56}$$

Since the constant current $I_S = 0.994$ A is switched to the two primary sides alternately by switching transistors M_1 and M_2 at an output frequency of $f = 30$ kHz at a 50% duty cycle, the input to the primary side is a square wave of $+I_S$ and $-I_S$. The Fourier analysis of the input square wave current I_S gives the peak value of the fundamental component as

$$I_{1m} = \frac{4I_S}{2\pi} \tag{6.57}$$

Therefore, the peak value of the primary side voltage is

$$V_{1m} = I_{1m} Z_1$$
$$= 0.633 \times 285.748 \angle - 0.884^\circ = 180.822 \angle - 0.884^\circ \tag{6.58}$$

Therefore, the peak value of the secondary voltage is

$$V_{3m} = aV_{1m}$$
$$= 2.646 \times 180.822 \angle - 0.884^\circ = 478.409 \angle - 0.884^\circ \tag{6.59}$$

6.6.1 Input-side parameters

(a) The peak value of the input current, $I_S = 0.994$ mA;
(b) The rms value of the primary input current, $I_1 = I_S = 0.994$ mA;

(c) Equation (6.57) gives the peak value of the primary side current

$$I_{1m} = \frac{4I_S}{2\pi}$$
$$= \frac{4 \times 0.994 \times 10^{-3}}{2\pi} = 0.633 \text{ A};$$

(d) Equation (6.58) gives the peak value of the primary side voltage $V_{1m} = I_{1m}Z_1 = 180.822$ V;
(e) The rms value of the primary input voltage $V_1 = \frac{V_{1m}}{\sqrt{2}} = \frac{180.822}{\sqrt{2}} = 127.86$ V.

6.6.2 Converter parameters

(a) The peak transistor current, $I_p = 994$ mA;
(b) The average transistor current $I_{av} = kI_S = 0.5 \times 994$ mA $= 0.0.497$ mA;
(c) The rms transistor current $I_{rms} = \sqrt{k}I_S = \sqrt{0.5} \times 994$ mA $= 0.703$ mA;
(d) Equation (6.58) gives the peak switch voltage $V_{peak} = V_{1m} = 180.822$ V.

6.6.3 Output-side parameters

(a) The rms output voltage, $V_{rms} = V_S = 100$ V;
(b) Equation (6.59) gives the peak value of the secondary voltage is $V_{3m} = aV_{1m} = 478.409$ V;
(c) The rms value of the secondary voltage is $V_3 = \frac{V_{3m}}{\sqrt{2}} = \frac{478.409}{\sqrt{2}} = 338.286$ V;

(d) Dividing V_{3m} by impedance Z_L in (6.53) gives the peak load current $I_{3m} = \frac{V_{3m}}{Z_L} = \frac{478.409}{\sqrt{1000^2+15.435^2}} = 0.478$ A;

(e) The rms value of the secondary load current, $I_3 = \frac{I_{3m}}{\sqrt{2}} = \frac{0.478}{\sqrt{2}} = 0.338$ A;

(f) The output load power $P_{out} = I_3^2 R = 0.338^2 \times 1000 = 114.41$ W.

The LTspice command (.four 30 kHz 29 1 *I(R)*) for Fourier analysis gives the Fourier components of the load current *I(R)* at the fundamental frequency of 25 kHz as follows:

Fourier components of *I(R)*

DC component: 0.0042385

Harmonic number	Frequency (Hz)	Fourier component	Normalized component	Phase (degrees)	Normalized phase (degrees)
1	3.000e+04	4.290e−01	1.000e+00	144.35	0.00
2	6.000e+04	1.180e−02	2.751e−02	−31.67	−176.02
3	9.000e+04	4.612e−02	1.075e−01	−58.44	−202.79

(Continues)

(*Continued*)

Fourier components of *I(R)*

DC component: 0.0042385

Harmonic number	Frequency (Hz)	Fourier component	Normalized component	Phase (degrees)	Normalized phase (degrees)
4	1.200e+05	1.203e−03	2.803e−03	75.93	−68.42
5	1.500e+05	1.831e−02	4.268e−02	−162.92	−307.27
6	1.800e+05	3.698e−03	8.619e−03	28.92	−115.43
7	2.100e+05	8.272e−03	1.928e−02	109.89	−34.46
8	2.400e+05	2.888e−03	6.732e−03	−20.93	−165.28
9	2.700e+05	3.330e−03	7.761e−03	−5.20	−149.55
10	3.000e+05	7.705e−04	1.796e−03	−69.73	−214.08
11	3.300e+05	3.570e−03	8.320e−03	−121.29	−265.64
12	3.600e+05	1.085e−03	2.528e−03	58.59	−85.76
13	3.900e+05	3.119e−03	7.269e−03	165.08	20.73
14	4.200e+05	1.765e−03	4.113e−03	9.75	−134.60
15	4.500e+05	1.446e−03	3.370e−03	90.69	−53.66
16	4.800e+05	1.213e−03	2.827e−03	−39.98	−184.33
17	5.100e+05	8.530e−04	1.988e−03	−70.32	−214.67
18	5.400e+05	4.034e−05	9.403e−05	−129.01	−273.36
19	5.700e+05	1.545e−03	3.601e−03	−153.52	−297.87
20	6.000e+05	9.363e−04	2.182e−03	43.85	−100.51
21	6.300e+05	1.232e−03	2.873e−03	149.77	5.42
22	6.600e+05	1.153e−03	2.687e−03	−5.83	−150.18
23	6.900e+05	3.388e−04	7.896e−04	87.69	−56.66
24	7.200e+05	6.230e−04	1.452e−03	−56.87	−201.22
25	7.500e+05	5.807e−04	1.353e−03	−119.47	−263.82
26	7.800e+05	2.283e−04	5.321e−04	85.32	−59.03
27	8.100e+05	8.833e−04	2.059e−03	−173.40	−317.75
28	8.400e+05	8.022e−04	1.870e−03	29.43	−114.93
29	8.700e+05	5.818e−04	1.356e−03	145.87	1.52

Total harmonic distortion: 12.211564%(12.226294%)

6.7 Class E-resonant pulse inverter

The schematic of a class E-resonant inverter is shown in Figure 6.9(a) [3]. The current and voltage waveforms are shown in Figure 6.9(b) and (c). When the DC supply voltage V_s is turned on, a resonant oscillation starts through L_e and C_e. The resonance current $i_s(t)$ flows through the DC source as given by [1,3,5]

$$i_s(t) = V_s\sqrt{\frac{C_e}{L_e}}\sin(\omega_e t) \tag{6.60}$$

where the resonant frequency $\omega_e = 1/\sqrt{L_e C_e}$.

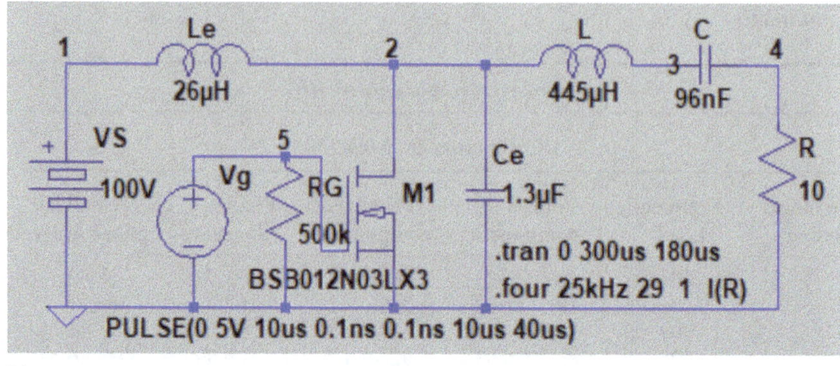

(a)

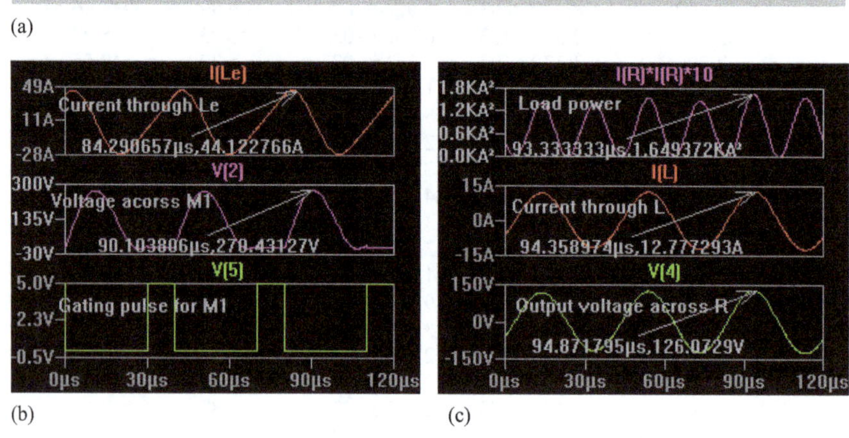

(b) (c)

Figure 6.9 Class E-resonant inverter: (a) schematic, (b) gate pulses, input current, and voltage, and (c) load voltage, current, and power

And the voltage across capacitor C_e can be described approximately as

$$v_{ce}(t) = V_s(1 - \cos(\omega_e t)) \tag{6.61}$$

The switch M_1 is turned on when the voltage across the capacitor becomes low and the current through inductor L_e rises. The gating pulse for M_1 can be adjusted to match with the capacitor voltage as shown in the pulse description. The voltage across the switch M_1 varies and serves as an input to the load consisting of the resonant RLC circuit. The location of the switch M_1 makes the circuit operate as Boost converter with a duly cycle k given by

$$k = \frac{t_{on}}{T_s} \tag{6.62}$$

where t_{on} is the on-time of the switch and T_s is the period of the output waveform. For a Boost converter, the average input voltage to the RLC load circuit is given by

$$V_i = \frac{V_S}{1 - k} \tag{6.63}$$

Thus, the voltage gain of the RLC circuit is frequency domain is given

$$G(j\omega) = \frac{V_o}{V_i} = \frac{R}{R + j\omega_s L - \frac{j}{\omega_s C}} \tag{6.64}$$

which gives peak value of the output voltage as given by

$$V_m = \frac{V_i}{\sqrt{1 + \left(\frac{\omega_s L}{R} - \frac{1}{\omega_s C}\right)^2}} \tag{6.65}$$

Let us consider following circuit parameters:
DC supply voltage $V_S = 100$ V, Load resistance $R = 10\ \Omega$
Output frequency $f_s = 25$ kHz $\quad \omega_s = 2 \times \pi \times f_s = 1.571 \times 10^5$ rad/s
On-time of the switch, $t_{on} = 10$ μs
The following guidelines are recommended for determining the optimum values of the inductors and capacitors.
Select series Quality factor Q_s for load RLC, $Q_s = 7$

$$\begin{aligned} L_e &= \frac{0.4001R}{\omega_s} \\ &= \frac{0.4001 \times 10}{1.571 \times 10^5} = 25.471\ \mu\text{H} \end{aligned} \tag{6.66}$$

$$\begin{aligned} C_e &= \frac{2.165}{R\omega_s} \\ &= \frac{2.165}{10 \times 1.571 \times 10^5} = 1.378\ \mu\text{F} \end{aligned} \tag{6.67}$$

$$\begin{aligned} \omega_e &= \sqrt{\frac{1}{L_e C_e}} \\ &= \sqrt{\frac{1}{25.471 \times 10^{-6} \times 1.378 \times 10^{-6}}} = 1.688 \times 10^5\ \text{rad/s} \end{aligned} \tag{6.68}$$

$$\begin{aligned} f_e &= \frac{1}{2\pi}\sqrt{\frac{1}{L_e C_e}} \\ &= \frac{1.688 \times 10^5}{2\pi} = 26.86\ \text{kHz} \end{aligned} \tag{6.69}$$

$$\begin{aligned} T_e &= \frac{1}{f_e} \\ &= \frac{1}{26.86\ \text{kHz}} = 37.228\ \mu\text{s} \end{aligned} \tag{6.70}$$

$$\begin{aligned} L &= \frac{Q_s R}{\omega_s} \\ &= \frac{7 \times 10}{1.571 \times 10^5} = 445.634\ \mu\text{H} \end{aligned} \tag{6.71}$$

$$\omega_s L - \frac{1}{\omega_s C} = 0.3533R$$

which gives

$$C = \frac{1}{(Q_s R - 0.3533R)\omega_s} = \frac{1}{(7 \times 10 - 0.3533 \times 10) \times 1.571 \times 10^5} = 95.78 \text{ nF} \tag{6.72}$$

Alternately,

$$C = \frac{1}{\omega_s(\omega_s L - 0.3533R)} = \frac{1}{1.571 \times 10^5 \times \left(1.571 \times 10^5 \times 445.634 \times 10^{-6} - 0.3533 \times 10\right)} = 95.78 \text{ nF}$$

Using the values of L and C, we obtain

$$\omega_s = \sqrt{\frac{1}{LC}} = \sqrt{\frac{1}{445.634 \times 10^{-6} \times 95.78 \times 10^{-9}}} = 1.531 \times 10^5 \text{ rad/s} \tag{6.73}$$

$$f_s = \frac{1}{2\pi}\sqrt{\frac{1}{LC}} = \frac{1.531 \times 10^5}{2\pi} = 24.36 \text{ kHz} \tag{6.74}$$

$$T_s = \frac{1}{f_s} = \frac{1}{24.36 \text{ kHz}} = 41.049 \text{ µs} \tag{6.75}$$

Equation (6.62) gives the duty cycle as

$$k = \frac{t_{on}}{T_s} = \frac{10 \text{ µs}}{40 \text{ µs}} = 0.25 \tag{6.76}$$

6.7.1 Output-side parameters

(a) Equation (6.63) gives the average input voltage to RLC load

$$V_i = \frac{V_S}{1-k} = \frac{100}{1 - 0.25} = 133.333 \text{ V}$$

(b) Equation (6.65) gives the peak output voltage

$$V_m = \frac{V_i}{\sqrt{1 + (\omega_s L/R - 1/(\omega_s C))^2}} = 125.718 \text{ V}$$

(c) The peak load current as

$$I_m = \frac{V_m}{\sqrt{R + j\omega L - \dfrac{j}{\omega C}}} = \frac{V_m}{\sqrt{R^2 + \left(\omega_s L - \dfrac{1}{\omega_s C}\right)^2}} = 11.854 \text{ A}$$

(d) The rms load current is $I_o = \frac{I_m}{\sqrt{2}} = 8.382$ A

(e) Power delivered to the load $P_{out} = I_o{}^2 R = 8.382^2 \times 10 = 702.556$ W

6.7.2 Input-side parameters

(a) The input power

$$P_{in} = P_{out} = 702.556 \text{ W}$$

(b) The average input current is

$$I_s = \frac{P_{in}}{V_S} = \frac{702.556}{100} = 7.026 \text{ A}$$

6.7.3 Converter parameters

(a) Equation (6.60) gives the peak transistor current

$$I_p = V_S\sqrt{\frac{C_e}{L_e}} = 23.262 \text{ A}$$

(b) The average transistor current

$$I_{av} = kI_s = 0.25 \times 7.026 = 1.756 \text{ A}$$

(c) The rms transistor current

$$I_{rms} = \sqrt{k}I_s = \sqrt{0.25} \times 7.026 = 3.513 \text{ A}$$

(d) Considering a safety factor of 2, the peak transistor voltage,

$$V_{peak} = 2V_s$$
$$= 2 \times 100 = 200 \text{ V}$$

The LTspice command (.four 25kHz 29 1 I(*R*)) for Fourier analysis gives the Fourier components of the load current *I*(*R*) at the fundamental frequency of 25 kHz as follows:

Fourier components of *I(R)*

DC component: −0.0398198

Harmonic Number	Frequency (Hz)	Fourier Component	Normalized Component	Phase (degrees)	Normalized Phase (degrees)
1	2.500e+04	1.264e+01	1.000e+00	−31.46	0.00
2	5.000e+04	3.480e−01	2.753e−02	166.75	198.21
3	7.500e+04	3.532e−02	2.794e−03	−169.97	−138.51
4	1.000e+05	8.444e−03	6.680e−04	77.15	108.61
5	1.250e+05	5.968e−03	4.721e−04	111.34	142.80
6	1.500e+05	1.384e−03	1.095e−04	22.80	54.26
7	1.750e+05	5.443e−03	4.306e−04	11.64	43.10
8	2.000e+05	4.250e−03	3.362e−04	21.82	53.28
9	2.250e+05	3.701e−03	2.928e−04	13.86	45.33
10	2.500e+05	3.450e−03	2.730e−04	5.47	36.93
11	2.750e+05	3.403e−03	2.692e−04	8.93	40.39
12	3.000e+05	3.914e−03	3.097e−04	7.19	38.65
13	3.250e+05	3.053e−03	2.416e−04	4.74	36.20
14	3.500e+05	2.615e−03	2.068e−04	8.81	40.27
15	3.750e+05	3.130e−03	2.476e−04	4.06	35.52
16	4.000e+05	2.772e−03	2.193e−04	3.32	34.78
17	4.250e+05	2.632e−03	2.083e−04	3.78	35.24
18	4.500e+05	2.524e−03	1.996e−04	4.25	35.71
19	4.750e+05	2.089e−03	1.653e−04	4.70	36.16
20	5.000e+05	2.607e−03	2.063e−04	−1.79	29.67
21	5.250e+05	2.305e−03	1.823e−04	1.07	32.53
22	5.500e+05	1.672e−03	1.323e−04	5.60	37.06
23	5.750e+05	2.138e−03	1.691e−04	−2.36	29.11
24	6.000e+05	2.019e−03	1.597e−04	−0.16	31.30
25	6.250e+05	1.633e−03	1.292e−04	3.07	34.54
26	6.500e+05	1.756e−03	1.389e−04	−3.22	28.24
27	6.750e+05	1.646e−03	1.302e−04	7.21	38.68
28	7.000e+05	1.580e−03	1.250e−04	3.72	35.19
29	7.250e+05	1.690e−03	1.337e−04	−6.87	24.59

THD: 2.770770% (2.771684%)

Note: A low THD of 2.8% indicates the quality of the output current.

6.8 Summary

Switching a DC voltage source to a resonant LC circuit causes a resonant oscillation of a sinusoidal current at a resonant frequency. The resonant frequency and the peak resonant current depend on the values of L and C. A small load resistance R is added to the resonant circuit for desired power dissipation. The addition of the resistance changes the resonant current to follow an exponentially decaying damped sinusoidal current and the resonant frequency is reduced to a damped resonant frequency. The switching the DC voltage source to an RLC circuit is done through two switching devices such as power transistors. The resonant oscillation continues for 180° and the transistors are turned off, when the resonant current falls to zero. One transistor switch causes a positive resonant pulse of current through the load resistance and another transistor causes a negative resonant pulse of current through the load resistance. There must be a dead–zone time between the turn–off time of one switch and the turn–on time of the other transistor switch. A transistor with an antiparallel diode will operate as a bidirectional switch, allowing the current flow in both directions. A resonant pulse inverter can operate as a fived out or a variable output voltage. We can vary the output voltage or the current in the half-bridge and the full-bridge inverters. The input to a parallel resonant inverter is a constant current source which is normally switched to a transformer with two primary windings and one secondary winding. The constant current source is normally varied normally varied by varying the DC input voltage to the converter through an inductor.

Problems

1. The half-bridge resonant pulse inverter is shown in Figure 6.2(a). The circuit parameters are: $C_1 = C_2 = C = 3$ μF, $L_1 = L_2 = L = 50$ μH, $R = 2$ Ω, and $Vs = 110$ V. Determine (a) the peak supply current I_{ps}, and (b) the resonant frequency f_r (kHz) and (c) the capacitor voltage range V_c.
2. The half-bridge resonant pulse inverter is shown in Figure 6.2(a). The circuit parameters are: $C_1 = C_2 = C = 3$ μF, $L_1 = L_2 = L = 50$ μH, $R = 2$ Ω, and $Vs = 110$ V. Calculate the performance parameters: (a) output-side parameters, (b) controller parameters, and (c) input-side parameters. Assume ideal transistor switches.
3. The half-bridge resonant pulse inverter is shown in Figure 6.2(a). The circuit parameters are: $C_1 = C_2 = C = 3$ μF, $L_1 = L_2 = L = 50$ μH, $R = 2$ Ω, and $Vs = 98$ V. Determine (a) the peak supply current I_{ps}, (b) the resonant frequency f_r (kHz), and (c) the capacitor voltage range V_c.
4. The half-bridge resonant pulse inverter is shown in Figure 6.2(a). The circuit parameters are: $C_1 = C_2 = C = 3$ μF, $L_1 = L_2 = L = 50$ μH, $R = 2$ Ω, and $Vs = 98$ V. Calculate the performance parameters: (a) Output-side parameters, (b) controller parameters, and (c) input-side parameters. Assume ideal transistor switches.

5. The half-bridge resonant pulse inverter is shown in Figure 6.2(a). The circuit parameters are: $C_1 = C_2 = C = 4$ μF, $L_1 = L_2 = L = 50$ μH, $R = 1.2$ Ω, and $Vs =$ 110 V. Determine (a) the peak supply current I_{ps}, (b) the resonant frequency f_r (kHz), and (c) the capacitor voltage range V_c.
6. The half-bridge resonant pulse inverter is shown in Figure 6.2(a). The circuit parameters are: $C_1 = C_2 = C = 4$ μF, $L_1 = L_2 = L = 50$ μH, $R = 1.2$ Ω, and $Vs =$ 110 V. Calculate the performance parameters: (a) output-side parameters, (b) controller parameters, and (c) input-side parameters. Assume ideal transistor switches.
7. The half-bridge resonant pulse inverter is shown in Figure 6.2(a). The circuit parameters are: $C_1 = C_2 = C = 4$ μF, $L_1 = L_2 = L = 50$ μH, $R = 1.2$ Ω, and $Vs =$ 98 V. Determine (a) the peak supply current I_{ps}, (b) the resonant frequency f_r (kHz), and (c) the capacitor voltage range V_c.
8. The half-bridge resonant pulse inverter is shown in Figure 6.2(a). The circuit parameters are: $C_1 = C_2 = C = 4$ μF, $L_1 = L_2 = L = 50$ μH, $R = 1.2$ Ω, and $Vs =$ 98 V. Calculate the performance parameters: (a) output-side parameters, (b) controller parameters, and (c) input-side parameters. Assume ideal transistor switches.
9. The half-bridge bidirectional resonant pulse inverter is shown in Figure 6.3 (a). The circuit parameters are: $C_1 = C_2 = C = 4$ μF, $L_1 = L_2 = L = 50$ μH, $R =$ 1.2 Ω, and $Vs =$ 110 V. Determine (a) the peak supply current I_{ps}, (b) the resonant frequency f_r (kHz), and (c) the capacitor voltage range V_c.
10. The half-bridge resonant pulse inverter is shown in Figure 6.3(a). The circuit parameters are: $C_1 = C_2 = C = 4$ μF, $L_1 = L_2 = L = 50$ μH, $R = 1.2$ Ω, and $Vs =$ 110 V. Calculate the performance parameters: (a) output-side parameters, (b) controller parameters, and (c) input-side parameters. Assume ideal transistor switches.
11. The half-bridge bidirectional resonant pulse inverter is shown in Figure 6.3 (a). The circuit parameters are: $C_1 = C_2 = C = 3$ μF, $L_1 = L_2 = L = 50$ μH, $R = 2$ Ω, and $Vs = 110$ V. Determine (a) the peak supply current I_{ps}, (b) the resonant frequency f_r (kHz), and (c) the capacitor voltage range V_c.
12. The half-bridge resonant pulse inverter is shown in Figure 6.3(a). The circuit parameters are: $C_1 = C_2 = C = 3$ μF, $L_1 = L_2 = L = 50$ μH, $R = 2$ Ω, and $Vs =$ 110 V. Calculate the performance parameters: (a) output-side parameters, (b) controller parameters, and (c) input-side parameters. Assume ideal transistor switches.
13. The half-bridge bidirectional resonant pulse inverter is shown in Figure 6.3 (a). The circuit parameters are: $C_1 = C_2 = C = 3$ μF, $L_1 = L_2 = L = 50$ μH, $R = 2$ Ω, and $Vs = 98$ V. Determine (a) the peak supply current I_{ps}, (b) the resonant frequency f_r (kHz), and (c) the capacitor voltage range V_c.
14. The half-bridge resonant pulse inverter is shown in Figure 6.3(a). The circuit parameters are: $C_1 = C_2 = C = 3$ μF, $L_1 = L_2 = L = 50$ μH, $R = 2$ Ω, and $Vs = 98$ V. Calculate the performance parameters: (a) output-side parameters, (b) controller parameters, and (c) input-side parameters. Assume ideal transistor switches.

15. The full-bridge bidirectional resonant pulse inverter is shown in Figure 6.5(a). for $T_o = T_r$. The circuit parameters are: $C_1 = C_2 = C = 4$ μF, $L_1 = L_2 = L = 50$ μH, $R = 1.2$ Ω, and $Vs = 110$ V. Determine (a) the peak supply current I_{ps}, (b) the resonant frequency f_r (kHz), and (c) the capacitor voltage range V_c.
16. The full-bridge bidirectional resonant pulse inverter is shown in Figure 6.5(a). for $T_o = T_r$. The circuit parameters are: $C_1 = C_2 = C = 4$ μF, $L_1 = L_2 = L = 50$ μH, $R = 1.2$ Ω, and $Vs = 110$ V. Calculate the performance parameters: (a) output-side parameters, (b) controller parameters, and (c) input-side parameters. Assume ideal transistor switch*es*.
17. The full-bridge bidirectional resonant pulse inverter is shown in Figure 6.5(a). for $T_o = T_r$. The circuit parameters are: $C_1 = C_2 = C = 3$ μF, $L_1 = L_2 = L = 50$ μH, $R = 2$ Ω, and $Vs = 110$ V. Determine (a) the peak supply current I_{ps}, (b) the resonant frequency f_r (kHz), and (c) the capacitor voltage range V_c.
18. The full-bridge bidirectional resonant pulse inverter is shown in Figure 6.5(a). for $T_o = T_r$. The circuit parameters are: $C_1 = C_2 = C = 3$ μF, $L_1 = L_2 = L = 50$ μH, $R = 2$ Ω, and $Vs = 110$ V. Calculate the performance parameters: (a) output-side parameters, (b) controller parameters, and (c) input-side parameters. Assume ideal transistor switch*es*.
19. The parallel resonant inverter is shown in Figure 6.7. The circuit parameters are: For $R = 1$ kΩ, Ls = 0.5 mH, $C = 0.3$ μF, $L= L_1 = L_2 = 0.4$ mH, $f= 25$ kHz, $V_s = 98$ V. Determine (a) the peak supply current I_{ps}, (b) the resonant frequency f_r (kHz), and (c) the capacitor voltage range V_c.
20. The parallel resonant inverter is shown in Figure 6.7. The circuit parameters are: For $R = 1$ kΩ, $Ls = 0.5$ mH, $C = 0.3$ μF, $L= L_1 = L_2 = 0.4$ mH, $f= 25$ kHz, $V_s = 98$ V. Calculate the performance parameters: (a) output-side parameters, (b) controller parameters, and (c) input-side parameters. Assume ideal transistor switch*es*.
21. The parallel resonant inverter is shown in Figure 6.7. The circuit parameters are: For $R = 1$ kΩ, Ls = 0.5 mH, $C = 0.3$ μF, $L= L_1 = L_2 = 0.4$ mH, $f= 25$ kHz, $V_s = 110$ V. Determine (a) the peak supply current I_{ps}, (b) the resonant frequency f_r (kHz), and (c) the capacitor voltage range V_c.
22. The parallel resonant inverter is shown in Figure 6.7. The circuit parameters are: For $R = 1$ kΩ, Ls = 0.5 mH, $C = 0.3$ μF, $L= L_1 = L_2 = 0.4$ mH, $f= 25$ kHz, $V_s = 110$ V. Calculate the performance parameters: (a) output-side parameters, (b) controller parameters, and (c) input-side parameters. Assume ideal transistor switch*es*.
23. The class E inverter as shown in Figure 6.9(a) operates at resonance and has $Vs = 48$ V and $R = 8$ Ω. The switching frequency is $fs = 25$ kHz. Assume a quality factor $Q = 7$. Calculate (a) the optimum values on inductor L, (b) the optimum value of capacitor C, (c) the optimum values on inductor L_e, (d) the optimum value of capacitor C_e, (e) the damping factor δ, and (*f*) the peak output voltage V_o for $V_i = 12$ V. Assume a quality factor $Q = 7$ and an ideal transistor switch.
24. The class E inverter as shown in Figure 6.9(a) operates at resonance and has $Vs = 48$ V and $R = 8$ Ω.

(a) Calculate (a) the optimum values on inductor L, (b) the optimum value of capacitor C, (c) the optimum values on inductor L_e, and (d) the optimum value of capacitor C_e.
(b) Calculate the performance parameters: (a) output-side parameters, (b) controller parameters, and (c) input-side parameters. Assume ideal transistor switches.

25. The class E inverter as shown in Figure 6.9(a) operates at resonance and has $Vs = 48$ V and $R = 4\ \Omega$.
(e) Calculate (a) the optimum values on inductor L, (b) the optimum value of capacitor C, (c) the optimum values on inductor L_e, and (d) the optimum value of capacitor C_e.
(ii) Calculate the performance parameters: (a) output-side parameters, (b) controller parameters, and (c) input-side parameters. Assume ideal transistor switches.

References

[1] M.H. Rashid, *Power Electronics: Circuits, Devices, and Applications*, 4th edition, Chapter 7. Englewood Cliffs, NJ: Prentice-Hall.
[2] M.H. Rashid, *SPICE and LTspice for Power Electronics and Electric Power*. Boca Raton, FL: CRC Press, 2024.
[3] M.H. Rashid, *Power Electronics Circuits, Devices and Applications*, 3rd ed. Englewood Cliffs, NJ: Prentice-Hall, 2003, Section 7.11 and Chapter 10.
[4] M.H. Rashid, *Introduction to PSpice Using OrCAD for Circuits and Electronics*, 3rd ed. Englewood Cliffs, NJ: Prentice-Hall, 2003, Chapter 6.
[5] M.K. Kazimierczuk and I. Jozwik, Class-E zero-voltage switching and zero-current switching rectifiers, *IEEE Transactions on Circuits and Systems*, 37(3), 1990, 436–444.

Chapter 7
Controlled rectifiers

7.1 Introduction

A rectifier converts an alternating current (AC) voltage to a direct current (DC) voltage and the average output voltage is fixed. However, a controlled rectifier also converts a fixed AC voltage to a variable DC voltage and uses thyristors commonly as switching devices. However, other bidirectional transistors can also be used. The average output voltage can be varied from zero to a maximum value of a diode rectifier. However, the input power factor depends on the output voltage. It uses one or more thyristors to make a unidirectional current flow from the positive terminal to the negative terminal of the output voltage. Thus, the voltage and the current on the input side are of AC types and the voltage and the current on the output side are of DC types.

7.2 Characteristics of silicon-controlled rectifiers

A silicon-controlled rectifier (SCR) is commonly known as a *thyristor* and it has three terminals: anode, cathode, and gate. A thyristor can be turned on by applying a pulse of short duration. Once the thyristor is on, the gate pulse has no effect, and it remains on until its current is reduced to zero. It is a latching device. The most common characteristics of a thyristor are [1–8]

- The device is normally off, and it can withstand both positive and negative voltages between the anode and cathode terminals.
- It can be turned on by applying a small triggering pulse of about 5–10 voltage and 100 μs width between the gate and the cathode terminal.
- When the device is turned on and the anode current reaches a certain magnitude known as the *Latching* current in the order of milli-amps, it latches in. The gate-pulse has no effect and it can be removed. An SCR is a latching device.
- When the anode current of a conducting device falls below a certain magnitude known as the *Holding* current in the order of milli-amps, it unlatches. It stops current flow and turns off completely.
- For complete recovery from the on-state to a fully off-state of its ability of withstanding voltages, there must be a minimum time allowed known as the *reverse recovery time* t_{off} before the device is turned on again so that the device can fully recover and operate normally regaining its ability of withstanding voltages. Otherwise, the device will not be turned off and it remains on.

- Once the device is turned on, the gate-pulse has no control over the current flow. For an AC supply, the anode current falls to zero due to the natural behavior of the supply voltage and the device is turned off naturally. For a DC source, there must be a commutation circuit to force a negative current flow to reduce anode current below the holding current and turn-off the device.

Figure 7.1(a) shows the schematic of a thyristor test circuit for a resistive load of $R = 10\ \Omega$. The thyristor is gated with a pulse on 6 V and a width of 100 μs. For a 60 Hz supply, the pulse is delayed by 60° from the zero crossing of the supply voltage. The gate signal v_g is applied between the gate and the cathode of the thyristor through a voltage-controlled voltage source E_1 providing an isolation between the power circuit and the low-level signal-generating circuit. For a frequency of 60 Hz, the period $T = 1/f = 1/60 = 16.67$ ms. The pulse is delayed by $\alpha = 60°$ from the zero crossing of the supply voltage and the corresponding time delay is $t_{delay} = \frac{60°}{360°} \times 16.67 \times 10^{-3} = 2.7778$ ms.

The LTspice statement for the gate pulse is: PULSE (0 5 V 2777.8 μs 0.1 ns 0.1 ns 100 μs 16.667 ms).

As shown in Figure 7.1(b), the thyristor starts conducting when the gate pulse starts, and the input voltage is the same as the output voltage. The thyristor stops conducting when the input voltage falls to zero due to the natural behavior of the sinusoidal input voltage. In practice, there will be a small drop of about 0.5–0.7 V due to the thyristor device. The thyristor conducts from 60° to 180° for a resistive load. For an RL load, the thyristor will continue to conduct until the anode current of the thyristor falls below the holding current. An inductor L is added to the test circuit in Figure 7.1(a) as shown in Figure 7.2(a). The inductor value is defined as a variable "LVAL" for $L = 0$, 6 mH, 16 mH, and 160 mH. The load currents for these inductor values are shown in Figure 7.2(b). As the value of the inductor increases, it takes a longer time for the load current to fall to zero. The angle when the current falls to zero is known as the *extinction angle*. The analysis of a controlled rectifiers depends on the type of load: (a) resistive load, and (b) RL load.

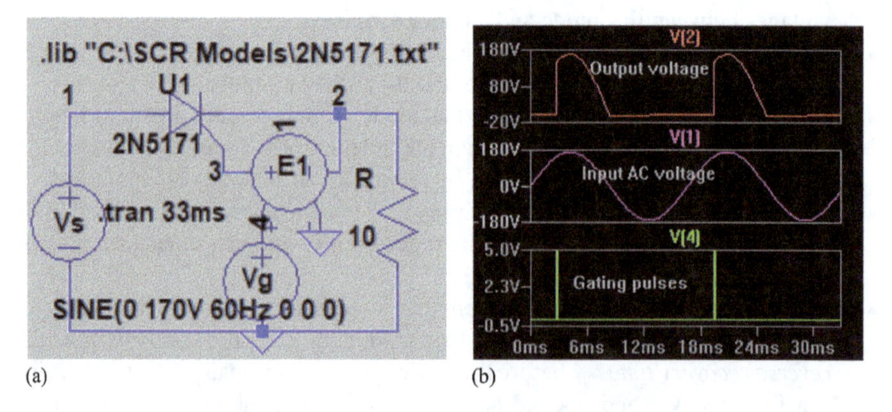

(a) (b)

Figure 7.1 Thyristor test circuit for a resistive load: (a) schematic and (b) waveforms

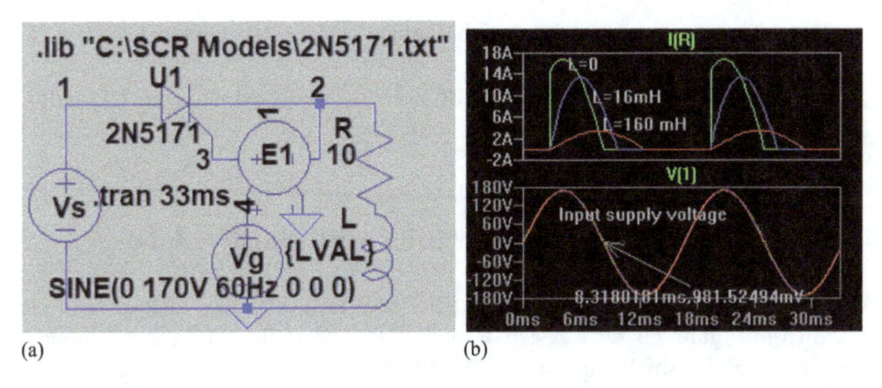

(a) (b)

Figure 7.2 Thyristor test circuit for an RL load: (a) schematic and (b) waveforms

Table 7.1 SPICE subcircuit model for 2N5171 [9]

```
* ============= SPICE Model =================
.SUBCKT 2N5171 10 30 20
* 10–A, 30–G, 20—K
.MODEL DGAT D (IS=1.0e−12 N=1 RS=0.001)
.MODEL DMOD D (IS=1.0e−12 N=0.001)
.MODEL DON D (IS=1.000e−012 N=1.000e+000 RS=3.534e−002 BV=7.200e+002)
.MODEL DBREAK D (IS=1.000e−012 N=9.404e+002 BV=7.200e+002)
V1 10 14 DC 0
DON1 14 222 DON
VV 222 22 DC 0
E1 22 20 poly(2) 10 20 3 20 0 0 0 0 1
DBRK1 14 27 DBREAK
DBRK2 20 27 DBREAK
RLEAK 14 20 6.000e+007
CRISE 14 20 1.000e−009
FC1 3 20 poly(2) VGD V1 −4.000e−002 1 8.000e−001
CON 3 20 4.000e−008 IC=1.5
DS1 3 31 DMOD
DS2 20 3 DMOD
VW 31 20 DC 1
DGATE 30 7 DGAT
VGD 7 20 DC 8.677e−001
.ENDS
*============= Model template =================
```

The LTspice statement for variable inductor is: .step param LVAL LIST 0 16 m 160 m.

The SPICE model of the Motorola SCR 2N5171 is listed in Table 7.1, and the location of the library file is included with the SPICE command: .lib "C:\SCR Models\2N5171.txt" [9]

7.3 Single-phase controlled rectifier with resistive load

A single-phase full-wave-controlled rectifier uses four thyristors instead of diodes as shown in Figure 7.3 for a resistive load of R. The addition of a small inductor $L = 100$ nH and a small capacitor $C = 440$ pF makes the circuit in a generalized form and these elements are often connected to serve as a filter. The AC input voltage is given by [10]

$$v_s(t) = V_m \sin(\omega t) = V_m \sin(2\pi f t) \tag{7.1}$$

where V_m = peak value of the AC input voltage, V; f = supply frequency, Hz; $\omega = 2\pi f$ = supply frequency, rad/s.

The peak supply voltage $V_{\mathrm{m}} = 169.7$ V. Let us consider a delay angle of $\alpha = 60°$,

Time delay $t_1 = \frac{60}{360} \times \frac{1000}{60\ \mathrm{Hz}} \times 1000 = 2777.8\ \mu\mathrm{s}$

Time delay $t_2 = \frac{240}{360} \times \frac{1000}{60\ \mathrm{Hz}} \times 1000 = 11{,}111.1\ \mu\mathrm{s}$

During the positive half-cycle of the input voltage, thyristors T_1 and T_2 are triggered at a delay angle of $\alpha = 60°$. The input voltage appears across the load from $\omega t = 60°$ to $180°$. During the negative half-cycle in the input voltage, thyristors T_3 and T_4 are triggered at a delay angle of $\alpha = 180° + 60° = 240°$ and the input voltage appears across the load from $\omega t = 240°$ to $360°$. The plots of the gating

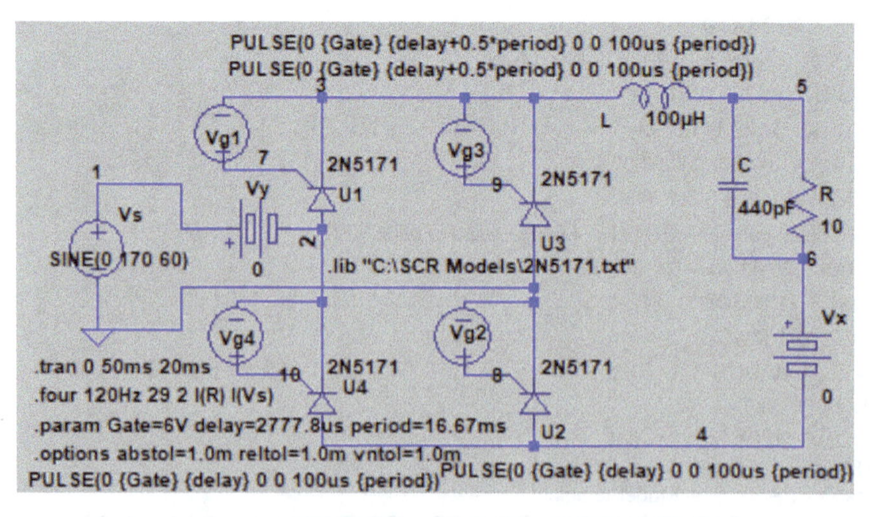

Figure 7.3 Schematic of single-phase controlled rectifier

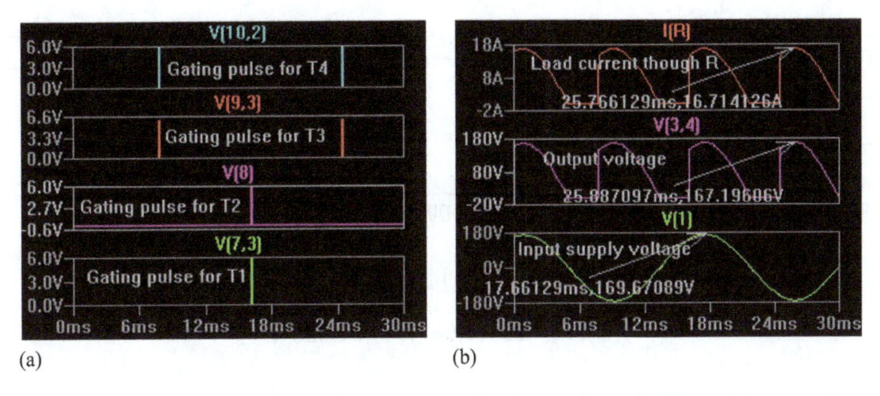

Figure 7.4 Single-phase controlled rectifier: (a) thyristor gating pulses and (b) output voltage and load current

pulses of 6 V, and width 100 μs are shown in Figure 7.4(a). The output voltage can be described as

$$\begin{aligned} v_o(t) &= V_m \sin(\omega t) \quad \text{for } \alpha < \omega t \leq \pi \\ &= V_m \sin(\omega t) \quad \text{for } (\pi + \alpha) < \omega t \leq 2\pi \end{aligned} \tag{7.2}$$

The plots of the input and output voltages are shown in Figure 7.4(b). While the thyristor is fully turned on and conducting, the output voltage is almost identical to the input voltage for a resistive load. Load, that is, $v_o = v_i$.

The gating pulses for T_1 and T_2 delayed by α are identical. Similarly, the gating pulses for T_3 and T_4 delayed by $(\pi+\alpha)$ are identical. As expected, the load current becomes zero when the input voltage falls to zero.

Since the thyristor conducts from $\omega t = \alpha$ to π, the average output voltage can be found from

$$\begin{aligned} V_{dc} &= \frac{1}{2\pi}\left[\int_{\alpha}^{\pi} V_m \sin(x)dx + \int_{\pi+\alpha}^{2\pi} V_m \sin(x)dx\right] \\ &= \frac{2}{2\pi}\int_{\alpha}^{\pi} V_m \sin(x)dx = \frac{V_m}{\pi}(1 + \cos\alpha) \end{aligned} \tag{7.3}$$

The rms output voltage can be found from

$$V_o = \sqrt{\frac{2}{2\pi}\int_{\alpha}^{\pi}(V_m \sin(x))^2 dx} = V_m\sqrt{\frac{\sin(2\alpha)}{4\pi} + \frac{1}{2} - \frac{\alpha}{2\pi}} \tag{7.4}$$

The performance parameters of controlled rectifiers can be divided into three types.

- Output-side parameters,
- Rectifier parameters, and
- Input-side parameters.

Let us consider an input voltage of $V_s = 120$ rms, $V_m = V_s = 120 \times \sqrt{2} = 169.7$ V, $R = 10\ \Omega$ and delay angle of $\alpha = 60°$. Converting to radian, $\alpha = \frac{60°}{180} \times \pi = 1.047$ rad

7.3.1 Output-side parameters

(a) Equation (7.3) gives the average output voltage is

$$V_{o(av)} = V_{dc} = \frac{V_m}{\pi}(1 + \cos\alpha)$$
$$= \frac{169.7}{\pi} \times (1 + \cos(1.047)) = 81.028 \text{ V}$$

(b) The average output load current is

$$I_{o(av)} = \frac{V_{o(av)}}{R}$$
$$= \frac{81.028}{10} = 8.103 \text{ A}$$

(c) The DC output power is

$$P_{dc} = V_{dc} I_{dc}$$
$$= 81.028 \times 8.103 = 656.561 \text{ W}$$

(d) Equation (7.4) gives the rms output voltage is

$$V_{o(rms)} = V_{rms} = V_m \sqrt{\frac{\sin(2\alpha)}{4\pi} + \frac{1}{2} - \frac{\alpha}{2\pi}}$$
$$= 169.7 \times \sqrt{\frac{\sin(2 \times 1.047)}{4 \times \pi} + \frac{1}{2} - \frac{1.047}{2 \times \pi}} = 107.633 \text{ V}$$

(e) The rms output current is

$$I_{o(rms)} = I_{rms} = \frac{V_{o(rms)}}{R}$$
$$= \frac{107.633}{10} = 10.763 \text{ A}$$

(f) The AC output power is

$$P_{ac} = V_{o(rms)} I_{o(rms)}$$
$$= 107.633 \times 10.763 = 1.158 \text{ kW}$$

(g) The output AC power

$$P_{out} = I_{rms}^2 R$$
$$= 10.763^2 \times 10 = 1.158 \text{ kW}$$

(h) The rms ripple content of the output voltage

$$V_{ac} = \sqrt{V_{o(rms)}^2 - V_{dc}^2}$$
$$= \sqrt{107.633^2 - 81.028^2} = 70.846 \text{ V}$$

(i) The rectification *efficiency or ratio*

$$\eta = \frac{P_{dc}}{P_o}$$
$$= \frac{656.561}{1158} = 56.674\%$$

(j) The *ripple factor*

$$RF = \frac{V_{ac}}{V_{dc}}$$
$$= \frac{70.846}{81.028} = 87.434\%$$

(k) The *form factor of the output voltage*

$$FF_{ov} = \frac{V_{rms}}{V_{dc}}$$
$$= \frac{107.633}{81.028} = 132.833\%$$

7.3.2 Controlled rectifier parameters

(a) The peak thyristor current

$$I_{T(peak)} = \frac{V_m}{R}$$
$$= \frac{169.7}{10} = 16.971 \text{ A}$$

(b) The load current is shared by two devices, the average thyristor current is

$$I_{T(av)} = \frac{I_{o(av)}}{2}$$
$$= \frac{8.103}{2} = 4.051 \text{ A}$$

(c) The load current is shared by two devices, the rms thyristor current is

$$I_{T(rms)} = \frac{I_{rms}}{\sqrt{2}}$$
$$= \frac{10.763}{\sqrt{2}} = 7.611 \text{ A}$$

(d) With a safely margin of two times, the peak voltage rating of thyristors is

$$V_{peak} = 2V_m$$
$$= 2 \times 169.7 = 339.411 \text{ V}$$

7.3.3 Input-side parameters

The rms input current is

$$I_s = I_{rms} = 10.763 \text{ A}$$

Assuming an ideal converter of no-power loss, the input power

$$P_{in} = P_{out} = 1.158 \text{ kW}$$

Input power factor

$$PF_i = \frac{P_{in}}{V_s I_s} = \frac{1158}{120 \times 10.763} = 0.897 \text{ (lagging)}$$

The transformer utilization factor

$$\text{TUF} = \frac{P_{dc}}{V_s I_s} = \frac{656.561}{120 \times 10.763} \times 100 = 50.8\%$$

The LTspice command (.four 120 Hz 29 1 $I(R)$) for Fourier analysis gives the Fourier components of the load current $I(R)$ at the fundamental frequency of 120 Hz as follows:

Fourier components of $I(R)$

DC component: 7.93869

Harmonic number	Frequency (Hz)	Fourier component	Normalized component	Phase (degrees)	Normalized phase (degrees)
1	1.200e+02	9.228e+00	1.000e+00	23.60	0.00
2	2.400e+02	1.828e+00	1.981e−01	45.41	21.81
3	3.600e+02	1.667e+00	1.806e−01	53.38	29.78
4	4.800e+02	1.311e+00	1.421e−01	93.25	69.65
5	6.000e+02	8.311e−01	9.007e−02	114.65	91.05
6	7.200e+02	7.922e−01	8.585e−02	130.90	107.31
7	8.400e+02	7.041e−01	7.631e−02	162.33	138.73
8	9.600e+02	5.380e−01	5.831e−02	−175.59	−199.19
9	1.080e+03	5.229e−01	5.667e−02	−157.31	−180.90
10	1.200e+03	4.806e−01	5.209e−02	−128.10	−151.70
11	1.320e+03	3.970e−01	4.303e−02	−106.39	−129.98
12	1.440e+03	3.903e−01	4.229e−02	−86.47	−110.06

(Continues)

(*Continued*)

Fourier components of *I*(*R*)					
DC component: 7.93869					
Harmonic number	**Frequency (Hz)**	**Fourier component**	**Normalized component**	**Phase (degrees)**	**Normalized phase (degrees)**
13	1.560e+03	3.637e−01	3.942e−02	−58.76	−82.35
14	1.680e+03	3.148e−01	3.411e−02	−37.00	−60.60
15	1.800e+03	3.104e−01	3.363e−02	−16.19	−39.79
16	1.920e+03	2.928e−01	3.174e−02	10.61	−12.98
17	2.040e+03	2.601e−01	2.818e−02	32.42	8.82
18	2.160e+03	2.577e−01	2.792e−02	53.75	30.15
19	2.280e+03	2.445e−01	2.649e−02	80.02	56.43
20	2.400e+03	2.214e−01	2.400e−02	101.80	78.21
21	2.520e+03	2.199e−01	2.383e−02	123.56	99.97
22	2.640e+03	2.094e−01	2.269e−02	149.40	125.81
23	2.760e+03	1.925e−01	2.086e−02	171.25	147.65
24	2.880e+03	1.915e−01	2.076e−02	−166.72	−190.31
25	3.000e+03	1.829e−01	1.983e−02	−141.17	−164.77
26	3.120e+03	1.701e−01	1.843e−02	−119.34	−142.93
27	3.240e+03	1.694e−01	1.836e−02	−97.04	−120.63
28	3.360e+03	1.622e−01	1.758e−02	−71.76	−95.35
29	3.480e+03	1.521e−01	1.648e−02	−49.91	−73.51
THD: 37.100038%(37.882164%)					

7.4 Single-phase controlled rectifier with RL load

A single-phase full-wave-controlled rectifier uses four thyristors as shown in Figure 7.5 to an RL load. During the positive half-cycle, thyristors T_1 and T_2 are turned on at a delay angle of $\omega t = \alpha$ and the load is connected to the input supply through T_1 and T_2. Due to the inductive load, thyristors T_1 and T_2 continue to conduct beyond $\omega t = \pi$ even though the input voltage is already negative. During the negative half-cycle of the input voltage, thyristors T_3 and T_4 are turned on. Turning on of thyristors T_3 and T_4 applies the supply voltage across thyristors T_1 and T_2 as reverse blocking voltage. T_1 and T_2 are turned off due to *line* or *natural commutation* and the load current is transferred from T_1 and T_2 to T_3 and T_4. Thus, with an inductive load, the load current will be delayed and will not fall to zero when the supply voltage falls to zero. Thus, T_1 and T_2 conduct during the first half of the period, and T_3 and T_4 conduct during the second half of the period. If the time constant of the RL load is greater than the period of the output voltage, the load current will continuous and will not fall to zero [10].

The peak supply voltage $V_{\mathrm{m}} = 169.7$ V, and a delay angle of $\alpha = 60°$,

Time delay $t_1 = \frac{60}{360} \times \frac{1000}{60\ \mathrm{Hz}} \times 1000 = 2777.8\ \mu\mathrm{s}$

Time delay $t_2 = \frac{240}{360} \times \frac{1000}{60\ \mathrm{Hz}} \times 1000 = 11{,}111.1\ \mu\mathrm{s}$

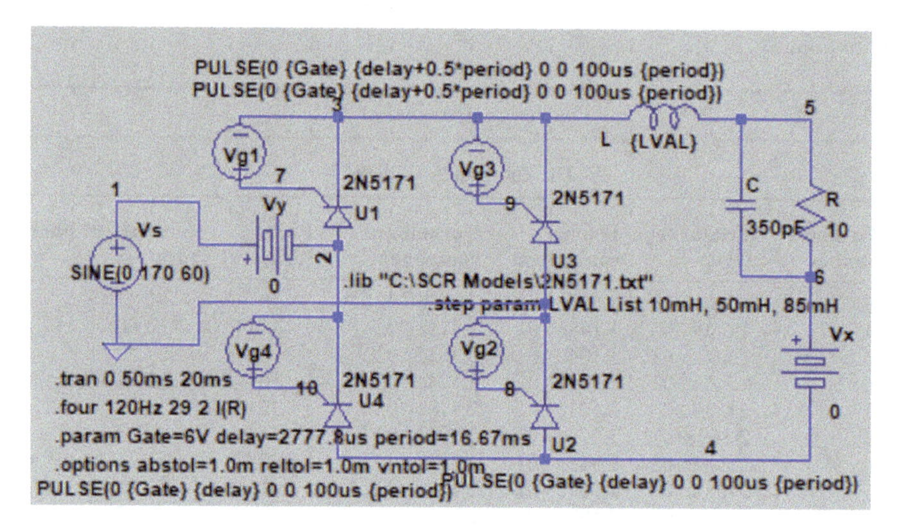

Figure 7.5 Schematic of single-phase controlled rectifier with an RL load [11]

During the positive half-cycle of the input voltage, thyristors T_1 and T_2 are triggered at a delay angle of $\alpha = 60°$. The input voltage appears across the load from $\omega t = 60°$ to $180° + 60°$. During the negative half-cycle in the input voltage, thyristors T_3 and T_4 are triggered at a delay angle of $\alpha = 180° + 60° = 240°$ and the input voltage appears across the load from $\omega t = 240°$ to $360° + 60°$. The output voltage can be described as

$$\begin{aligned} v_o(t) &= V_m \sin(\omega t) \quad \text{for } \alpha < \omega t \leq (\pi + \alpha) \\ &= V_m \sin(\omega t) \quad \text{for } (\pi + \alpha) < \omega t \leq (2\pi + \alpha) \end{aligned} \tag{7.5}$$

Depending on the value of α, the average output voltage could be either positive or negative. The average output voltage can be found from

$$V_{dc} = \frac{2}{2\pi}\int_{\alpha}^{\pi+\alpha} V_m \sin(x)dx = \frac{2V_m}{\pi}\cos\alpha \tag{7.6}$$

which gives the average load current as

$$I_{dc} = \frac{V_{dc}}{R} = \frac{2V_m}{\pi R}\cos\alpha \tag{7.7}$$

The rms output voltage can be found from

$$V_o = \sqrt{\frac{2}{2\pi}\int_{\alpha}^{\pi+\alpha} (V_m \sin(x))^2 dx} = \frac{V_m}{\sqrt{2}} \tag{7.8}$$

Let us consider an input voltage of $V_s = 120$ rms, $V_m = V_s = 120 \times \sqrt{2} = 169.7$ V, $R = 10\ \Omega$, $E = 0$, and delay angle of $\alpha = 60°$. Converting to radian, $\alpha = \frac{60°}{180} \times \pi = 1.047$ rad.

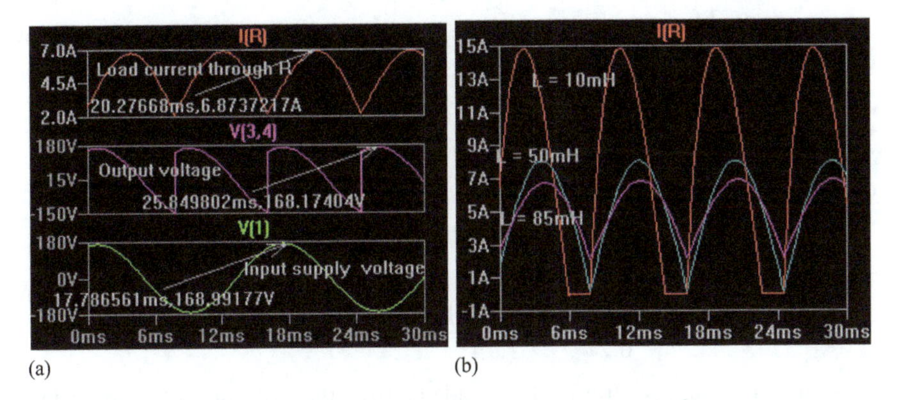

(a) (b)

Figure 7.6 Single-phase controlled rectifier with an RL load: (a) output voltage and load current and (b) effects of load inductances

The plots of the input and output voltages and the load current are shown in Figure 7.6(b). As expected, increasing the value of inductor reduces the ripple content and makes the load current continuous. By making the load highly inductive, the load current will have practically negligible ripple contents.

7.4.1 Load inductor

The frequency f_o of the output voltage is twice the supply frequency f. There are two ripples on the output current: one positive ripple and one negative ripple around the average value. We can find the inductor value to maintain continuous load current. That is,

$$\frac{L}{R} \geq T_o = \frac{1}{2 \times f} \tag{7.9}$$

which give the value of L as

$$L \geq \frac{R}{2 \times 2 \times f} = \frac{10}{2 \times 2 \times 60} = 41.67 \text{ mH}$$

For a desired value of the ripple factor of the output voltage $RF_v = 6\% = 0.06$, we can find the ripple voltage as

$$\Delta V = RF_v V_{dc}$$
$$= 0.06 \times 54.02 = 3.241 \text{ V}.$$

Similarly, for a desired value of the ripple factor of the output current $RF_i = 5\% = 0.05$, we can find the ripple current as

$$\Delta I = RF_i I_{dc}$$
$$= 0.05 \times 5.402 = 0.27 \text{ A}$$

We can find the inductor L from the ripple voltage ΔV and the ripple current ΔI as

$$L\frac{\Delta I}{\Delta T} = \Delta V$$

which gives the inductor value L as

$$\begin{aligned} L &\geq \frac{\Delta V \times \Delta T}{\Delta I} = \frac{\Delta V}{\Delta I} \times \frac{T}{2 \times 2} \\ &= \frac{3.241}{0.27} \times \frac{16.67 \times 10^{-3}}{4} = 50\text{mH} \end{aligned} \tag{7.10}$$

Using the values from (7.9) and (7.10), the average of two values $L = \frac{41.67 + 50}{2} = 45.83$ mH. Let us $L = 45$ mH. Figure 7.6(b) shows the effects of load inductances on the load current.

7.4.2 Circuit model

The operation of the converter can be divided into two identical modes: mode 1 when T_1 and T_2 conduct, and mode 2 when T_3 and T_4 conduct. The output currents during these modes are similar and we need to consider only one mode to find the output current i_L [9–11].

Mode 1 is valid for $\alpha \leq \omega t \leq (\pi + \alpha)$. If $v_s = V_m \sin(\omega t)$ is the input voltage, the load current i_L during mode 1 can be found from [9–11]

$$L\frac{di_L}{dt} + Ri_L + E = V_m \sin \omega t \quad \text{for } i_L \geq 0 \tag{7.11}$$

where E is a battery voltage connected in series to the load side.

The solution of (7.11) is of the form

$$i_L = \frac{V_s}{Z} \sin(\omega t - \theta) + A_1 e^{-(R/L)t} - \frac{E}{R} \text{ for } i_L \geq 0 \tag{7.12}$$

where the load impedance, $Z = \sqrt{R^2 + (\omega L)^2}$, and the load angle $\theta = \tan^{-1}(\omega L/R)$.

Constant A_1, which can be determined from the initial condition: at $\omega t = \alpha$, $i_L = I_{L0}$

$$A_1 = \left[I_{L0} + \frac{E}{R} - \frac{V_m}{Z} \sin(\alpha - \theta)\right] e^{(R/L)(\alpha/\omega)} \tag{7.13}$$

Substitution of A_1 in (7.12) gives i_L as

$$i_L = \frac{V_s}{Z} \sin(\omega t - \theta) - \frac{E}{R} + \left[I_{L0} + \frac{E}{R} - \frac{V_m}{Z} \sin(\alpha - \theta)\right] e^{(R/L)(\alpha/\omega - t)} \tag{7.14}$$

At the end of mode 1 in the steady-state condition $i_L(\omega t = \pi + \alpha) = I_{L1} = I_{L0}$. Applying this condition to (7.14) and solving for I_{L0}, we get

$$I_{L1} = I_{L0} = \frac{V_s}{Z}\left[\frac{-\sin(\alpha-\theta) - \sin(\alpha-\theta)e^{-(R/L)(\pi/\omega)}}{1-e^{-(R/L)(\pi/\omega)}}\right] - \frac{E}{R} \quad \text{for } I_{L0} \geq 0 \tag{7.15}$$

The critical value of α at which I_{L0} becomes zero can be solved for known values of θ, R, L, E, and V_s by an iterative method. That is,

$$0 = \frac{V_s}{Z}\left[\frac{-\sin(\alpha_c-\theta)(1+e^{-(R/L)(\pi/\omega)})}{1-e^{-(R/L)(\pi/\omega)}}\right] - \frac{E}{R} \tag{7.16}$$

which, for $E = 0$, gives $\sin(\alpha_c - \theta) = 0$. That is

$$\begin{aligned} \alpha_c &= \theta \\ &= \tan^{-1}\left(\frac{\omega L}{R}\right) = \tan^{-1}\left(\frac{2 \times \pi \times 60 \times 85 \times 10^{-3}}{10}\right) = 72.67^\circ \end{aligned} \tag{7.17}$$

Using Mathcad software, the rms current of a thyristor can be found from

$$\begin{aligned} I_R &= \sqrt{\frac{1}{\pi}\int_{\alpha}^{\pi+\alpha} i_L^2 d(\omega t)} \\ &= 4.17\text{ A} \end{aligned} \tag{7.18}$$

Using Mathcad software, the rms load current can be found from

$$\begin{aligned} I_R &= \sqrt{2} I_R \\ &= \sqrt{2} \times 4.17 = 5.89\text{ A} \end{aligned} \tag{7.19}$$

Using Mathcad software, the average current of a thyristor can be found from

$$\begin{aligned} I_A &= \frac{1}{\pi}\int_{\alpha}^{\pi+\alpha} i_L d(\omega t) \\ &= 3.85\text{ A} \end{aligned} \tag{7.20}$$

7.4.3 Converter parameters

The performance parameters of controlled rectifiers can be divided into three types.

- Output-side parameters,
- Rectifier parameters, and
- Input-side parameters.

7.4.3.1 Output-side parameters

(a) Equation (7.6) gives the average output voltage is

$$\begin{aligned} V_{o(av)} &= V_{dc} = \frac{2V_m}{\pi}\cos\alpha \\ &= \frac{2 \times 169.7}{\pi} \times \cos(1.047) = 54.02\text{ V} \end{aligned}$$

(b) The average output current is

$$I_{o(av)} = \frac{V_{o(av)}}{R}$$
$$= \frac{54.02}{10} = 5.402 \text{ A}$$

(c) The DC output power is

$$P_{dc} = V_{dc}I_{dc}$$
$$= 54.02 \times 5.402 = 291.81 \text{ W}$$

(d) Equation (7.8) gives the rms output voltage is

$$V_{o(rms)} = V_{rms} = \frac{V_m}{\sqrt{2}} = 120 \text{ V}$$

(e) Equation (7.19) gives the rms output current is

$$I_{o(rms)} = \sqrt{2}I_R$$
$$= \sqrt{2} \times 4.17 = 5.89 \text{ A}$$

(f) The output AC power

$$P_{out} = I_{o(rms)}^2 R$$
$$= 5.89^2 \times 10 = 347.49 \text{ W}$$

(g) The rms ripple content of the output voltage

$$V_{ac} = \sqrt{V_{o(rms)}^2 - V_{dc}^2}$$
$$= \sqrt{120^2 - 54.02^2} = 107.15 \text{ V}$$

(h) The rectification *efficiency or ratio*

$$\eta = \frac{P_{dc}}{P_o}$$
$$= \frac{291.81}{347.49} = 83.97\%$$

(i) The *ripple factor*

$$RF = \frac{V_{ac}}{V_{dc}}$$
$$= \frac{107.15}{54.02} = 198.36\%$$

(j) The *form factor of the output voltage*

$$FF_{ov} = \frac{V_{rms}}{V_{dc}}$$
$$= \frac{120}{54.02} = 222.14\%$$

7.4.3.2 Controlled rectifier parameters

(a) The peak thyristor current

$$I_{T(peak)} = \frac{V_m}{R} = \frac{169.7}{10} = 16.971 \text{ A}$$

(b) The load current is shared by two devices, the average thyristor current is

$$I_{T(av)} = \frac{I_{o(av)}}{2} = \frac{5.402}{2} = 2.7 \text{ A}$$

(c) The load current is shared by two devices, the rms thyristor current is

$$I_{T(rms)} = \frac{I_{rms}}{\sqrt{2}} = \frac{5.89}{\sqrt{2}} = 4.17 \text{ A}$$

(d) With a safely margin of two times, the peak voltage rating of thyristors is

$$V_{peak} = 2V_m = 2 \times 169.7 = 339.411 \text{ V}$$

7.4.3.3 Input-side parameters

(a) The rms input current is

$$I_s = I_{rms} = 5.89 \text{ A}$$

(b) Assuming ideal no-loss converter, the input power

$$P_{in} = P_{out} = 347.49 \text{ W}$$

(c) The input power factor

$$PF_i = \frac{P_{in}}{V_s I_s} = \frac{347.49}{120 \times 5.89} = 0.49 \text{ (lagging)}$$

(d) The *transformer utilization factor*

$$\text{TUF} = \frac{P_{dc}}{V_s I_s} = \frac{291.81}{120 \times 5.89} \times 100 = 41.25\%$$

7.4.3.4 Load ripple current

For a single-phase full-wave-controlled rectifier, the lowest order harmonic frequency is $\omega_o = 2\omega$ and the impedance of the RL load at the dominant frequency is $Z_L = R + j2\omega L$.

The approximate value of the ripple current due to the dominant harmonic at $\omega_o = 2\omega$ is given by

$$I_{ripple} = \frac{V_{ac}}{\sqrt{R^2 + (2\omega L)^2}} = \frac{107.15}{35.37} = 3.03 \text{ A}$$

7.4.4 LTspice Fourier analysis

The LTspice command (.four 120 Hz 29 1 *I*(*R*)) for Fourier analysis gives the Fourier components of the load current *I*(*R*) at the fundamental frequency of 120 Hz as follows:

Fourier components of *I*(*R*)

DC component: 5.16489

Harmonic number	Frequency (Hz)	Fourier component	Normalized component	Phase (degrees)	Normalized phase (degrees)
1	1.200e+02	2.012e+00	1.000e+00	−73.32	0.00
2	2.400e+02	3.904e−01	1.940e−01	−46.40	26.93
3	3.600e+02	1.681e−01	8.352e−02	−22.68	50.65
4	4.800e+02	9.466e−02	4.704e−02	2.99	76.32
5	6.000e+02	5.843e−02	2.904e−02	27.21	100.54
6	7.200e+02	4.067e−02	2.021e−02	48.50	121.82
7	8.400e+02	3.024e−02	1.503e−02	75.52	148.84
8	9.600e+02	2.217e−02	1.102e−02	97.92	171.24
9	1.080e+03	1.834e−02	9.116e−03	117.89	191.22
10	1.200e+03	1.545e−02	7.677e−03	147.46	220.78
11	1.320e+03	1.189e−02	5.907e−03	169.09	242.42
12	1.440e+03	1.042e−02	5.177e−03	−171.07	−97.75
13	1.560e+03	9.178e−03	4.561e−03	−143.31	−69.98
14	1.680e+03	7.164e−03	3.560e−03	−118.72	−45.39
15	1.800e+03	6.613e−03	3.287e−03	−101.68	−28.36
16	1.920e+03	6.128e−03	3.045e−03	−71.65	1.68
17	2.040e+03	4.594e−03	2.283e−03	−47.94	25.38
18	2.160e+03	4.704e−03	2.338e−03	−31.02	42.31
19	2.280e+03	4.330e−03	2.152e−03	1.37	74.69
20	2.400e+03	3.409e−03	1.694e−03	21.59	94.92
21	2.520e+03	3.457e−03	1.718e−03	39.65	112.97
22	2.640e+03	3.228e−03	1.604e−03	69.74	143.06

(Continues)

(*Continued*)

Fourier components of *I*(*R*)					
DC component: 5.16489					
Harmonic number	**Frequency (Hz)**	**Fourier component**	**Normalized component**	**Phase (degrees)**	**Normalized phase (degrees)**
23	2.760e+03	2.520e−03	1.252e−03	93.26	166.59
24	2.880e+03	2.579e−03	1.282e−03	109.40	182.73
25	3.000e+03	2.632e−03	1.308e−03	142.69	216.01°
26	3.120e+03	1.891e−03	9.399e−04	164.18	237.50
27	3.240e+03	2.114e−03	1.051e−03	177.88	251.21
28	3.360e+03	2.042e−03	1.015e−03	−143.33	−70.00
29	3.480e+03	1.512e−03	7.516e−04	−125.56	−52.24
THD: 22.071220% (22.083575%)					

Note: THD is reduced from 37.10 % to 22.07% due to the load inductor.

7.4.5 LC-filter

Capacitor C provides a low impedance path for the ripple currents to flow through it rather than the load resistance R and lowers the Total Harmonic Distortion (THD). The frequency f_o of the output voltage is two times the supply frequency f. There are two ripples on the output current: one positive ripple and one negative ripple around the average value. We can find the capacitor value to maintain continuous load voltage. That is,

$$RC \geq T_o = \frac{1}{2 \times 2 \times f} \tag{7.21}$$

which give the value of C as

$$C \geq \frac{1}{2 \times 2fR} = \frac{1}{2 \times 2 \times 60 \times 10} = 416.67\ \mu\text{F}$$

Using the voltage and current ripples in (7.10), we can find the capacitor C from the ripple voltage ΔV and the ripple current ΔI as

$$C\frac{\Delta V}{\Delta T} = \Delta I \tag{7.22}$$

which gives the capacitor value C as

$$\begin{aligned} C &= \frac{\Delta I \times \Delta T}{\Delta V} = \frac{\Delta I}{\Delta V} \times \frac{T}{2 \times 2} \\ &= \frac{0.27}{3.241} \times \frac{16.67 \times 10^{-3}}{2 \times 2} = 347.22\ \mu\text{F} \end{aligned} \tag{7.23}$$

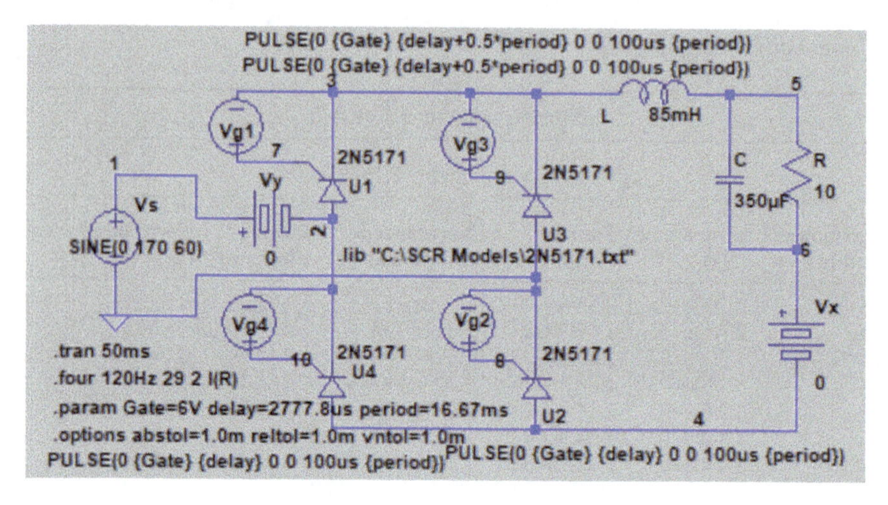

Figure 7.7 Schematic of single-phase controlled rectifier with an LC filter

Using the values from (7.21) and (7.23), let us take the average of two values $C = \frac{416.67 + 347.22}{2} = 381.94$ mH. Let us select $C = 380$ mH.

Figure 7.7 shows the full-wave-controlled rectifier with an LC filter.

7.4.6 Circuit model

$R = 10\ \Omega$, $C = 380\ \mu$F, and $L = 45$ mH (from Section 7.4.1), $f = 60$ Hz

$$\omega_o = 2\pi f_o = 2\pi \times 2 \times 60 = 753.98 \text{ rad/s}$$

The nth harmonic impedance of the load with the LC-filter is given by

$$Z(n) = jn\omega L + \frac{R \| \frac{1}{jn\omega C}}{1 + jn\omega RC} = \frac{R + jn\omega L - (n\omega)RLC}{1 + jn\omega RC} \tag{7.24}$$

For a single-phase full-wave-controlled rectifier, $n = 2$ and (7.24) gives $|Z(2)| = 23.97$.

The approximate value of the ripple current through the RLC load due to the dominant harmonic at$\omega_o = 2\omega$ is given by

$$I_{ripple} = \frac{V_{ac}}{|Z(n)|}$$
$$= \frac{107.15}{23.97} = 3.95 \text{ A}$$

The Fourier components of the output voltage are given by [1]

$$V_{dc} = a_o = \frac{2}{2\pi}\int_{\alpha}^{\pi+\alpha} V_m \sin(\omega t) d(\omega t) = \frac{2V_m}{\pi}\cos(\alpha) \tag{7.25}$$

$$a(n) = \frac{2}{2\pi}\int_{\alpha}^{\pi+\alpha} V_m \sin(\omega t)\sin(n\omega t)d(\omega t) \quad \text{for } n = 2,4,\ldots \\ = 0 \quad \text{for } n = 3,5,\ldots \tag{7.26}$$

$$b(n) = \frac{2}{2\pi}\int_{\alpha}^{\pi+\alpha} V_m \sin(\omega t)\cos(n\omega t)d(\omega t) \quad \text{for } n = 2,4,\ldots \\ = 0 \quad \text{for } n = 3,5,\ldots \tag{7.27}$$

Therefore, the Fourier series of the output voltage is given by

$$v_o(t) = V_{dc} + \sum_{n=2,4\ldots}^{\infty} \sqrt{a^2(n) + b^2(n)}\sin(n\omega t) \tag{7.28}$$

Diving the voltage expression in (7.28) by $Z(n)$ in (7.24) gives the Fourier components of the load currents as [1]

$$i_o(t) = \frac{2V_m}{\pi R}\cos(\alpha) + \sum_{n=2,4\ldots}^{\infty} \frac{\sqrt{a^2(n) + b^2(n)}}{Z(n)}\sin(n\omega t - \theta_n) \tag{7.29}$$

where $\theta_n = \tan^{-1}\left(\frac{a_n}{b_n}\right)$ is the impedance angle of the nth harmonic component, rad.

Equation (7.29) gives the peak magnitude of the nth harmonic currents

$$I_m(n) = \frac{\sqrt{a^2(n) + b^2(n)}}{Z(n)} \quad \text{for } n = 2,4,\ldots \tag{7.30}$$

Using Mathcad software, (7.30) gives the peak load current through RLC load for $n = 2$, $n = 4$, $n = 6$, $n = 8$, 10, and $n = 12$ as

$$I_m(2) = 0.76 \quad I_m(4) = 0.07 \quad I_m(6) = 0.08$$
$$I_m(8) = 0.02 \quad I_m(10) = 0.01 \quad I_m(12) = 0.02$$

Equation (5.30) gives the peak values of the nth harmonic currents through the RLC load. Since the load resistance R forms a parallel circuit with the filter capacitance C. The peak load current through R can be found from the current division rule as

$$I_{mR}(n) = I_m(n) \times \left|\frac{X_c}{R + X_c}\right| = \frac{\sqrt{a^2(n) + b^2(n)}}{|Z(n)|}\left|\frac{X_c}{R + X_c}\right| \tag{7.31}$$

where $X_c = \frac{1}{jn\omega_o C}$ is the impedance of the filter capacitor C, and $I_{mR}(n)$ is the peak value of the nth harmonic current through the load resistor R.

Using Mathcad software, (7.31) gives the peak load current through load R for $n = 2$, $n = 4$, $n = 6$, $n = 8$, $n = 10$, and $n = 12$ as

$$I_{mR}(2) = 0.25 \quad I_{mR}(4) = 0.01 \quad I_{mR}(6) = 0$$
$$I_{mR}(8) = 0 \quad I_{mR}(10) = 0 \quad I_{mR}(12) = 0$$

Dividing the peak values by $\sqrt{2}$ gives the rms values for $n = 2$, $n = 4$, $n = 6$, $n = 8$, $n = 10$, and $n = 12$ as

$$I_{rms}(2) = \frac{I_{mR}(2)}{\sqrt{2}} = 0.18 \qquad I_{rms}(4) = \frac{I_{mR}(4)}{\sqrt{2}} = 0 \qquad I_{rms}(6) = \frac{I_{mR}(6)}{\sqrt{2}} = 0$$

$$I_{rms}(8) = \frac{I_{mR}(8)}{\sqrt{2}} = 0 \qquad I_{rms}(10) = \frac{I_{mR}(10)}{\sqrt{2}} = 0 \quad I_{rms}(12) = \frac{I_{mR}(12)}{\sqrt{2}} = 0$$

The total rms value of the ripple current can be determined by adding the square of the individual rms ripple current and then taking the square root as given by

$$\begin{aligned} I_{ripple} &= \sqrt{\sum_{n=2,4}^{30} |I_{rms}(n)|^2} \\ &= 0.19 \text{ A} \end{aligned} \tag{7.32}$$

The total rms value of the load current can be determined by adding the square of the average (DC) current and the rms ripple current as

$$\begin{aligned} I_o &= \sqrt{I_{dc}^2 + I_{ripple}^2} \\ &= \sqrt{5.402^2 + 0.19^2} = 5.41 \text{ A} \end{aligned} \tag{7.33}$$

The plots of the input and output voltages and the load current are shown in Figure 7.8(a). Figure 7.8(b) shows the effects of filter capacitances on the load voltage. As expected, increasing the value of capacitance reduces the ripple on the output voltage.

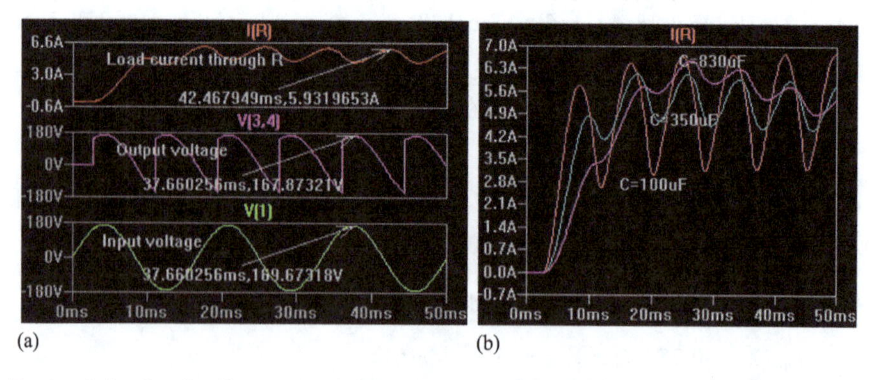

Figure 7.8 Single-phase controlled rectifier with an LC filter: (a) output voltage and load current, (b) effects on capacitance om load current

7.4.7 LTspice Fourier analysis

The LTspice command (.four 120 Hz 29 1 *I*(*R*)) for Fourier analysis gives the Fourier components of the load current *I*(*R*) at the fundamental frequency of 120 Hz as follows:

Fourier components of *I*(*R*)

DC component: 5.19982

Harmonic number	Frequency (Hz)	Fourier component	Normalized component	Phase (degrees)	Normalized Phase (degrees)
1	1.200e+02	7.660e−01	1.000e+00	64.94	0.00
2	2.400e+02	7.741e−02	1.011e−01	−53.98	−118.93
3	3.600e+02	1.728e−02	2.256e−02	178.01	113.06
4	4.800e+02	1.049e−02	1.369e−02	49.19	−15.76
5	6.000e+02	5.884e−03	7.682e−03	−44.77	−109.71
6	7.200e+02	4.525e−04	5.907e−04	−171.17	−236.12
7	8.400e+02	3.113e−03	4.064e−03	26.91	−38.04
8	9.600e+02	2.157e−03	2.816e−03	−17.62	−82.57
9	1.080e+03	4.202e−04	5.485e−04	29.39	−35.55
10	1.200e+03	1.418e−03	1.851e−03	21.44	−43.51
11	1.320e+03	1.220e−03	1.592e−03	−20.07	−85.02
12	1.440e+03	6.045e−04	7.891e−04	−6.00	−70.94
13	1.560e+03	1.180e−03	1.541e−03	5.41	−59.53
14	1.680e+03	8.676e−04	1.133e−03	−4.27	−69.21
15	1.800e+03	5.898e−04	7.699e−04	−4.51	−69.45
16	1.920e+03	8.209e−04	1.072e−03	8.66	−56.29
17	2.040e+03	6.490e−04	8.472e−04	−2.34	−67.29
18	2.160e+03	5.174e−04	6.755e−04	−4.60	−69.55
19	2.280e+03	6.667e−04	8.704e−04	3.63	−61.31
20	2.400e+03	6.290e−04	8.211e−04	−4.13	−69.07
21	2.520e+03	5.215e−04	6.808e−04	1.23	−63.72
22	2.640e+03	5.505e−04	7.186e−04	7.80	−57.15
23	2.760e+03	5.244e−04	6.846e−04	−0.63	−65.58
24	2.880e+03	4.194e−04	5.475e−04	2.21	−62.74
25	3.000e+03	4.576e−04	5.973e−04	3.98	−60.96
26	3.120e+03	4.728e−04	6.172e−04	−3.25	−68.19
27	3.240e+03	3.750e−04	4.895e−04	−0.55	−65.50
28	3.360e+03	4.428e−04	5.780e−04	3.14	−61.80
29	3.480e+03	4.188e−04	5.467e−04	−1.24	−66.19

THD: 10.492823%(10.952328%)

Note: THD is reduced from 22.07% to 10.49 % due to the addition of a capacitor filter across *R*.

7.5 Three-phase controlled rectifier with resistive load

A three-phase full-wave-controlled rectifier uses six thyristors instead of six diodes as shown in Figure 7.9 to a resistive load of *R*. The addition of a small inductor

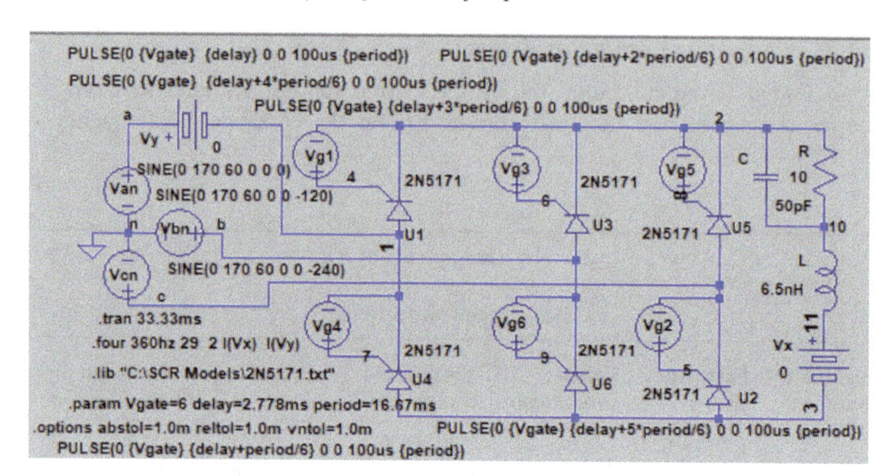

Figure 7.9 Three-phase controlled rectifier for α = 30°

$L = 6.5$ nH and a small capacitor $C = 50$ pF makes the circuit in a generalized form and these elements are often connected to serve as a filter. If the AC line-to-neutral voltages are defined as [10,11]

$$\begin{aligned} v_{an} &= V_m \sin(\omega t) \\ v_{bn} &= V_m \sin\left(\omega t - \frac{2\pi}{3}\right) \\ v_{cn} &= V_m \sin\left(\omega t - \frac{4\pi}{3}\right) = V_m \sin\left(\omega t + \frac{2\pi}{3}\right) \end{aligned} \tag{7.35}$$

the corresponding line-to-line voltages are

$$\begin{aligned} v_{ab} &= v_{an} - v_{bn} = \sqrt{3}V_m \sin\left(\omega t + \frac{\pi}{6}\right) \\ v_{bc} &= v_{bn} - v_{cn} = \sqrt{3}V_m \sin\left(\omega t - \frac{\pi}{2}\right) \\ v_{ca} &= v_{cn} - v_{an} = \sqrt{3}V_m \sin\left(\omega t - \frac{3\pi}{2}\right) = V_m \sin\left(\omega t + \frac{\pi}{2}\right) \end{aligned} \tag{7.35}$$

where V_m = peak value of the AC phase input voltage, V; f = supply frequency, Hz; $\omega = 2\pi f$ = supply frequency, rad/s.

The peak phase supply voltage $V_{\rm m} = 169.7$ V. Let us consider a delay angle of $\alpha = 30°$,

Time delay $t_1, t_2 = \frac{30}{360} \times \frac{1000}{60\ \text{Hz}} \times 1000 = 1.389$ ms

Time delay $t_3, t_4 = \frac{150}{360} \times \frac{1000}{60\ \text{Hz}} \times 1000 = 11.111$ ms

Time delay $t_5, t_6 = \frac{270}{360} \times \frac{1000}{60\ \text{Hz}} \times 1000 = 16.667$ ms

Each thyristor is turned on for an interval of $\pi/3$. The frequency of the output ripple voltage is $6f$. At $\omega t = \pi/6 + \alpha$, thyristor T_6 is already conducting and thyristor T_1 is turned on. During interval $(\pi/6+\alpha) \leq \omega t \leq (\pi/2+\alpha)$, thyristors T_1 and T_6 conduct and the line-to-line voltage $v_{ab}(= v_{an} - v_{bn})$ AC appears across the load. At $\omega t = (\pi/2+\alpha)$, thyristor T_2 is turned on and thyristor T_6 is reverse biased immediately. T_6 is turned off due to natural commutation. During interval $(\pi/2+\alpha) \leq \omega t \leq (5\pi/2+\alpha)$, thyristors T_1 and T_2 conduct and the line-to-line voltage $v_{ac}(= v_{an} - v_{cn})$ appears across the load. The thyristors are numbered, as shown in Figure 7.9 according to the conduction sequence, and the firing sequence is 12, 23, 34, 45, 56, and 61. The line-line voltages appear across the output for an internal of $\pi/3$ from $(\pi/6+\alpha)$ to $(\pi/2+\alpha)$. The average output voltage over an interval of $\pi/3$ can be found from

$$V_{dc} = \frac{3}{\pi}\int_{\pi/6+\alpha}^{\pi/2+\alpha} v_{ab}(\omega t)d(\omega t) = \frac{3}{\pi}\int_{\pi/6+\alpha}^{\pi/2+\alpha} \sqrt{3}V_m \sin(\omega t + \frac{\pi}{6})d(\omega t)$$
$$= \frac{3\sqrt{3}V_m}{\pi} \cos\alpha \quad \text{for } 0 \leq \alpha \leq \frac{\pi}{3} \tag{7.36}$$

The rms value of the output voltage is found from

$$V_{rms} = \sqrt{\frac{3}{\pi}\int_{\pi/6+\alpha}^{\pi/2+\alpha} (v_{ab}(\omega t))^2 d(\omega t)} = \sqrt{\frac{3}{\pi}\int_{\pi/6+\alpha}^{\pi/2+\alpha} (\sqrt{3}V_m \sin(\omega t + \frac{\pi}{6}))^2 dx}$$
$$= \sqrt{3}V_m\sqrt{\frac{1}{2} + \frac{3\sqrt{3}}{4\pi}\cos 2\alpha} \quad \text{for } 0 \leq \alpha \leq \frac{\pi}{3} \tag{7.37}$$

For $\alpha > \pi/3$, the instantaneous output voltage has a negative part. Because the current through thyristors cannot be negative, the load current is always positive. Thus, with a resistive load, the instantaneous load voltage cannot be negative, and the full converter behaves as a semi converter, not a full-wave converter. The plots of the gating pulses of 6 V, with a width 100 μs are shown in Figure 7.10(a). The

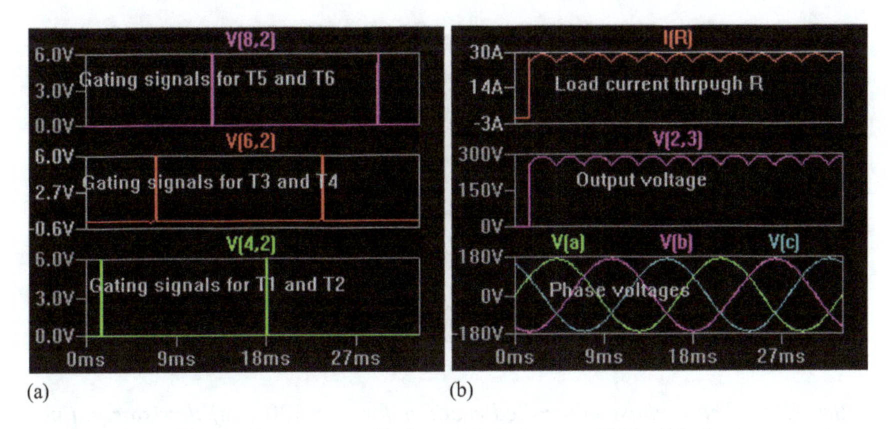

Figure 7.10 Three-phase controlled rectifier for α = 30°: (a) thyristor gating pulses and (b) input and output voltages and load current

output voltage and the load current are shown in Figure 7.10(b) for $\alpha = 30°$. As expected the output voltage and the current have less ripple and the frequency of the output ripples are six times the supply frequency of 60 Hz.

The plots of the gating pulses of 6 V and width 100 μs are shown in Figure 7.11 (a) for $\alpha = 120°$. The gating pulses for T_1 and T_2 delayed by α are identical. The gating pulses for T_3 and T_4 delayed by $(2\pi/3+\alpha)$ are identical. Similarly, the gating pulses for T_5 and T_6 delayed by $(4\pi/3+\alpha)$ are identical. The delay angle is $\alpha > \pi/3$. The output voltage and the load current as shown Figure 7.11(b) are discontinuous for $\alpha = 120°$. As expected, the load current becomes to zero when the input voltage falls to zero.

The performance parameters of controlled rectifiers can be divided into three types:

- Output-side parameters,
- Rectifier parameters, and
- Input-side parameters.

Let us consider an input phase voltage of $V_s = 120$ rms, $V_m = \sqrt{2}$ $V_s = 120 \times \sqrt{2} = 169.7$ V, $R = 10\ \Omega$ and delay angle of $\alpha = 30°$. Converting to radian, $\alpha = \frac{30°}{180} \times \pi = 0.524$ rad

7.5.1 Output-side parameters

(a) Equation (7.36) gives the average output voltage is

$$V_{o(av)} = V_{dc} = \frac{3\sqrt{3}V_m}{\pi} \cos\ \alpha$$

$$= \frac{3 \times \sqrt{3} \times 169.7}{\pi} \times (1 + \cos(0.524)) = 243.085\ \text{V}$$

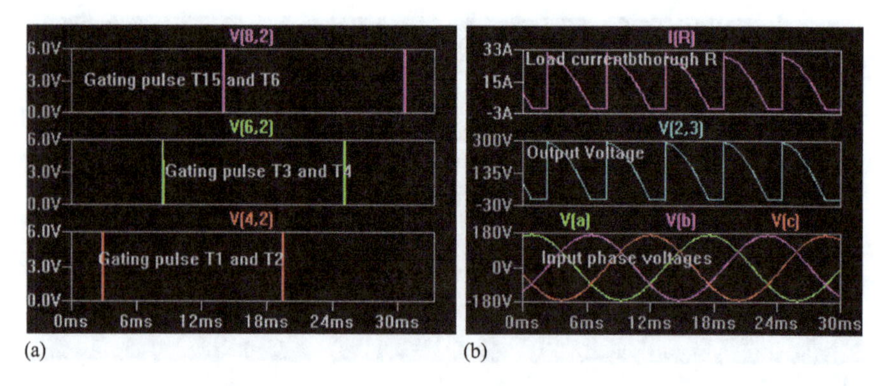

Figure 7.11 Three-phase controlled rectifier for α = 120°: (a) thyristor gating pulses and (c) input and output voltages and load current

(b) The average output current is

$$I_{o(av)} = \frac{V_{o(av)}}{R}$$
$$= \frac{243.085}{10} = 24.3085 \text{ A}$$

(c) The DC output power is

$$P_{dc} = V_{dc} I_{dc}$$
$$= 243.085 \times 24.3085 = 5.909 \text{ kW}$$

(d) Equation (7.37) gives the rms output voltage is

$$V_{o(rms)} = V_{rms} = \sqrt{3} V_m \sqrt{\frac{1}{2} + \frac{3\sqrt{3}}{4\pi} \cos 2\alpha}$$
$$= \sqrt{3} \times 169.7 \times \sqrt{\frac{1}{2} + \frac{3\sqrt{3}}{4\pi} \cos (2 \times 0.524)} = 247.109 \text{ V}$$

(e) The rms output current is

$$I_{o(rms)} = I_{rms} = \frac{V_{o(rms)}}{R}$$
$$= \frac{247.109}{10} = 24.7109 \text{ A}$$

(f) The AC output power is

$$P_{ac} = V_{o(rms)} I_{o(rms)}$$
$$= 247.109 \times 24.7109 = 6.106 \text{ kW}$$

(g) The output power

$$P_{out} = I_{rms}^2 R$$
$$= 24.7109^2 \times 10 = 6.106 \text{ kW}$$

Note: For a resistive load, $P_{ac} = P_{out}$.

(h) The rms ripple content of the output voltage

$$V_{ac} = \sqrt{V_{o(rms)}^2 - V_{dc}^2}$$
$$= \sqrt{247.109^2 - 243.085^2} = 44.413 \text{ V}$$

(i) The rectification *efficiency or ratio*

$$\eta = \frac{P_{dc}}{P_o}$$
$$= \frac{5.909}{6.106} = 96.77\%$$

(j) The *ripple factor of the output voltage*

$$RF_{ov} = \frac{V_{ac}}{V_{dc}}$$
$$= \frac{44.413}{243.085} = 18.271\%$$

(k) The *form factor of the output voltage*

$$FF_{ov} = \frac{V_{rms}}{V_{dc}}$$
$$= \frac{247.109}{243.085} = 101.655\%$$

7.5.2 Controlled rectifier parameters

(a) The line–line voltage gives the peak thyristor current

$$I_{T(peak)} = \frac{\sqrt{3}V_m}{R}$$
$$= \frac{\sqrt{3}\times 169.7}{10} = 29.394 \text{ A}$$

(b) The load current is shared by three devices, the average thyristor current

$$I_{T(av)} = \frac{I_{o(av)}}{3}$$
$$= \frac{24.3085}{3} = 8.103 \text{ A}$$

(c) Rms load current is shared by three devices, the rms thyristor current

$$I_{T(rms)} = I_{rms}\sqrt{\frac{1}{3}}$$
$$= 24.711 \times \sqrt{\frac{1}{3}} = 14.267 \text{ A}$$

(d) With a safely margin of two times, the peak thyristor voltage rating is

$$V_{peak} = 2\sqrt{3}V_m$$
$$= 2 \times \sqrt{3} \times 169.7 = 587.878 \text{ V}$$

7.5.3 Input-side parameters

(a) The rms input current is

$$I_s = I_{rms}\sqrt{\frac{4}{6}}$$
$$= 24.711 \times \sqrt{\frac{4}{6}} = 20.176 \text{ A}$$

(b) The input power

$$P_{in} = P_{out} = 6.106 \text{ kW}$$

(c) Input power factor

$$PF_i = \frac{P_{in}}{3V_sI_s} = \frac{6.106 \text{ kW}}{3 \times 120 \times 20.176} = 0.841 \text{ (lagging)}$$

(d) The *transformer utilization factor*

$$\text{TUF} = \frac{P_{dc}}{3V_sI_s} = \frac{5.909 \text{ kW}}{3 \times 120 \times 20.176} \times 100 = 81.4\%$$

7.5.4 LTspice Fourier analysis

The LTspice command (.four 360 Hz 29 1 *I*(*R*)) for Fourier analysis gives the Fourier components of the load current *I*(*R*) at the fundamental frequency of 360 Hz as follows:

Fourier components of *I*(*vx*)

DC component: 27.7581

Harmonic number	Frequency (Hz)	Fourier component	Normalized component	Phase (degrees)	Normalized phase (degrees)
1	3.600e+02	1.596e+00	1.000e+00	89.99	0.00
2	7.200e+02	3.917e−01	2.454e−01	−90.11	−180.10
3	1.080e+03	1.744e−01	1.093e−01	89.65	−0.34
4	1.440e+03	9.750e−02	6.109e−02	−90.28	−180.27
5	1.800e+03	6.200e−02	3.884e−02	89.87	−0.12
6	2.160e+03	4.312e−02	2.702e−02	−89.98	−179.97
7	2.520e+03	3.160e−02	1.980e−02	90.81	0.82
8	2.880e+03	2.431e−02	1.523e−02	−88.63	−178.62
9	3.240e+03	1.945e−02	1.219e−02	91.87	1.88
10	3.600e+03	1.594e−02	9.984e−03	−88.01	−178.00
11	3.960e+03	1.335e−02	8.362e−03	91.23	1.24
12	4.320e+03	1.128e−02	7.068e−03	−89.66	−179.64
13	4.680e+03	9.557e−03	5.987e−03	89.23	−0.76
14	5.040e+03	8.168e−03	5.117e−03	−91.77	−181.76
15	5.400e+03	6.988e−03	4.378e−03	87.89	−2.10
16	5.760e+03	6.060e−03	3.797e−03	−92.28	−182.27
17	6.120e+03	5.330e−03	3.339e−03	87.99	−2.00

(Continues)

(*Continued*)

Fourier components of *I*(*vx*)					
DC component: 27.7581					
Harmonic number	**Frequency (Hz)**	**Fourier component**	**Normalized component**	**Phase (degrees)**	**Normalized phase (degrees)**
18	6.480e+03	4.757e−03	2.980e−03	−91.47	−181.46
19	6.840e+03	4.317e−03	2.705e−03	88.52	−1.47
20	7.200e+03	3.948e−03	2.473e−03	−91.29	−181.28
21	7.560e+03	3.627e−03	2.272e−03	88.28	−1.71
22	7.920e+03	3.326e−03	2.084e−03	−92.04	−182.03
23	8.280e+03	3.055e−03	1.914e−03	88.03	−1.96
24	8.640e+03	2.807e−03	1.759e−03	−91.75	−181.74
25	9.000e+03	2.610e−03	1.635e−03	89.23	−0.76
26	9.360e+03	2.450e−03	1.535e−03	−89.92	−179.91
27	9.720e+03	2.323e−03	1.455e−03	90.81	0.82
28	1.008e+04	2.221e−03	1.391e−03	−88.98	−178.97
29	1.044e+04	2.114e−03	1.325e−03	90.68	0.69
THD: 28.155555% (28.160452%)					

7.6 Three-phase controlled rectifier with RL load

A three-phase full-wave-controlled rectifier uses six thyristors is shown Figure 7.12 with an RL load. If the time constant of the RL load is greater than the period of the output voltage, the load current will continuous and will not fall to zero. The line-line voltages appear across the output for an internal of $\pi/3$ from $(\pi/6 + \alpha)$ to $(\pi/2 + \alpha)$. Depending on the value of α, the average output voltage could be either positive or negative.

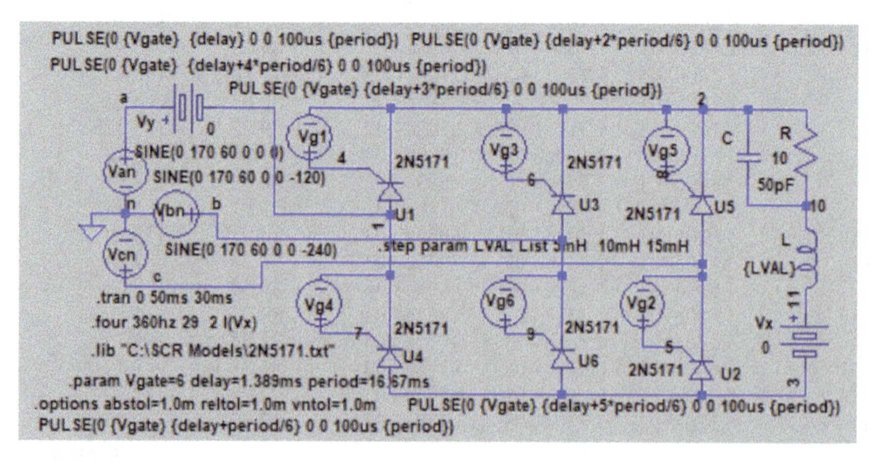

Figure 7.12 Three-phase controlled rectifier with RL load for $\alpha = 60°$

The average output voltage can be found from

$$V_{dc} = \frac{3}{\pi}\int_{\pi/6+\alpha}^{\pi/2+\alpha} v_{ab}(\omega t)d(\omega t) = \frac{3}{\pi}\int_{\pi/6+\alpha}^{\pi/2+\alpha} \sqrt{3}V_m \sin(\omega t + \frac{\pi}{6})d(\omega t)$$
$$= \frac{3\sqrt{3}V_m}{\pi} \cos\alpha \quad \text{for } 0 \le \alpha \le \pi \tag{7.38}$$

which gives the average load current as

$$I_{dc} = \frac{V_{dc}}{R} = \frac{3\sqrt{3}V_m}{\pi R} \cos\alpha \tag{7.39}$$

The rms value of the output voltage is found from

$$V_o = \sqrt{\frac{3}{\pi}\int_{\pi/6+\alpha}^{\pi/2+\alpha} (v_{ab}(\omega t))^2 d(\omega t)} = \sqrt{\frac{3}{\pi}\int_{\pi/6+\alpha}^{\pi/2+\alpha} (\sqrt{3}V_m \sin(\omega t + \frac{\pi}{6}))^2 dx}$$
$$= \sqrt{3}V_m\sqrt{\frac{1}{2} + \frac{3\sqrt{3}}{4\pi}\cos 2\alpha} \quad \text{for } 0 \le \alpha \le \pi \tag{7.40}$$

which gives the rms load current as

$$I_o = \frac{V_o}{\sqrt{R^2 + (6\omega L)^2}} = \frac{\sqrt{3}V_m}{\sqrt{R^2 + (6\omega L)^2}}\sqrt{\frac{1}{2} + \frac{3\sqrt{3}}{4\pi}\cos 2\alpha} \tag{7.41}$$

where L is the load inductance.

Let us consider an input phase voltage of $V_s = 120$ rms, $V_m = V_s = 120 \times \sqrt{2} = 169.7$ V, $R = 10\ \Omega$, $E = 0$, and delay angle of $\alpha = 60°$. Converting to radian, $\alpha = \frac{60°}{180} \times \pi = 1.047$ rad.

The plots of the input and output voltages and the load currents are shown in Figure 7.13(a). As expected, the ripples on the load current decreases with the values of load inductance L as shown in Figure 7.13(b).

7.6.1 Load inductor

Frequency f_o of the output voltage is six times the supply frequency f. There are two ripples on the output current: one positive ripple and one negative around the average value. We can find the inductor value to maintain continuous load current from the following relation given by

$$\frac{L}{R} \ge \frac{T_o}{2} = \frac{1}{2 \times 6 \times f} \tag{7.42}$$

which gives the value of L as

$$L \ge \frac{R}{2 \times 6 \times f} = \frac{10}{2 \times 6 \times 60} = 13.89 \text{ mH}$$

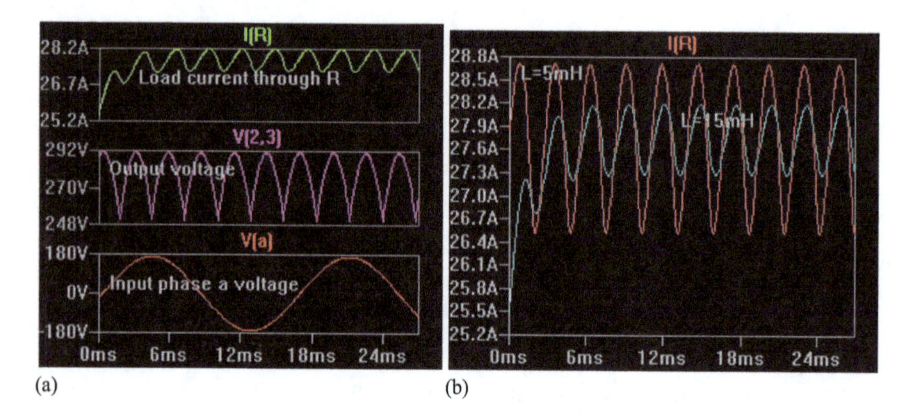

Figure 7.13 Out voltage and currents for $\alpha = 60°$: (a) output voltage and load current and (b) effects of load inductances

For a desired value of the ripple factor of the output voltage $RF_v = 6\% = 0.06$, we can find the ripple voltage as

$$\begin{aligned}\Delta V &= RF_v\, V_{dc} \\ &= 0.06 \times 140.35 = 8.42 \text{ V}\end{aligned}$$

Similarly, for a desired value of the ripple factor of the output current $RF_i = 5\% = 0.05$, we can find the ripple current as

$$\begin{aligned}\Delta I &= RF_i\, I_{dc} \\ &= 0.05 \times 14.03 = 0.7 \text{ A}\end{aligned}$$

We can find the inductor L from the ripple voltage ΔV and the ripple current ΔI as

$$L\frac{\Delta I}{\Delta T} = \Delta V$$

which gives the inductor value L as

$$\begin{aligned}L &\geq \frac{\Delta V \times \Delta T}{\Delta I} = \frac{\Delta V}{\Delta I} \times \frac{T}{2 \times 6} \\ &= \frac{8.42}{0.7} \times \frac{16.67 \times 10^{-3}}{12} = 16.67 \text{ mH}\end{aligned} \tag{7.43}$$

Using the values from (7.42) and (7.43), the average of two values is $L = \frac{13.89 + 16.67}{2} = 15.33$ mH. Let us select $L = 15$ mH. Figure 7.13(b) shows the effects of load inductances on the load current.

7.6.2 Circuit model

The operation of the converter can be divided into two identical modes: mode 1 when T_1 and T_2 conduct, and mode 2 when T_3 and T_4 conduct. The output currents during these modes are similar and we need to consider only one mode to find the output current i_L.

Mode 1 is valid for $\left(\frac{\pi}{6}+\alpha\right) \le \omega t \le \left(\frac{\pi}{2}+\alpha\right)$. If line-to-line voltage appears across the load, we obtain the instantaneous output voltage as

$$\begin{aligned} v_{ab} &= \sqrt{2}\, V_{ab} \sin\left(\omega t + \frac{\pi}{6}\right) \quad \text{for} \left(\frac{\pi}{6}+\alpha\right) \le \omega t \le \left(\frac{\pi}{2}+\alpha\right) \\ &= \sqrt{2}\, V_{ab} \sin\, \omega t' \quad \text{for} \left(\frac{\pi}{3}+\alpha\right) \le \omega t' \le \left(\frac{2\pi}{3}+\alpha\right) \end{aligned} \tag{7.44}$$

where $\omega t' = \omega t + \pi/6$ and V_{ab} is the line-to-line (rms) input voltage. Choosing v_{ab} as the time reference voltage and $v_{ab} = \sqrt{2} V_{ab} \sin\, \omega t'$ as the instantaneous input voltage, the instantaneous load current can be found from

$$L\frac{di_L}{dt} + Ri_L + E = \sqrt{2} V_{ab} \sin\, \omega t' \quad \text{for} \left(\frac{\pi}{3}+\alpha\right) \le \omega t' \le \left(\frac{2\pi}{3}+\alpha\right) \tag{7.45}$$

where E is a battery voltage connected in series to the load side. Applying (7.14), the load current i_L during mode 1 can be found from

$$\begin{aligned} i_L &= \frac{\sqrt{2}\, V_{ab}}{Z} \sin\left(\omega t' - \theta\right) - \frac{E}{R} \\ &\quad + \left[I_{L1} + \frac{E}{R} - \frac{\sqrt{2} V_{ab}}{Z} \sin\left(\frac{\pi}{3}+\alpha-\theta\right)\right] e^{-(R/L)(\pi/3+\alpha)/\omega - t')} \quad \text{for} \quad i_L \ge 0 \end{aligned} \tag{7.46}$$

where $Z = \sqrt{R^2 + (\omega L)^2}$ and $\theta = \tan^{-1}(\omega L/R)$.

At the end of mode 1 under a steady-state condition $i_L(\omega t' = 2\pi/3 + \alpha) = i_L(\omega t' = \pi/3 + \alpha) = I_{L1}$. Applying this condition to (7.35) gives the value of I_{L1} as

$$\begin{aligned} I_{L1} &= I_{L0} \\ &= \frac{\sqrt{2}\, V_{ab}}{Z} \frac{\sin\left(\frac{2\pi}{3}+\alpha-\theta\right) - \sin\left(\frac{\pi}{3}+\alpha-\theta\right) e^{-(R/L)(\pi/3\omega)}}{1 - e^{-(R/L)(\pi/\omega)}} - \frac{E}{R} \quad \text{for} \quad I_{L1} \ge 0 \end{aligned} \tag{7.47}$$

The critical value of α at which I_{L0} becomes zero can be solved for known values of θ, R, L, E, and V_{ab} by an iterative method. That is,

$$0 = \frac{\sqrt{2}\, V_{ab}}{Z} \frac{\sin\left(\frac{2\pi}{3}+\alpha-\theta\right) - \sin\left(\frac{\pi}{3}+\alpha-\theta\right) e^{-(R/L)(\pi/3\omega)}}{1 - e^{-(R/L)(\pi/\omega)}} - \frac{E}{R} \tag{7.48}$$

which for $E = 0$ and the iterative Mathcad software function of Gues statement gives

$$\alpha_c = 1.27 \text{ rad}, 72.67°$$

Using Mathcad software, the rms current of a thyristor can be found from

$$I_R = \sqrt{\frac{1}{\pi}\int_{\frac{\pi}{3}+\alpha}^{\frac{2\pi}{3}+\alpha} i_L^2 d(\omega t)} \tag{7.49}$$
$$= 8.17 \text{ A}$$

Using Mathcad software, the rms load current can be found from

$$\begin{aligned} I_R &= \sqrt{3} I_R \\ &= \sqrt{3} \times 8.17 = 14.14 \text{ A} \end{aligned} \tag{7.50}$$

Using Mathcad software, the average current of a thyristor can be found from

$$I_A = \frac{1}{\pi}\int_{\frac{\pi}{3}+\alpha}^{\frac{2\pi}{3}+\alpha} i_L d(\omega t) \tag{7.51}$$
$$= 4.68 \text{ A}$$

7.6.3 *Converter parameters*

The performance parameters of controlled rectifiers can be divided into three types:

- Output-side parameters,
- Rectifier parameters, and
- Input-side parameters.

7.6.3.1 Output-side parameters

(a) Equation (7.38) gives the average output voltage is

$$\begin{aligned} V_{o(av)} &= V_{dc} = \frac{3\sqrt{3}V_m}{\pi} \cos \alpha \\ &= \frac{3 \times \sqrt{3} \times 169.7}{\pi} \times \cos(1.05) = 140.35 \text{ V} \end{aligned}$$

(b) The average output current is

$$\begin{aligned} I_{o(av)} &= \frac{V_{o(av)}}{R} \\ &= \frac{140.35}{10} = 14.03 \text{ A} \end{aligned}$$

(c) The DC output power is

$$P_{dc} = V_{dc} I_{dc}$$
$$= 140.35 \times 14.03 = 1.97 \text{ kW}$$

(d) Equation (7.40) gives the rms output voltage is

$$V_{o(rms)} = V_{rms} = \sqrt{3} V_m \sqrt{\frac{1}{2} + \frac{3\sqrt{3}}{4\pi} \cos 2\alpha}$$
$$= \sqrt{3} \times 169.7 \times \sqrt{\frac{1}{2} + \frac{3\sqrt{3}}{4\pi} \cos (2 \times 1.05)} = 159.18 \text{ V}$$

(e) The rms output current is

$$I_{o(rms)} = \frac{V_{o(rms)}}{R}$$
$$= \frac{159.18}{10} = 15.92 \text{ A}$$

(f) The output AC power

$$P_{out} = I^2_{o(rms)} R$$
$$= 15.92^2 \times 10 = 2.53 \text{ kW}$$

(g) The rms ripple content of the output voltage

$$V_{ac} = \sqrt{V^2_{o(rms)} - V^2_{dc}}$$
$$= \sqrt{159.18^2 - 140.35^2} = 75.1 \text{ V}$$

(h) The rectification *efficiency or ratio*

$$\eta = \frac{P_{dc}}{P_o}$$
$$= \frac{1.97 \text{ kW}}{2.53 \text{ kW}} = 77.74\%$$

(i) The *ripple factor of the output voltage*

$$RF_{ov} = \frac{V_{ac}}{V_{dc}}$$
$$= \frac{75.1}{140.35} = 53.51\%$$

(j) The *form factor of the output voltage*

$$FF_{ov} = \frac{V_{rms}}{V_{dc}}$$
$$= \frac{120}{54.02} = 222.14\%$$

7.6.3.2 Controlled rectifier parameters

(a) The peak thyristor current

$$I_{T(peak)} = \frac{\sqrt{3}V_m}{R}$$
$$= \frac{\sqrt{3} \times 169.7}{10} = 29.391 \text{ A}$$

(b) The load current is shared by three devices, and the average thyristor current is

$$I_{T(av)} = \frac{I_{o(av)}}{3}$$
$$= \frac{14.03}{3} = 4.68 \text{ A}$$

(c) The rms load current is shared three devices, and the rms thyristor current is

$$I_{T(rms)} = \frac{I_{rms}}{\sqrt{3}}$$
$$= \frac{15.92}{\sqrt{3}} = 9.19 \text{ A}$$

(d) With a safe margin of two times, the peak voltage rating of thyristors is

$$V_{peak} = 2\sqrt{3}V_m$$
$$= 2 \times \sqrt{3} \times 169.7 = 587.88 \text{ V}$$

7.6.3.3 Input-side parameters

The rms input current is

$$I_s = \frac{I_{rms}}{\sqrt{3}}$$
$$= 9.19 \text{ A}$$

The input power

$$P_{in} = P_{out}$$
$$= 2.53 \text{ kW}$$

Note: For an ideal converter with no-power loses, $P_{in} = P_{out}$.

The input power factor

$$PF_i = \frac{P_{in}}{3V_sI_s}$$
$$= \frac{2.53 \text{ kW}}{3 \times 120 \times 9.19} = 0.77 \text{ (lagging)}$$

The transformer utilization factor

$$\text{TUF} = \frac{P_{dc}}{3V_sI_s}$$
$$= \frac{1.97 \text{ kW}}{3 \times 120 \times 9.19} \times 100 = 59.54\%$$

7.6.3.4 Load Ripple Current

For a three-phase full-wave-controlled rectifier, the lowest order harmonic frequency is $\omega_o = 6\omega$ and the impedance of the RL load at the dominant frequency is $Z_L = R + j6\omega L$.

The approximate value of the ripple current due to the dominant harmonic at $\omega_o = 6\omega$ is given by

$$I_{ripple} = \frac{V_{ac}}{\sqrt{R^2 + (6\omega L)^2}}$$
$$= \frac{75.1}{35.37} = 2.12 \text{ A}$$

7.6.4 LTspice Fourier analysis

The LTspice command (.four 120 Hz 29 1 $I(R)$) for Fourier analysis gives the Fourier components of the load current $I(R)$ at the fundamental frequency of 360 Hz as follows:

Harmonic number	Frequency (Hz)	Fourier component	Normalized component	Phase (degrees)	Normalized phase (degrees)
1	3.600e+02	4.542e−01	1.000e+00	−26.66	0.00
2	7.200e+02	5.661e−02	1.246e−01	102.43	129.09
3	1.080e+03	1.724e−02	3.796e−02	−125.02	−98.36
4	1.440e+03	7.689e−03	1.693e−02	2.62	29.28
5	1.800e+03	2.999e−03	6.603e−03	159.11	185.77
6	2.160e+03	2.446e−03	5.385e−03	−45.20	−18.54
7	2.520e+03	1.911e−03	4.207e−03	55.06	81.73
8	2.880e+03	1.035e−03	2.277e−03	−177.18	−150.51
9	3.240e+03	6.253e−04	1.377e−03	−35.27	−8.61
10	3.600e+03	4.044e−04	8.903e−04	75.83	102.49
11	3.960e+03	3.498e−04	7.701e−04	−91.24	−64.58
12	4.320e+03	3.330e−04	7.330e−04	26.54	53.21
13	4.680e+03	1.694e−04	3.729e−04	122.60	149.27
14	5.040e+03	1.861e−04	4.098e−04	−53.04	−26.37
15	5.400e+03	1.114e−04	2.453e−04	61.01	87.67
16	5.760e+03	9.680e−05	2.131e−04	−139.44	−112.77
17	6.120e+03	1.847e−04	4.065e−04	−7.61	19.05
18	6.480e+03	4.833e−05	1.064e−04	116.14	142.80

(Continues)

(*Continued*)

Harmonic number	Frequency (Hz)	Fourier component	Normalized component	Phase (degrees)	Normalized phase (degrees)
19	6.840e+03	5.605e−05	1.234e−04	−69.00	−42.34
20	7.200e+03	5.455e−05	1.201e−04	50.89	77.56
21	7.560e+03	3.971e−05	8.743e−05	−97.25	−70.59
22	7.920e+03	6.800e−05	1.497e−04	−47.52	−20.85
23	8.280e+03	6.316e−05	1.390e−04	78.29	104.95
24	8.640e+03	6.722e−05	1.480e−04	−118.37	−91.71
25	9.000e+03	9.101e−05	2.003e−04	−8.65	18.01
26	9.360e+03	6.768e−05	1.490e−04	147.82	174.48
27	9.720e+03	5.548e−05	1.221e−04	−79.05	−52.39
28	1.008e+04	3.877e−05	8.535e−05	3.24	29.91
29	1.044e+04	1.313e−05	2.890e−05	152.69	179.35
		THD: 13.175524% (13.176868%)			

Note: THD is reduced from 28.160452% to 13.17% due to the load inductor.

7.6.5 LC-filter

The capacitor C provides a low impedance path for the ripple currents to flow through it and lowers the THD. Since frequency f_o of the output voltage is six times the supply frequency f, we can find the capacitor value to maintain continuous load voltage. That is,

$$RC \geq T_o = \frac{1}{2 \times 6f} \tag{7.52}$$

which gives the value of C as

$$C \geq \frac{1}{2 \times 6fR} = \frac{1}{2 \times 6 \times 60 \times 10} = 138.89\ \mu\text{F}$$

Using the voltage and current ripples in (7.22), we can find the capacitor C from the ripple voltage ΔV and the ripple current ΔI as

$$C\frac{\Delta V}{\Delta T} = \Delta I$$

which gives the capacitor value C as

$$\begin{aligned} C &= \frac{\Delta I \times \Delta T}{\Delta V} = \frac{\Delta I}{\Delta V} \times \frac{T}{2 \times 6} \\ &= \frac{0.7}{8.42} \times \frac{16.67 \times 10^{-3}}{2 \times 6} = 115.74\ \mu\text{F} \end{aligned} \tag{7.53}$$

Using the values from (7.21) and (7.23), the average of two values $C = \frac{138.89 + 115.74}{2} = 127.31\ \mu\text{H}$.

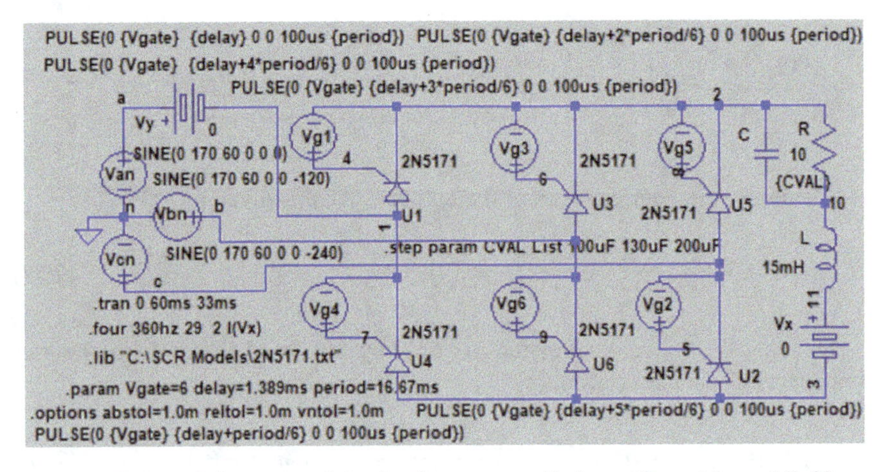

Figure 7.14 Schematic of single-phase controlled rectifier with an LC filter

Let us select $C = 130\ \mu$F and $L = 15$ mH. Figure 7.14 shows the three-phase controlled rectifier with an LC filter. The nth harmonic impedance of the load with the LC-filter is given by

$$Z(n) = jn\omega L + \frac{R \| \frac{1}{jn\omega C}}{1 + jn\omega RC} = \frac{R + jn\omega L - (n\omega)RLC}{1 + jn\omega RC} \tag{7.54}$$

For a three-phase full-wave-controlled rectifier, $n = 6$ and (7.54) gives $|Z(6)| = 23.97$.

The approximate value of the ripple current through the RLC load due to the dominant harmonic at $\omega_o = 6\omega$ is given by

$$\begin{aligned} I_{ripple} &= \frac{V_{ac}}{|Z(n)|} \\ &= \frac{75.1}{30.9} = 2.43 \text{ A} \end{aligned}$$

The Fourier components of the output voltage are given by [1]

$$V_{dc} = a_o = \frac{3\sqrt{3}}{2\pi} \int_{\pi/3+\alpha}^{2\pi/3+\alpha} \sin(\omega t) d(\omega t) = \frac{3\sqrt{3} V_m}{\pi R} \cos \alpha \tag{7.54}$$

$$\begin{aligned} a(n) &= \frac{3\sqrt{3}}{2\pi} \int_{\pi/3+\alpha}^{2\pi/3+\alpha} V_m \sin(\omega t) \sin(n\omega t) d(\omega t) \quad \text{for } n = 6, 12, 18, \ldots \\ &= 0 \quad \text{for } n = 3, 5, \ldots \end{aligned} \tag{7.55}$$

$$b(n) == \frac{3\sqrt{3}}{2\pi}\int_{\pi/3+\alpha}^{2\pi/3+\alpha} V_m \sin(\omega t)\cos(n\omega t)d(\omega t) \quad \text{for } n = 6, 12, 18, \ldots$$
$$= 0 \quad \text{for } n = 3, 5, \ldots \tag{7.56}$$

Therefore, the Fourier series of the output voltage is given by

$$v_o(t) = V_{dc} + \sum_{n=2,4,\ldots}^{\infty} \sqrt{a^2(n) + b^2(n)}\sin(n\omega t) \tag{7.57}$$

Dividing the voltage expression in (7.57) by $Z(n)$ in (7.54) gives the Fourier components of the load currents as [1]

$$i_o(t) = \frac{2V_m}{\pi R}\cos(\alpha) + \sum_{n=2,4,\ldots}^{\infty} \frac{\sqrt{a^2(n) + b^2(n)}}{Z(n)}\sin(n\omega t - \theta_n) \tag{7.58}$$

where $\theta_n = \tan^{-1}\left(\frac{a_n}{b_n}\right)$ is the impedance angle of the nth harmonic component, rad.

Equation (7.58) gives the peak magnitude of the nth harmonic currents

$$I_m(n) = \frac{\sqrt{a^2(n) + b^2(n)}}{Z(n)} \quad \text{for } n = 2, 4, \ldots \tag{7.59}$$

Using Mathcad software, (7.30) gives the peak load current through RLC load for $n = 2$, $n = 4$, $n = 6$, $n = 8$, 10, and $n = 12$ as

$$I_m(2) = 0.76 \quad I_m(4) = 0.07 \quad I_m(6) = 0.08$$
$$I_m(8) = 0.02 \quad I_m(10) = 0.01 \quad I_m(12) = 0.02$$

Equation (7.30) gives the peak values of the nth harmonic currents through the RLC load. Since the load resistance R forms a parallel circuit with the filter capacitance C. The peak load current through R can be found from the current division rule as

$$I_{mR}(n) = I_m(n) \times \left|\frac{X_c}{R + X_c}\right| = \frac{\sqrt{a^2(n) + b^2(n)}}{|Z(n)|}\left|\frac{X_c}{R + X_c}\right| \tag{7.60}$$

where $X_c = \frac{1}{jn\omega_o C}$ is the impedance of the filter capacitor C, and $I_{mR}(n)$ is the peak value of the nth harmonic current through the load resistor R.

Using Mathcad software, (7.31) gives the peak load current through load R for $n = 2$, $n = 4$, $n = 6$, $n = 8$, $n = 10$, and $n = 12$ as

$$I_{mR}(2) = 0.25 \quad I_{mR}(4) = 0.01 \quad I_{mR}(6) = 0$$
$$I_{mR}(8) = 0 \quad I_{mR}(10) = 0 \quad I_{mR}(12) = 0$$

Dividing the peak values by $\sqrt{2}$ gives the rms values for $n = 2$, $n = 4$, $n = 6$, $n = 8$, $n = 10$, and $n = 12$ as

$$I_{rms}(2) = \frac{I_{mR}(2)}{\sqrt{2}} = 0.18 \quad I_{rms}(4) = \frac{I_{mR}(4)}{\sqrt{2}} = 0 \quad I_{rms}(6) = \frac{I_{mR}(6)}{\sqrt{2}} = 0$$

$$I_{rms}(8) = \frac{I_{mR}(8)}{\sqrt{2}} = 0 \quad I_{rms}(10) = \frac{I_{mR}(10)}{\sqrt{2}} = 0 \quad I_{rms}(12) = \frac{I_{mR}(12)}{\sqrt{2}} = 0$$

The total rms value of the ripple current can be determined by adding the square of the individual rms ripple current and then taking the square root as given by

$$\begin{aligned} I_{ripple} &= \sqrt{\sum_{n=2,4}^{30} |I_{rms}(n)|^2} \\ &= 0.19\ \text{A} \end{aligned} \tag{7.61}$$

The total rms value of the load current can be determined by adding the square of the average (DC) current and the rms ripple current as

$$\begin{aligned} I_o &= \sqrt{I_{dc}^2 + I_{ripple}^2} \\ &= \sqrt{5.402^2 + 0.19^2} = 5.41\ \text{A} \end{aligned} \tag{7.62}$$

A three-phase full-wave-controlled rectifier uses six thyristors is shown Figure 7.14 with an RLC load. The plots of the input and output voltages and the load current are shown in Figure 7.15(a). Figure 7.15(b) shows the effects of filter capacitances on the load voltage.

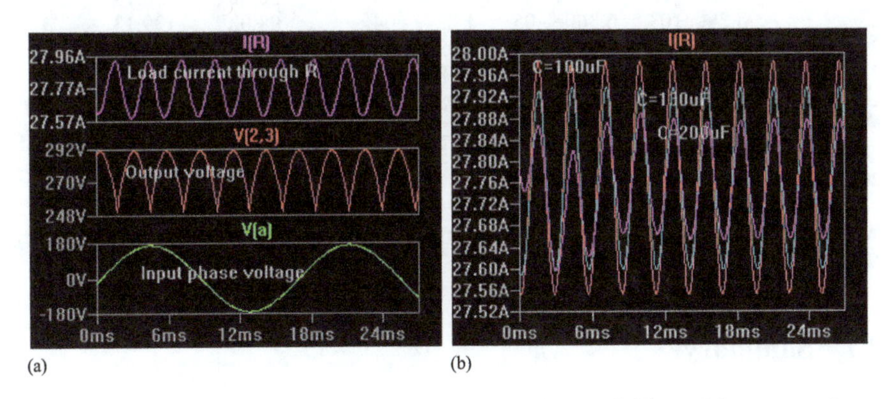

Figure 7.15 Single-phase controlled rectifier with an LC filter: (a) output voltage and load current and (b) effects of capacitance on load current

7.6.6 LTspice Fourier analysis

The LTspice command (.four 360 Hz 29 1 *I*(*R*)) for Fourier analysis gives the Fourier components of the load current *I*(*R*) at the fundamental frequency of 360 Hz as follows:

Fourier components of *I*(*vx*)

DC component: 27.7586

Harmonic number	Frequency (Hz)	Fourier component	Normalized component	Phase (degrees)	Normalized phase (degrees)
1	3.600e+02	5.194e−01	1.000e+00	−41.12	0.00
2	7.200e+02	5.939e−02	1.143e−01	93.96	135.07
3	1.080e+03	1.743e−02	3.356e−02	−129.64	−88.52
4	1.440e+03	7.274e−03	1.400e−02	8.10	49.22
5	1.800e+03	3.790e−03	7.297e−03	143.48	184.59
6	2.160e+03	2.172e−03	4.183e−03	−78.67	−37.56
7	2.520e+03	1.374e−03	2.645e−03	59.11	100.23
8	2.880e+03	8.914e−04	1.716e−03	−167.75	−126.63
9	3.240e+03	6.662e−04	1.283e−03	−24.67	16.45
10	3.600e+03	4.742e−04	9.129e−04	107.68	148.80
11	3.960e+03	3.337e−04	6.424e−04	−113.19	−72.07
12	4.320e+03	2.936e−04	5.653e−04	28.81	69.93
13	4.680e+03	1.804e−04	3.473e−04	159.37	200.49
14	5.040e+03	1.760e−04	3.389e−04	−64.90	−23.79
15	5.400e+03	1.534e−04	2.952e−04	69.04	110.15
16	5.760e+03	8.531e−05	1.642e−04	−155.10	−113.99
17	6.120e+03	9.233e−05	1.778e−04	−19.11	22.00
18	6.480e+03	8.380e−05	1.613e−04	131.17	172.28
19	6.840e+03	5.194e−05	9.999e−05	−95.75	−54.63
20	7.200e+03	5.180e−05	9.973e−05	43.07	84.18
21	7.560e+03	5.340e−05	1.028e−04	−170.85	−129.74
22	7.920e+03	2.655e−05	5.112e−05	−30.92	10.19
23	8.280e+03	6.360e−05	1.224e−04	98.02	139.13
24	8.640e+03	5.391e−05	1.038e−04	−153.34	−112.22
25	9.000e+03	1.863e−05	3.586e−05	−19.41	21.70
26	9.360e+03	5.609e−05	1.080e−04	135.59	176.70
27	9.720e+03	2.462e−05	4.739e−05	−93.89	−52.78
28	1.008e+04	1.828e−05	3.520e−05	68.86	109.98
29	1.044e+04	3.538e−05	6.811e−05	168.32	209.44

THD: 12.032953%(12.033203%)

Note: THD is reduced from 13.17% to 12.03 % due to the addition of a capacitor filter across *R*.

7.7 Summary

A controlled rectifier converts a fixed AC voltage to a variable DC voltage and commonly uses thyristors as switching devices. It uses one or more thyristors to make a unidirectional current flow from the positive terminal to the negative terminal of the output voltage. Thus, the voltage and the current on the input side are of AC types and the voltage and current on the output side are of DC types. The average output voltage can be varied from zero to the maximum value of a diode rectifier by varying the delay angle. However, the input power factor depends on the output voltage. L and LC filters are often connected to a resistive load to limit the ripple contents of the output voltage and current. For a resistive load, the load current can be discontinuous depending on a higher value of the delay angle. For a highly inductive load, the load current becomes continuous, and the conducting pair of thyristors continues to conduct until the next pair of thyristors is turned on. The output voltage of a three-phase controlled rectifiers is much higher than that of a single-phase controlled rectifier. A thyristor can be turned on by applying a pulse for a short duration. Once the thyristor is on, the gate pulse has no effect, and it remains on until its current is falls to zero. It is a latching device. The thyristors of the controlled rectifiers are turned off due to the natural behavior of the AC sinusoidal input voltage.

Problems

1. The single-phase full-wave converter as shown in Figure 7.3 has a resistive load of 5 Ω. The input voltage is $V_s = 120$ V at (rms) 60 Hz. The delay angle is $\omega t = \alpha = \pi/6$. Calculate the performance parameters. (a) the average output current $I_{dc,}$ (b) the rms output current $I_{rms,}$ (c) the average thyristor current $I_{A,}$ (d) the rms thyristor current I_R, (e) the rectification efficiency, (f) the TUF, and (g) the input PF. Assume ideal thyristor switches.
2. The single-phase full converter as shown in Figure 7.3 has a resistive load of 5 Ω. The input voltage is $V_s = 120$ V at (rms) 60 Hz. The delay angle is $\omega t = \alpha = \pi/6$. Calculate the performance parameters. (a) Output-side parameters, (b) controller parameters, and (c) input-side parameters. Assume ideal thyristor switches.
3. The single-phase full converter as shown in Figure 7.3 has a resistive load of 5 Ω. The input voltage is $V_s = 120$ V at (rms) 60 Hz. The delay angle is $\omega t = \alpha = \pi/3$. Calculate the performance parameters. (a) the average output current $I_{dc,}$ (b) the rms output current I_{rms}, (c) the average thyristor current $I_{A,}$ (d) the rms thyristor current I_R, (e) the rectification efficiency, (f) the TUF, and (g) the input PF. Assume ideal thyristor switches.
4. The single-phase full converter as shown in Figure 7.3 has a resistive load of 5 Ω. The input voltage is $V_s = 120$ V at (rms) 60 Hz. The delay angle is $\omega t = \alpha = \pi/3$. Calculate the performance parameters. (b) Output-side parameters, (c) controller parameters, and (d) input-side parameters. Assume ideal thyristor switches.
5. The single-phase full converter as shown in Figure 7.3 has a resistive load of 5 Ω. The input voltage is $V_s = 120$ V at (rms) 60 Hz. The delay angle is $\omega t = \alpha = \pi/2$. Calculate the performance parameters. (a) the average output

current I_{dc}, (b) the rms output current I_{rms}, (c) the average thyristor current I_A, (d) the rms thyristor current I_R, (e) the rectification efficiency, (f) the TUF, and (g) the input PF. Assume ideal thyristor switches.

6. The single-phase full converter as shown in Figure 7.3 has a resistive load of 5 Ω. The input voltage is $V_s = 120$ V at (rms) 60 Hz. The delay angle is $\omega t = \alpha = \pi/2$. Calculate the performance parameters. (b) Output-side parameters, (c) controller parameters, and (d) input-side parameters. Assume ideal thyristor switches.
7. The single-phase full converter as shown in Figure 7.5 has an RL load of $R = 5$ Ω. The input voltage is $V_s = 120$ V at (rms) 60 Hz. The delay angle is $\omega t = \alpha = \pi/6$. Calculate the performance parameters. (a) the value of inductance L for a 6% ripple of the average load current, (b) the average output current I_{dc}, (c) the rms output current I_{rms}, (d) the average thyristor current I_A, (e) the rms thyristor current I_R, (f) the rectification efficiency, (g) the TUF, and (h) the input PF. Assume ideal thyristor switches.
8. The single-phase full converter as shown in Figure 7.5 has an RL load of $R = 5$ Ω. The input voltage is $V_s = 120$ V at (rms) 60 Hz. The delay angle is $\omega t = \alpha = \pi/6$. Calculate the performance parameters. (a) the value of inductance L for a 6% ripple of the average load current, (b) output-side parameters, (c) controller parameters, and (d) input-side parameters. Assume ideal thyristor switches.
9. The single-phase full converter as shown in Figure 7.5 has an RL load of $R = 5$ Ω. The input voltage is $V_s = 120$ V at (rms) 60 Hz. The delay angle is $\omega t = \alpha = \pi/3$. Calculate the performance parameters. (a) the value of inductance L for a 6% ripple of the average load current, (b) the average output current I_{dc}, (c) the rms output current I_{rms}, (d) the average thyristor current I_A, (e) the rms thyristor current I_R, (f) the rectification efficiency, (g) the TUF, and (h) the input PF. Assume ideal thyristor switches.
10. The single-phase full converter as shown in Figure 7.5 has an RL load of $R = 5$ Ω. The input voltage is $V_s = 120$ V at (rms) 60 Hz. The delay angle is $\omega t = \alpha = \pi/3$. Calculate the performance parameters. (a) the value of inductance L for a 6% ripple of the average load current, (b) output-side parameters, (c) controller parameters, and (d) input-side parameters. Assume ideal thyristor switches.
11. The single-phase full converter as shown in Figure 7.5 has an RL load of $R = 5$ Ω. The input voltage is $V_s = 120$ V at (rms) 60 Hz. The delay angle is $\omega t = \alpha = \pi/3$. Calculate the performance parameters. (a) the value of inductance L for a 6% ripple of the average load current, (b) the average output current I_{dc}, (c) the rms output current I_{rms}, (d) the average thyristor current I_A, (e) the rms thyristor current I_R, (f) the rectification efficiency, (g) the TUF, and (h) the input PF. Assume ideal thyristor switches.
12. The single-phase full converter as shown in Figure 7.5 has an RL load of $R = 5$ Ω. The input voltage is $V_s = 120$ V at (rms) 60 Hz. The delay angle is $\omega t = \alpha = \pi/3$. Calculate the performance parameters. (a) the value of inductance L for a 6% ripple of the average load current, (b) output-side parameters,

(c) controller parameters, and (d) input-side parameters. Assume ideal thyristor switches.

13. The single-phase full converter as shown in Figure 7.5 has an RL load of $R = 5\ \Omega$. The input voltage is $V_s = 120$ V at (rms) 60 Hz. The delay angle is $\omega t = \alpha = \pi/2$. Calculate the performance parameters. (a) the value of inductance L for a 6% ripple of the average load current, (b) the average output current $I_{dc,}$ (c) the rms output current $I_{rms,}$ (d) the average thyristor current $I_{A,}$ (e) the rms thyristor current I_R, (f) the rectification efficiency, (g) the TUF, and (h) the input PF. Assume ideal thyristor switches.
14. The single-phase full converter as shown in Figure 7.5 has an RL load of $R = 5\ \Omega$. The input voltage is $V_s = 120$ V at (rms) 60 Hz. The delay angle is $\omega t = \alpha = \pi/2$. Calculate the performance parameters. (a) the value of inductance L for a 6% ripple of the average load current, (b) output-side parameters, (c) controller parameters, and (d) input-side parameters. Assume ideal thyristor switches.
15. The single-phase full converter as shown in Figure 7.7 has an RLC-load of $R = 5\ \Omega$. The input voltage is $V_s = 120$ V at (rms) 60 Hz. The delay angle is $\omega t = \alpha = \pi/6$. Calculate the performance parameters. (a) the value of inductance L for a 6% ripple of the average load current, and the value of capacitance C for a 5% ripple of the average output voltage, (b) the average output current $I_{dc,}$ (c) the rms output current $I_{rms,}$ (d) the average thyristor current $I_{A,}$ (e) the rms thyristor current I_R, (f) the rectification efficiency, (g) the TUF, and (h) the input PF. Assume ideal thyristor switches.
16. The single-phase full converter as shown in Figure 7.7 has an RCL-load of $R = 5\ \Omega$. The input voltage is $V_s = 120$ V at (rms) 60 Hz. The delay angle is $\omega t = \alpha = \pi/6$. Calculate the performance parameters. (a) the value of inductance L for a 6% ripple of the average load current, and the value of capacitance C for a 5% ripple of the average output voltage, (b) output-side parameters, (c) controller parameters, and (d) input-side parameters. Assume ideal thyristor switches.
17. The single-phase full converter as shown in Figure 7.7 has an RLC-load of $R = 5\ \Omega$. The input voltage is $V_s = 120$ V at (rms) 60 Hz. The delay angle is $\omega t = \alpha = \pi/3$. Calculate the performance parameters. (a) the value of inductance L for a 6% ripple of the average load current, and the value of capacitance C for a 5% ripple of the average output voltage, (b) the average output current $I_{dc,}$ (c) the rms output current $I_{rms,}$ (d) the average thyristor current $I_{A,}$ (e) the rms thyristor current I_R, (f) the rectification efficiency, (g) the TUF, and (h) the input PF. Assume ideal thyristor switches.
18. The single-phase full converter as shown in Figure 7.7 has an RCL-load of $R = 5\ \Omega$. The input voltage is $V_s = 120$ V at (rms) 60 Hz. The delay angle is $\omega t = \alpha = \pi/3$. Calculate the performance parameters. (a) the value of inductance L for a 6% ripple of the average load current, and the value of capacitance C for a 5% ripple of the average output voltage, (b) output-side parameters, (c) controller parameters, and (d) input-side parameters. Assume ideal thyristor switches.

19. The single-phase full converter as shown in Figure 7.7 has an RLC-load of $R = 5\ \Omega$. The input voltage is $V_s = 120$ V at (rms) 60 Hz. The delay angle is $\omega t = \alpha = \pi/2$. Calculate the performance parameters. (a) the value of inductance L for a 6% ripple of the average load current, and the value of capacitance C for a 5% ripple of the average output voltage, (b) the average output current I_{dc}, (c) the rms output current I_{rms}, (d) the average thyristor current I_A, (e) the rms thyristor current I_R, (f) the rectification efficiency, (g) the TUF, and (h) the input PF. Assume ideal thyristor switches.
20. The single-phase full converter as shown in Figure 7.7 has an RCL-load of $R = 5\ \Omega$. The input voltage is $V_s = 120$ V at (rms) 60 Hz. The delay angle is $\omega t = \alpha = \pi/2$. Calculate the performance parameters. (a) the value of inductance L for a 6% ripple of the average load current, and the value of capacitance C for a 5% ripple of the average output voltage, (b) output-side parameters, (c) controller parameters, and (d) input-side parameters. Assume ideal thyristor switches.
21. The three-phase full-wave converter as shown in Figure 7.9 has a resistive load of $5\ \Omega$. The input phase voltage is $V_s = 120$ V at (rms) 60 Hz. The delay angle is $\omega t = \alpha = \pi/6$. Calculate the performance parameters. (a) the average output current I_{dc}, (b) the rms output current I_{rms}, (c) the average thyristor current I_A, (d) the rms thyristor current I_R, (e) the rectification efficiency, (f) the TUF, and (g) the input PF. Assume ideal thyristor switches.
22. The three-phase full converter as shown in Figure 7.9 has a resistive load of $5\ \Omega$. The input voltage is $V_s = 120$ V at (rms) 60 Hz. The delay angle is $\omega t = \alpha = \pi/6$. Calculate the performance parameters. (a) Output-side parameters, (b) controller parameters, and (c) input-side parameters. Assume ideal thyristor switches.
23. The three-phase full converter as shown in Figure 7.12 has an RL load of $R = 5\ \Omega$. The input phase voltage is $V_s = 120$ V at (rms) 60 Hz. The delay angle is $\omega t = \alpha = \pi/3$. Calculate the performance parameters. (a) the value of inductance L for a 6% ripple of the average load current, (b) the average output current I_{dc}, (c) the rms output current I_{rms}, (d) the average thyristor current I_A, (e) the rms thyristor current I_R, (f) the rectification efficiency, (g) the TUF, and (h) the input PF. Assume ideal thyristor switches.
24. The three-phase full converter as shown in Figure 7.12 has an RL load of $R = 5\ \Omega$. The input phase voltage is $V_s = 120$ V at (rms) 60 Hz. The delay angle is $\omega t = \alpha = \pi/3$. Calculate the performance parameters. (a) the value of inductance L for a 6% ripple of the average load current, (b) output-side parameters, (c) controller parameters, and (d) input-side parameters. Assume ideal thyristor switches.
25. The three-phase full converter as shown in Figure 7.14 has an RLC-load of $R = 5\ \Omega$. The input phase voltage is $V_s = 120$ V at (rms) 60 Hz. The delay angle is $\omega t = \alpha = \pi/3$. Calculate the performance parameters. (a) the value of inductance L for a 6% ripple of the average load current, and the value of capacitance C for a 5% ripple of the average output voltage, (b) the average output current I_{dc}, (c) the rms output current I_{rms}, (d) the average thyristor

current $I_{A,}$ (e) the rms thyristor current I_R, (f) the rectification efficiency, (g) the TUF, and (h) the input PF. Assume ideal thyristor switches.

26. The three-phase full converter as shown in Figure 7.14 has an RLC-load of $R = 5\ \Omega$. The input phase voltage is $V_s = 120$ V at (rms) 60 Hz. The delay angle is $\omega t = \alpha = \pi/3$. Calculate the performance parameters. (a) the value of inductance L for a 6% ripple of the average load current, and the value of capacitance C for a 5% ripple of the average output voltage, (b) output-side parameters, (c) controller parameters, and (d) input-side parameters. Assume ideal thyristor switches.

References

[1] L.J. Giacoletto, Simple SCR and TRIAC PSPICE computer models, *IEEE Transactions on Industrial Electronics*, 36(3), 1989, 451–455.

[2] G.L. Arsov, Comments on a nonideal macromodel of thyristor for transient analysis in power electronics, *IEEE Transactions on Industrial Electronics*, 39(2), 1992, 175–176.

[3] F.J. Garcia, F. Arizti, and F.J. Aranceta, A nonideal macromodel of thyristor for transient analysis in power electronics, *IEEE Transactions on Industrial Electronics*, 37(6), 1990, 514–520.

[4] R.L. Avant, and F.C.Y. Lee, The J3 SCR model applied to resonant converter simulation, *IEEE Transactions on Industrial Electronics*, 32(1), 1985, 1–12.

[5] E.Y. Ho, and P.C. Sen, Effect of gate drive on GTO thyristor characteristics, *IEEE Transactions on Industrial Electronics*, 33(3), 1986, 325–331.

[6] M.A.I. El-Amin, GTO PSPICE model and its applications, *The Fourth Saudi Engineering Conference*, November 1995, Vol. III, pp. 271–277.

[7] G. Busatto, F. Iannuzzo, and L. Fratelli, Proceedings of Symposium on Power Electronics Electrical Drives, Advanced Machine Power Quality, *The Fourth Saudi Engineering Conference*, June 1998, Sorrento, Italy, Col. 1, pp. P2/5–10.

[8] A.A. Zekry, G.T. Sayah, and F.A. Soliman, SPICE model of thyristors with amplifying gate and emitter-shorts. IET Power Electronics, 7(3), 2014, 724–735.

[9] https://pdf1.alldatasheet.com/datasheet-pdf/download/91463/MOTOROLA/2N5171.html

[10] M.H. Rashid, *Power Electronics: Circuits, Devices, and Applications*, 4th ed., Englewood Cliffs, NJ: Prentice-Hall, 2014, Chapter 5.

[11] M.H. Rashid, *SPICE and LTspice for Power Electronics and Electric Power*, Boca Raton, FL: CRC Press, 2024.

Chapter 8
AC voltage controllers

8.1 Introduction

An AC voltage controller converts an alternating current (AC) voltage to a variable AC voltage. The input voltage and the output current of AC voltage controllers pass through the value zero in every cycle. A thyristor can be turned on by applying a pulse of short duration, and it is turned off by natural commutation due to the natural behavior of the input voltage and the current. An AC voltage controller uses anti-parallel thyristor – one thyristor allows current flow in one direction and the other thyristor allows current flow in the other direction [1,2].

8.2 Single-phase AC voltage controller with resistive load

A single-phase full-wave AC voltage controller uses two thyristors as shown in Figure 8.1 [3,4] to a resistive load of R. Vx is added to measure the load current, $I(Vx)$ in LTspice. One thyristor allows current flow in one direction and the other thyristor allows current flow in the other direction. As a result, it also creates an alternating current on the output side. The addition of a small inductor $L = 100$ nH makes the circuit in a generalized form and an inductor is often connected to serve as a filter. R_s and C_s form a snubber circuit to suppress transient voltages across the thyristors. The voltage-controlled voltage sources E_1 and E_2 are connected to isolate the gating signals from the power circuit. The AC input voltage is given by

$$v_s(t) = V_m \sin(\omega t) = V_m \sin(2\pi f t) \tag{8.1}$$

where V_m= peak value of the AC input voltage, V; f = supply frequency, Hz; $\omega = 2\pi f$= supply frequency, rad/s.

For an AC supply of $V_s = 120$ V (rms), the peak supply voltage is

$$V_m = \sqrt{2}V_s = \sqrt{2} \times 120 = 169.7 \text{ V}.$$

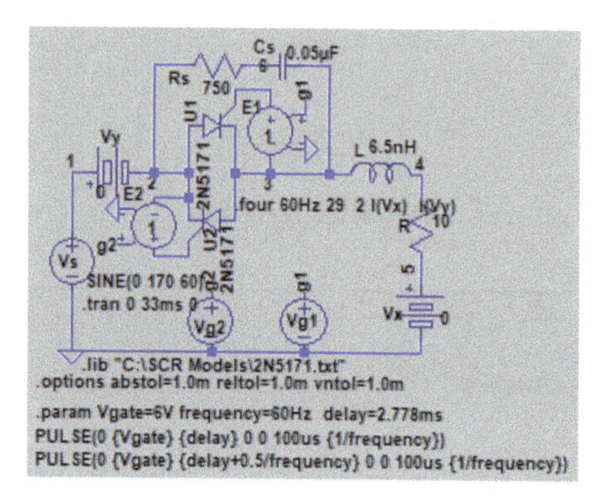

Figure 8.1 Schematic of single-phase AC voltage controller

Let us consider a delay angle of $\alpha = 60°$,
Time delay for triggering thyristor T_1, $t_1 = \frac{60}{360} \times \frac{1000}{60\ \text{Hz}} \times 1000 = 2777.8\ \mu\text{s}$

Time delay for triggering thyristor T_2 with an extra delay of π, $t_2 = \frac{(180+60)}{360} \times \frac{1000}{60\ \text{Hz}} \times 1000 = 11111.1\ \mu\text{s}$

During the positive half-cycle of the input voltage, thyristor T_1 is triggered at a delay angle of $\alpha = 60°$. The input voltage appears across the load from $\omega t = 60°$ to 180°. During the negative half-cycle of the input voltage, thyristors T_2 is triggered at a delay angle of $\alpha = 180° + 60° = 240°$ and the input voltage appears across the load from $\omega t = 240°$ to 360°. The output voltage can be described as

$$\begin{aligned} v_o(t) &= V_m \sin(\omega t) \quad \text{for } \alpha < \omega t \leq \pi \\ &= V_m \sin(\omega t) \quad \text{for } (\pi + \alpha) < \omega t \leq 2\pi \end{aligned} \tag{8.2}$$

8.3 Thyristor gating pulses output voltage and load current

The plots of the gating pulses of 6 V and width of 100 µs are shown in Figure 8.2 (a). The gating pulse for T_1 is delayed by α and the gating pulse for T_2 is delayed by $(\pi + \alpha)$. As expected, the load current becomes zero when the input voltage falls to zero. Thyristor conducts from $\omega t = \alpha$ to π, and from $(\pi + \alpha)$ to (2π). The plots of the input and output voltages are shown in Figure 8.2(b). The rms output voltage

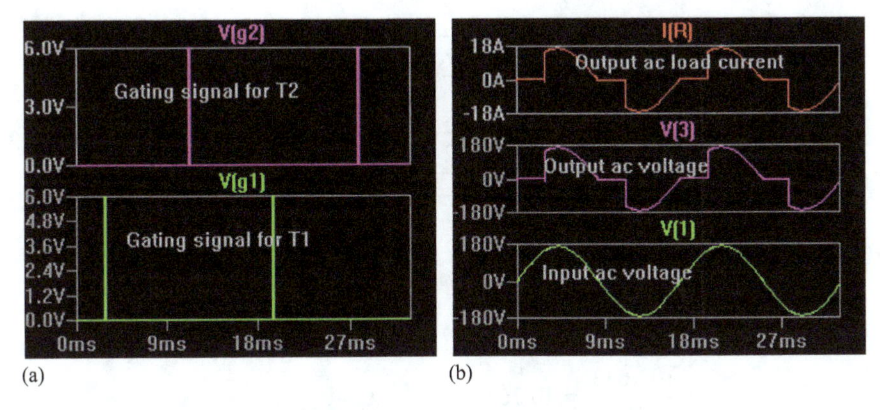

Figure 8.2 Waveforms of single-phase voltage controller: (a) Thyristor Gating pulses and (b) Output voltage and load current

can be found from [1]

$$V_o = \sqrt{\frac{2}{2\pi}\int_{\alpha}^{\pi}(\sqrt{2}V_s \sin(x))^2 dx} = \sqrt{\frac{V_s^2}{\pi}\int_{\alpha}^{\pi}(1 - \cos(2x))dx}$$
$$= V_s\sqrt{\frac{1}{\pi}(\pi - \alpha + \frac{\sin(2\alpha)}{2}} \tag{8.3}$$

The rms current through the load can be found from

$$I_o = \frac{V_o}{R} = \frac{V_s}{R}\sqrt{\frac{1}{\pi}(\pi - \alpha + \frac{\sin(2\alpha)}{2}} \tag{8.4}$$

The average current through the thyristor can be found from

$$I_A = \frac{1}{2\pi R}\int_{\alpha}^{\pi}\sqrt{2}V_s \sin(x)dx = \frac{\sqrt{2}V_s}{2\pi R}(1 + \cos(\alpha)) \tag{8.5}$$

The performance parameters of an AC voltage controller can be divided into three types.

- Output-side parameters,
- Controller parameters, and
- Input-side parameters.

Let us consider an input voltage of $V_s = 120$ V (rms), $V_m = \sqrt{2}V_s = 120 \times \sqrt{2} = 169.7$ V, $R = 10\ \Omega$ and delay angle of $\alpha = 60°$. Converting to radian, $\alpha = \frac{60°}{180} \times \pi = 1.047$ rad.

8.3.1 Output-side parameters

(a) Equation (8.3) gives the rms output voltage as

$$V_o = V_{rms} = V_s\sqrt{\frac{1}{\pi}\left(\pi - \alpha + \frac{\sin(2\alpha)}{2}\right)}$$

$$= 120 \times \sqrt{\frac{1}{\pi}\left(\pi - 1.047 + \frac{\sin(2 \times 1.047)}{2}\right)} = 107.633 \text{ V}$$

(b) The rms output current is

$$I_{o(rms)} = \frac{V_{o(rms)}}{R}$$

$$= \frac{107.633}{10} = 10.763 \text{ A}$$

(c) The AC output power is

$$P_{ac} = V_{o(rms)} I_{o(rms)}$$

$$= 107.633 \times 10.763 = 1.158 \text{ kW}$$

(d) The output power is

$$P_{out} = I_{rms}^2 R$$

$$= 10.763^2 \times 10 = 1.158 \text{ kW}$$

Note: For a resistive load, $P_{ac} = P_{out}$.

8.3.2 Controlled rectifier parameters

(a) The peak thyristor current is

$$I_{T(peak)} = \frac{V_m}{R}$$

$$= \frac{169.7}{10} = 16.971 \text{ A}$$

(b) Equation (8.5) gives the average thyristor current is

$$I_A = \frac{\sqrt{2}V_s}{2\pi R}(1 + \cos(\alpha))$$

$$= \frac{\sqrt{2} \times 120}{2 \times \pi \times 10}(1 + \cos(1.047)) = 4.051$$

(c) The load current is shared by two devices, the rms thyristor current is

$$I_{T(rms)} = \frac{I_{rms}}{\sqrt{2}}$$

$$= \frac{10.763}{\sqrt{2}} = 7.611 \text{ A}$$

(d) With a safely margin of two times, the peak voltage rating of thyristors is

$$V_{peak} = 2V_m$$
$$= 2 \times 169.7 = 339.411 \text{ V}$$

8.3.3 Input-side parameters

The rms input current is

$$I_s = I_{rms}$$
$$= 10.763 \text{ A}$$

For an ideal no-power loss converter, the input power is

$$P_{in} = P_{out}$$
$$= 1.158 \text{ kW}$$

The input power factor is

$$PF_i = \frac{P_{in}}{V_s I_s}$$
$$= \frac{1158}{120 \times 10.763} = 0.897 \text{ (lagging)}$$

8.3.4 LTspice Fourier analysis

The LTspice command (.four 600 Hz 29 1 *I*(*R*)) for Fourier Analysis gives the Fourier components of the load current *I*(*R*) at the fundamental frequency of 60 Hz as follows:

Fourier components of *I*(*vx*)

DC component: 0.00989838

Harmonic number	Frequency (Hz)	Fourier component	Normalized component	Phase (degrees)	Normalized phase (degrees)
1	6.000e+01	1.414e+01	1.000e+00	−16.71	0.00°
2	1.200e+02	2.119e−02	1.498e−03	99.94	116.65°
3	1.800e+02	4.081e+00	2.886e−01	149.21	165.93
4	2.400e+02	2.025e−02	1.432e−03	112.13	128.85
5	3.000e+02	2.366e+00	1.673e−01	60.72	77.44
6	3.600e+02	1.791e−02	1.267e−03	120.57	137.28
7	4.200e+02	1.152e+00	8.143e−02	−61.22	−44.50
8	4.800e+02	1.927e−02	1.363e−03	119.96	136.68
9	5.400e+02	1.093e+00	7.729e−02	168.02	184.73
10	6.000e+02	2.261e−02	1.599e−03	135.57	152.28
11	6.600e+02	9.407e−01	6.653e−02	60.84	77.56
12	7.200e+02	2.079e−02	1.471e−03	156.38	173.10
13	7.800e+02	6.522e−01	4.612e−02	−62.11	−45.40
14	8.400e+02	1.690e−02	1.195e−03	163.95	180.66
15	9.000e+02	6.555e−01	4.635e−02	172.40	189.12
16	9.600e+02	1.722e−02	1.218e−03	165.77	182.48

(Continues)

(*Continued*)

Fourier components of *I(vx)*					
DC component: 0.00989838					
Harmonic number	**Frequency (Hz)**	**Fourier component**	**Normalized component**	**Phase (degrees)**	**Normalized phase (degrees)**
17	1.020e+03	5.776e−01	4.084e−02	60.73	77.45
18	1.080e+03	1.722e−02	1.218e−03	177.55	194.26
19	1.140e+03	4.606e−01	3.257e−02	−63.32	−46.60
20	1.200e+03	1.737e−02	1.228e−03	−174.38	−157.66
21	1.260e+03	4.686e−01	3.314e−02	174.75	191.47
22	1.320e+03	1.685e−02	1.192e−03	−162.26	−145.55
23	1.380e+03	4.123e−01	2.915e−02	60.11	76.83
24	1.440e+03	1.612e−02	1.140e−03	−149.92	−133.21
25	1.500e+03	3.617e−01	2.558e−02	−64.24	−47.52
26	1.560e+03	1.514e−02	1.071e−03	−142.14	−125.42
27	1.620e+03	3.606e−01	2.550e−02	176.52	193.24
28	1.680e+03	1.463e−02	1.035e−03	−129.78	−113.06
29	1.740e+03	3.200e−01	2.263e−02	58.75	75.46
Total harmonic distortion: 37.298336% (36.494201%)					

8.4 Single-phase AC voltage controller with RL load

A single-phase full-wave AC voltage controller uses two thyristors as shown in Figure 8.3 [1,2] with an RL load. One thyristor allows current flow in one direction

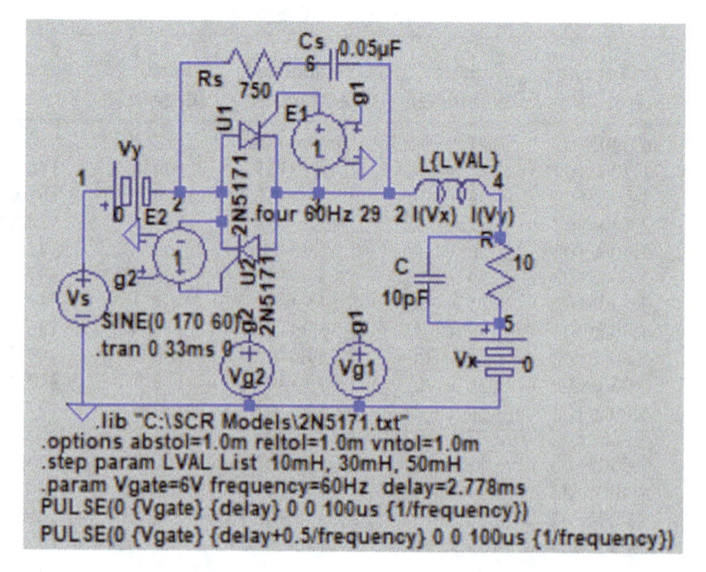

Figure 8.3 Schematic of single-phase controller with an RL load

and the other thyristor allows current flow in the other direction. During the positive half-cycle, thyristor T_1 is turned on at a delay angle of $\omega t = \alpha$ and the load is connected to the input supply through T_1. Due to the inductive load, thyristor T_1 continues to conduct beyond $\omega t = \pi$ even though the input voltage is already negative and thyristor T_1 continues to conduct until the load current becomes zero at an angle $\omega t = \beta$ known *extinction angle.*

During the negative half-cycle of the input voltage, thyristor T_2 is turned on and the input voltage appears across the load. Thus, with an inductive load, the load current will be delayed and will not fall to zero when the supply voltage falls to zero. If the time constant of the RL load is greater than the period of the output voltage, the load current will continues and will not fall to zero.

The peak supply voltage V_{m} = 169.7 V, and a delay angle of α = 60° or 1.05 rad

Time delay for triggering thyristor T_1,

$$t_1 = \frac{60}{360} \times \frac{1000}{60\ \mathrm{Hz}} \times 1000 = 2777.8\ \mu\mathrm{s}$$

Time delay for triggering thyristor T_2,

$$t_2 = \frac{(180 + 60)}{360} \times \frac{1000}{60\ \mathrm{Hz}} \times 1000 = 11111.1\ \mu\mathrm{s}$$

During the positive half-cycle of the input voltage, thyristor T_1 is triggered at a delay angle of $\alpha = 60°$. The input voltage appears across the load from $\omega t = 60°$ to $(180° + 60°)$. During the negative half-cycle of the input voltage, thyristor T_2 is triggered at a delay angle of α = 180° + 60° = 240° and the input voltage appears across the load from $\omega t = 240°$ to $360° + 60°$. The output voltage can be described as

$$\begin{aligned} v_o(t) &= V_m \sin(\omega t) \quad \text{for } \alpha < \omega t \leq (\pi + \alpha) \\ &= V_m \sin(\omega t) \quad \text{for } (\pi + \alpha) < \omega t \leq (2\pi + \alpha) \end{aligned} \tag{8.6}$$

The plots of the gating pulses of 6 V, and width 100 μs are shown in Figure 8.2 (a). The gating pulse for T_1 is delayed by α and the gating pulse for T_2 is delayed by $(\pi + \alpha)$. As expected, the load current does not become zero when the input voltage falls to zero. Thyristor T_1 conducts from $\omega t = \alpha$ until the current extinguishes to zero at $\omega t = \beta$, and thyristor T_2 conducts from $(\pi + \alpha)$ until the load current extinguishes at $\omega t = \pi + \beta$. The plots of the input and output voltages are shown in Figure 8.4(a). The rms output voltage can be found in [1]

$$\begin{aligned} V_o &= \sqrt{\frac{2}{2\pi}\int_{\alpha}^{\beta}(\sqrt{2}V_s \sin(x))^2 dx} = \sqrt{\frac{4V_s^2}{4\pi}\int_{\alpha}^{\beta}(1 - \cos(2x))dx} \\ &= V_s\sqrt{\frac{1}{\pi}(\beta - \alpha + \frac{\sin(2\alpha)}{2} - \frac{\sin(2\beta)}{2}} \end{aligned} \tag{8.7}$$

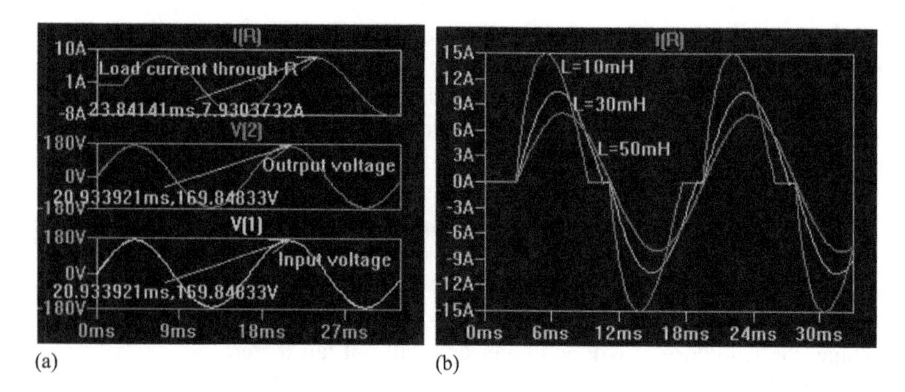

(a) (b)

Figure 8.4 Waveforms of single-phase controller with an RL load: (a) output voltage and load current and (b) effects of load inductances

8.4.1 Circuit model

With an inductive load, the AC sinusoidal voltage is connected through thyristor T_1 to an RL load. The load current i_1 during mode 1 when thyristor T_1 conducts can be found from [1]

$$L\frac{di_1}{dt} + Ri_1 = \sqrt{2}V_s \sin(\omega t) \tag{8.8}$$

The solution of (8.8) is of the form

$$i_1 = \frac{\sqrt{2}V_s}{Z} \sin(\omega t - \theta) + A_1 e^{-(R/L)t} \tag{8.9}$$

where the load impedance, $Z = \sqrt{R^2 + (\omega L)^2}$, and the load angle $\theta = \tan^{-1}(\omega L/R)$.

The constant A_1 in (8.9) can be found from the initial condition at $\omega t = \alpha$, $i_1 = 0$. Applying the initial condition, (8.9) gives

$$A_1 = -\frac{\sqrt{2}V_s}{Z} \sin(\alpha - \theta) e^{(R/L)(\alpha/\omega)} \tag{8.10}$$

Substituting A_1 from (8.10) to (8.9) gives the load current during mode 1 as

$$i_1 = \frac{\sqrt{2}V_s}{Z}\left[\sin(\omega t - \theta) - \sin(\alpha - \theta) e^{(R/L)(\alpha/\omega - t)}\right] \tag{8.11}$$

which indicates that the load current and the load voltage can be sinusoidal if the delay angle α is less than the load impedance angle θ. If α is greater than θ, the load current would be discontinuous.

The angle β when the load current falls to zero and thyristor T_1 is turned off can be found from the condition $i_1(\omega t = \beta) = 0$ in (8.11) and is given by the relation

$$\sin(\beta - \theta) = \sin(\alpha - \theta)e^{(R/L)(\alpha-\beta)/\omega} \tag{8.12}$$

The extinction angle β can be determined from transcendental equation (8.11) by an iterative method of solution.

If $\alpha = \theta$, (8.12) gives

$$\sin(\beta - \theta) = \sin(\beta - \alpha) = 0 \tag{8.13}$$

And

$$\beta - \alpha = \pi \tag{8.14}$$

The rms value of the load current can be determined from (8.11) as given by

$$I_R = \sqrt{\frac{1}{2\pi}\int_{\alpha}^{\beta} i_1^2(\omega t)d(\omega t)} \tag{8.15}$$

The average value of the thyristor current can be determined from (8.9) as given by

$$I_A = \frac{1}{2\pi}\int_{\alpha}^{\beta} i_1(\omega t)d(\omega t) \tag{8.16}$$

8.4.2 *Load inductor*

The frequency f_o of the output voltage is the same as the supply frequency f. We can find the inductor value to maintain continuous load current. That is,

$$\frac{L}{R} \geq T_o = \frac{1}{f} \tag{8.17}$$

which give the value of L as

$$L \geq \frac{R}{f} = \frac{10}{60} = 166.67 \text{ mH}$$

For a continuous load current,

$$\tan^{-1}\left(\frac{\omega L}{R}\right) \geq \alpha \tag{8.18}$$

which gives the minimum value of inductor as

$$\begin{aligned} L &\geq \frac{R\tan(\alpha)}{\omega} \\ &\geq \frac{10 \times \tan(1.05)}{376.99} = 45.94 \text{ mH} \end{aligned} \tag{8.19}$$

If α is greater than θ, the load current would be discontinuous. Let us choose $L = 50$ mH.

Let us consider an input voltage of $V_s = 120$ V (rms), $V_m = \sqrt{2}V_s = 120 \times \sqrt{2} = 169.7$ V, $R = 10\ \Omega$, $V_x = 0$, and delay angle of $\alpha = 60°$. Converting to radian, $\alpha = \frac{60°}{180} \times \pi = 1.047$ rad.

The plots of the input and output voltages and the load current are shown in Figure 8.4(a). The effects of load inductances are shown in Figure 8.4(b). As expected, the load current becomes discontinuous for $L = 10$ mH and close to sinusoidal for 30 mH.

Equation (8.14) gives the approximate value of the extinction angle

$$\begin{aligned}\beta &= \pi + \alpha \\ &= \pi + 1.047 \quad = 4.19 \text{ rad } (240°)\end{aligned} \tag{8.20}$$

Rewriting (8.12) gives

$$\sin(\beta - \theta) - \sin(\alpha - \theta)e^{(R/L)(\alpha-\beta)/\omega} = 0 \tag{8.21}$$

which, for $R = 10\ \Omega$, $L = 50$ mH and $\alpha = 1.047$, gives by the Mathcad iterative method of solution as

$$\beta = 4.23\text{rad } (242.43°).$$

8.4.3 Converter parameters

The performance parameters of AC voltage controller can be divided into three types.

- Output-side parameters,
- Controller parameters, and
- Input-side parameters.

8.4.3.1 Output-side parameters

(a) Equation (8.7) gives the rms output voltage is

$$V_{o(rms)} = V_s\sqrt{\frac{1}{\pi}\left(\beta - \alpha + \frac{\sin(2\alpha)}{2} - \frac{\sin(2\beta)}{2}\right)}$$

$$= 120 \times \sqrt{\frac{1}{\pi}\left(4.23 - 1.047 + \frac{\sin(2 \times 1.047)}{2} - \frac{\sin(2 \times 4.23)}{2}\right)} = 121.24\text{ V}$$

(b) Since the rms output current is contributed by two thyristors, (8.15) gives the rms output current,

$$\begin{aligned}I_o = I_{o(rms)} &= \sqrt{\frac{2}{2\pi}\int_{\alpha}^{\beta} i_1^2(\omega t)d(\omega t)} \\ &= 5.74\ \text{A}\end{aligned} \tag{8.22}$$

(c) The output power

$$P_{out} = I_{o(rms)}^2 R$$
$$= 5.74^2 \times 10 = 329.63 \text{ W}$$

8.4.3.2 Controller parameters

(a) The peak thyristor current

$$I_{T(peak)} = \frac{V_m}{R}$$
$$= \frac{169.7}{10} = 16.971 \text{ A}$$

(b) Equation (8.16) gives the average thyristor current is

$$I_A = \frac{1}{2\pi}\int_{\alpha}^{\beta} i_1(\omega t)d(\omega t)$$
$$= 2.6 \text{ A}$$

(c) Equation (8.15) gives the rms thyristor current is

$$I_R = \sqrt{\frac{1}{2\pi}\int_{\alpha}^{\beta} i_1^2(\omega t)d(\omega t)}$$
$$= 4.17 \text{ A}$$

(d) With a safely margin of two times, the peak voltage rating of thyristors is

$$V_{peak} = 2V_m$$
$$= 2 \times 169.7 = 339.411 \text{ V}$$

8.4.3.3 Input-side parameters

The rms input current is

$$I_s = I_{rms}$$
$$= 5.74 \text{ A}$$

The input power is

$$P_{in} = P_{out}$$
$$= 329.63 \text{ W}$$

Note: For an ideal no-loss converter, $P_{in} = P_{out}$.

The input power factor,

$$PF_i = \frac{P_{in}}{V_s I_s}$$
$$= \frac{329.63}{120 \times 5.74} = 0.48 \text{ (lagging)}$$

The transformer utilization factor,

$$\mathrm{TUF} = \frac{P_{dc}}{V_s I_s}$$

$$= \frac{291.81}{120 \times 5.74} \times 100 = 41.25\%$$

8.4.3.4 LTspice Fourier analysis

The LTspice command (.four 60 Hz 29 1 *I*(*R*)) for Fourier analysis gives the Fourier components of the load current *I*(*R*) at the fundamental frequency of 60 Hz as follows:

***N*-Period=2**					
Fourier components of *I(vx)*					
DC component: 0.142659					
Harmonic number	**Frequency (HZ)**	**Fourier component**	**Normalized component**	**Phase (degrees)**	**Normalized phase (degrees)**
1	6.000e+01	1.312e+01	1.000e+00	−33.76	0.00
2	1.200e+02	2.837e−01	2.162e−02	86.17	119.92
3	1.800e+02	2.969e+00	2.262e−01	91.34	125.10
4	2.400e+02	2.623e−01	1.998e−02	82.71	116.46
5	3.000e+02	1.353e+00	1.031e−01	0.52	34.27
6	3.600e+02	2.295e−01	1.749e−02	80.10	113.85
7	4.200e+02	3.565e−01	2.717e−02	−118.21	−84.46
8	4.800e+02	1.896e−01	1.444e−02	79.05	112.81
9	5.400e+02	3.341e−01	2.546e−02	96.69	130.45
10	6.000e+02	1.474e−01	1.124e−02	80.78	114.54
11	6.600e+02	1.461e−01	1.113e−02	−2.20	31.56
12	7.200e+02	1.089e−01	8.300e−03	87.37	121.13
13	7.800e+02	1.807e−01	1.377e−02	−178.67	−144.91
14	8.400e+02	8.022e−02	6.113e−03	101.48	135.24
15	9.000e+02	2.156e−01	1.643e−02	111.26	145.02
16	9.600e+02	6.591e−02	5.022e−03	122.40	156.16
17	1.020e+03	5.297e−02	4.036e−03	60.73	94.49
18	1.080e+03	6.313e−02	4.811e−03	142.12	175.88
19	1.140e+03	9.974e−02	7.600e−03	169.69	203.44
20	1.200e+03	6.323e−02	4.818e−03	155.76	189.51
21	1.260e+03	9.589e−02	7.307e−03	115.55	149.31
22	1.320e+03	6.063e−02	4.620e−03	165.41	199.17
23	1.380e+03	1.731e−02	1.319e−03	−75.57	−41.81
24	1.440e+03	5.458e−02	4.159e−03	174.55	208.31
25	1.500e+03	8.329e−02	6.347e−03	−154.20	−120.44
26	1.560e+03	4.699e−02	3.581e−03	−173.81	−140.06
27	1.620e+03	4.688e−02	3.572e−03	168.65	202.41
28	1.680e+03	4.064e−02	3.097e−03	−158.00	−124.25
29	1.740e+03	4.527e−02	3.450e−03	−111.02	−77.27
Total harmonic distortion: 25.637749% (24.130706%)					

Note: THD is reduced from 37.29% to 25.64 % due to the addition of an inductor L.

8.5 Three-phase AC voltage controller with RL

A three-phase AC voltage controller uses six thyristors as shown in Figure 8.5 [1,2] with a wye-connected resistive load. Two thyristors are connected in pairs in opposite directions to form a single-phase voltage controller and allows current flow in both the direction. One thyristor allows current flow in one direction and the other thyristor allows current flow in the other direction. The firing sequence of thyristors is T_1, T_2, T_3, T_4, T_5, T_6, T_1. If the AC line-to-neutral input voltages of a wye-connected source are defined as [5]

$$\begin{aligned} v_{AN} &= \sqrt{2}V_s \sin(\omega t) \\ v_{BN} &= \sqrt{2}V_s \sin\left(\omega t - \frac{2\pi}{3}\right) \\ v_{CN} &= \sqrt{2}V_s \sin\left(\omega t - \frac{4\pi}{3}\right) \end{aligned} \tag{8.23}$$

The corresponding line-to-line input voltages are

$$\begin{aligned} v_{AB} &= v_{AN} - v_{BN} = \sqrt{6}V_s \sin\left(\omega t + \frac{\pi}{6}\right) \\ v_{BC} &= v_{BN} - v_{CN} = \sqrt{6}V_S \sin\left(\omega t - \frac{\pi}{2}\right) \\ v_{CA} &= v_{CN} - v_{AN} = \sqrt{6}V_s \sin\left(\omega t - \frac{7\pi}{6}\right) \end{aligned} \tag{8.24}$$

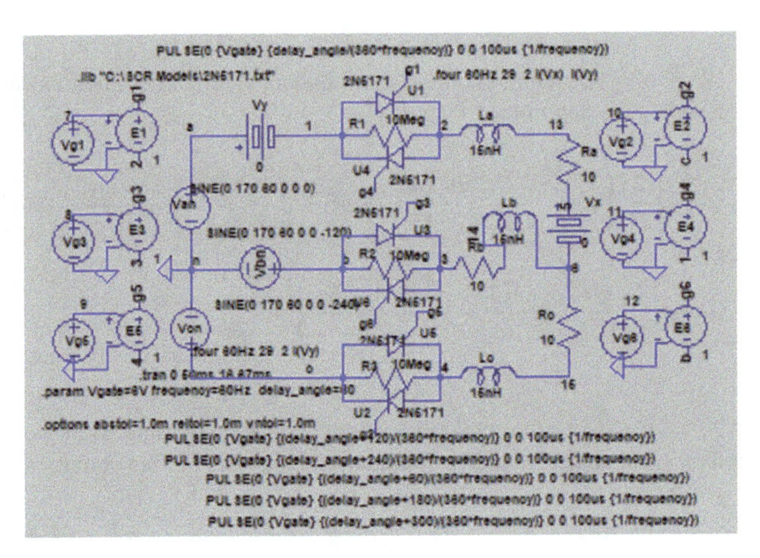

Figure 8.5 Schematic of three-phase AC voltage controller for $\alpha = 60°$

where V_s= rms value of the AC phase input voltage, V; f= supply frequency, Hz; $\omega = 2\pi f$= supply frequency, rad/s.

At least one of the forward current conducting thyristors (T_1,T_3,T_5) and one of the reversing current conducting thyristors (T_2,T_4,T_6) must conduct to produce an output phase current. The number of conducting thyristors and the rms value of the output voltage V_o depends on the delay angle α.

The range of delay angle is

$$0 \le \alpha \le 150^\circ \tag{8.25}$$

For $0 \le \alpha \le 60^\circ$: Once T_1 is turned on, three thyristors conduct. A thyristor turns off when its current attempts to reverse. The rms value of the output voltage is determined from [1]

$$\begin{aligned} V_o &= \left[\frac{1}{2\pi}\int_0^{2\pi} v_{an}^2 d(\omega t)\right]^{1/2} \\ &= \sqrt{6}V_s\left\{\frac{2}{2\pi}\left[\int_\alpha^{\pi/3}\frac{\sin^2(\omega t)}{3}d(\omega t) + \int_{\pi/4}^{\pi/2+\alpha}\frac{\sin^2(\omega t)}{4}d(\omega t) + \int_{\pi/3+\alpha}^{2\pi/3}\frac{\sin^2(\omega t)}{3}d(\omega t)\right.\right. \\ &\quad \left.\left. + \int_{\pi/2}^{\pi/2+\alpha}\frac{\sin^2(\omega t)}{4}d(\omega t) + \int_{2\pi/3+\alpha}^{\pi}\frac{\sin^2(\omega t)}{3}d(\omega t)\right]\right\}^{1/2} \\ &= \sqrt{6}V_s\left[\frac{1}{\pi}\left(\frac{\pi}{6} - \frac{\alpha}{4} + \frac{\sin(2\alpha)}{8}\right]^{1/2} \end{aligned} \tag{8.26}$$

For $60^\circ \le \alpha \le 90^\circ$: Only two thyristors conduct at any time. The rms value of the output voltage is determined from [1]

$$\begin{aligned} V_o &= \sqrt{6}V_s\left[\frac{2}{2\pi}\left(\int_{\pi/2-\pi/3+\alpha}^{5\pi/6-\pi/3+\alpha}\frac{\sin^2(\omega t)}{4}d(\omega t) + \int_{\pi/2-\pi/3+\alpha}^{5\pi/6-\pi/3+\alpha}\frac{\sin^2(\omega t)}{4}d(\omega t)\right)\right]^{1/2} \\ &= \sqrt{6}V_m\left[\frac{1}{\pi}\left(\frac{\pi}{12} + \frac{3\sin(2\alpha)}{16} + \frac{\sqrt{3}\cos(2\alpha)}{16}\right]^{1/2} \end{aligned} \tag{8.27}$$

For $90^\circ \le \alpha \le 150^\circ$: Only two thyristors conduct at any time. However, there are periods when no other thyristors conduct. The rms value of the output voltage is determined from [1]

$$V_o = \sqrt{6}V_s\left[\frac{2}{2\pi}\left(\int_{\pi/2-\pi/3+\alpha}^{\pi}\frac{\sin^2(\omega t)}{4}d(\omega t)+\int_{\pi/2-\pi/3+\alpha}^{\pi}\frac{\sin^2(\omega t)}{4}d(\omega t)\right)\right]^{1/2}$$

$$= \sqrt{6}V_m\left[\frac{1}{\pi}(\frac{5\pi}{24}-\frac{\alpha}{4}+\frac{\sin(2\alpha)}{16}+\frac{\sqrt{3}\cos(2\alpha)}{16}\right]^{1/2} \tag{8.28}$$

For an input phase voltage of V_s = 120 V. The peak supply voltage is $V_m = \sqrt{2}V_s = \sqrt{2} \times 120 = 169.7$ V. Let us consider a delay angle of $\alpha = 60°$,

Time delay $t_1 = \frac{60}{360} \times \frac{1000}{60\text{ Hz}} \times 1000 = 2.778$ ms

Time delay $t_2 = \frac{120}{360} \times \frac{1000}{60\text{ Hz}} \times 1000 = 5.556$ ms

Time delay $t_3 = \frac{180}{360} \times \frac{1000}{60\text{ Hz}} \times 1000 = 8.333$ ms

Time delay $t_4 = \frac{240}{360} \times \frac{1000}{60\text{ Hz}} \times 1000 = 11.111$ ms

Time delay $t_5 = \frac{300}{360} \times \frac{1000}{60\text{ Hz}} \times 1000 = 13.889$ ms

Time delay $t_6 = \frac{360}{360} \times \frac{1000}{60\text{ Hz}} \times 1000 = 16.667$ ms

The gating pulses for the thyristors are shown in Figure 8.6(a). The input phase voltages, the line phase voltage, and phase current are shown in Figure 8.6(b) for $\alpha = 60°$. The load current is distorted.

8.5.1 Controller parameters

The performance parameters of AC voltage controller can be divided into three types

- Output-side parameters,

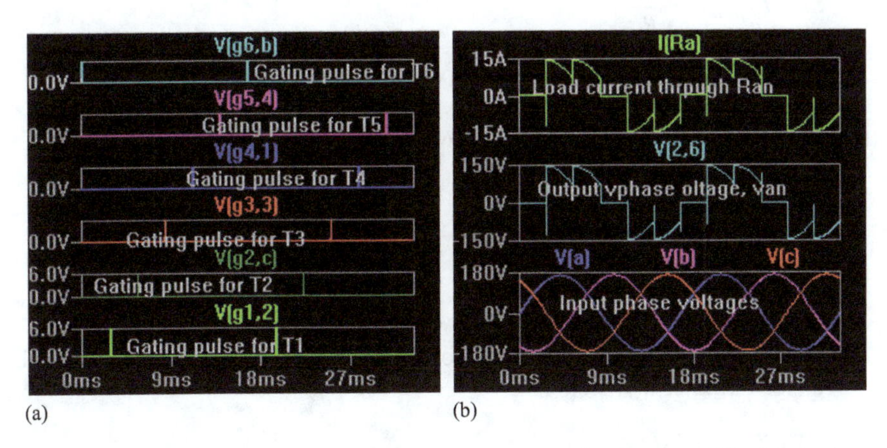

Figure 8.6 Waveforms of three-phase voltage controller for α = 60°: (a) thyristor gating pulses and (b) input and output voltages and load current

- Controller parameters, and
- Input-side parameters.

Let us consider an input voltage of $V_s = 120$ rms, $V_m = \sqrt{2}V_s = 120\times\sqrt{2} = 169.7$ V, $R = 10\ \Omega$ and delay angle of $\alpha = 60°$. Converting to radian, $\alpha = \frac{60°}{180} \times \pi = 1.047$ rad.

8.5.2 Output-side parameters

(a) Equation (8.27) gives the rms output voltage is

$$V_o = V_{o(rms)} = \sqrt{6}V_s\left[\frac{1}{\pi}(\frac{\pi}{12}+\frac{3\sin(2\alpha)}{16}+\frac{\sqrt{3}\cos(2\alpha)}{16}\right]^{1/2}$$

$$= \sqrt{6}\times 120\times\sqrt{\frac{1}{\pi}(\frac{\pi}{12}+\frac{3\sin(2\times 1.047)}{16}+\frac{\sqrt{3}\cos(2\times 1.047)}{16}} = 100.957\ \text{V}$$

(b) The rms output phase current is

$$I_{o(rms)} = \frac{V_{o(rms)}}{R}$$

$$= \frac{100.957}{10} = 10.096\ \text{A}$$

(c) The output power for three phases

$$P_{out} = 3I_{rms}^2R$$

$$= 3\times 10.096^2\times 10 = 3.058\ \text{kW}$$

8.5.3 Controller parameters

(a) The peak thyristor current

$$I_{T(peak)} = \frac{V_m}{R}$$

$$= \frac{169.7}{10} = 16.971\ \text{A}$$

(b) The load current is shared by two devices, the rms thyristor current is

$$I_{T(rms)} = \frac{I_{rms}}{\sqrt{2}}$$

$$= \frac{10.096}{\sqrt{2}} = 7.139\ \text{A}$$

(c) With a safely margin of two times of the peak line-to-line voltage, the peak voltage rating of thyristors is

$$V_{peak} = 2\sqrt{3}V_m$$

$$= 2\times\sqrt{3}\times 169.7 = 588.313\ \text{V}$$

8.5.4 *Input-side parameters*

The rms input line current is

$$I_s = I_{o(rms)}$$
$$= 10.096 \text{ A}$$

Assuming an ideal no loss converter, the input power

$$P_{in} = P_{out}$$
$$= 3.058 \text{ kW}$$

The input power factor

$$PF_i = \frac{P_{in}}{3V_sI_s}$$
$$= \frac{3058}{3 \times 120 \times 10.096} = 0.841 \text{ (lagging)}$$

8.5.4.1 LTspice Fourier analysis

The LTspice command (.four 600Hz 29 1 *I*(*R*)) for Fourier analysis gives the Fourier components of the load current *I*(*R*) at the fundamental frequency of 60Hz as follows:

Fourier components of *I(vx)*

DC component: 6.01951e−005

Harmonic number	Frequency (HZ)	Fourier component	Normalized component	Phase (degrees)	Normalized phase (degrees)
1	6.000e+01	1.333e+01	1.000e+00	−26.83	0.00
2	1.200e+02	1.262e−03	9.471e−05	76.77	103.60
3	1.800e+02	7.313e−04	5.488e−05	111.52	138.34
4	2.400e+02	1.994e−03	1.496e−04	78.69	105.52
5	3.000e+02	3.488e+00	2.618e−01	60.27	87.10
6	3.600e+02	7.850e−04	5.891e−05	120.04	146.87
7	4.200e+02	1.739e+00	1.305e−01	−60.13	−33.30
8	4.800e+02	1.298e−03	9.743e−05	101.81	128.63
9	5.400e+02	2.697e−04	2.024e−05	67.59	94.42
10	6.000e+02	1.639e−03	1.230e−04	85.08	111.91
11	6.600e+02	1.396e+00	1.048e−01	60.39	87.22
12	7.200e+02	6.625e−04	4.972e−05	147.25	174.08
13	7.800e+02	9.940e−01	7.459e−02	−60.02	−33.19
14	8.400e+02	8.314e−04	6.239e−05	95.44	122.26
15	9.000e+02	1.809e−03	1.358e−04	105.57	132.40
16	9.600e+02	1.532e−03	1.150e−04	62.05	88.88
17	1.020e+03	8.715e−01	6.540e−02	60.48	87.31
18	1.080e+03	1.205e−03	9.046e−05	92.44	119.27

(Continues)

(*Continued*)

Fourier components of *I(vx)*					
DC component: 6.01951e−005					
Harmonic number	**Frequency (HZ)**	**Fourier component**	**Normalized component**	**Phase (degrees)**	**Normalized phase (degrees)**
19	1.140e+03	6.956e−01	5.220e−02	−59.89	−33.06
20	1.200e+03	1.310e−03	9.827e−05	130.31	157.14
21	1.260e+03	1.282e−03	9.618e−05	98.60	125.43
22	1.320e+03	1.378e−03	1.034e−04	58.66	85.49
23	1.380e+03	6.345e−01	4.761e−02	60.59	87.41
24	1.440e+03	9.598e−04	7.203e−05	106.42	133.25
25	1.500e+03	5.347e−01	4.012e−02	−59.80	−32.97
26	1.560e+03	8.758e−04	6.572e−05	116.63	143.45
27	1.620e+03	1.436e−03	1.078e−04	109.94	136.77
28	1.680e+03	1.704e−03	1.279e−04	58.69	85.51
29	1.740e+03	4.983e−01	3.739e−02	60.68	87.51
		THD: 33.817473% (35.546486%)			

8.6 Three-phase AC voltage controller with RL load

The load current as shown in Figure 8.6(b) is distorted and contains harmonic contents. An inductor connected in series with the load resistance serves as a filter and reduces the distortion. The schematic is shown in Figure 8.7. The indicator value is shown as a variable "LVAL" in LTspice so that we can observe the effects of the load inductor values on the load current.

8.6.1 Load inductor

The frequency f_o of the output voltage is the same as the supply frequency f. We can find the inductor value to maintain continuous load current. Equation (8.17) gives

$$\frac{L}{R} \geq T_o = \frac{1}{f}$$

which give the value of L as

$$L \geq \frac{R}{f} = \frac{10}{60} = 166.67 \text{ mH}$$

Equation (8.21) gives the condition for the extinction of the load current when the thyristor is turned off as

$$\sin(\beta - \theta) = \sin(\alpha - \theta)e^{(R/L)(\alpha - \beta)/\omega}$$

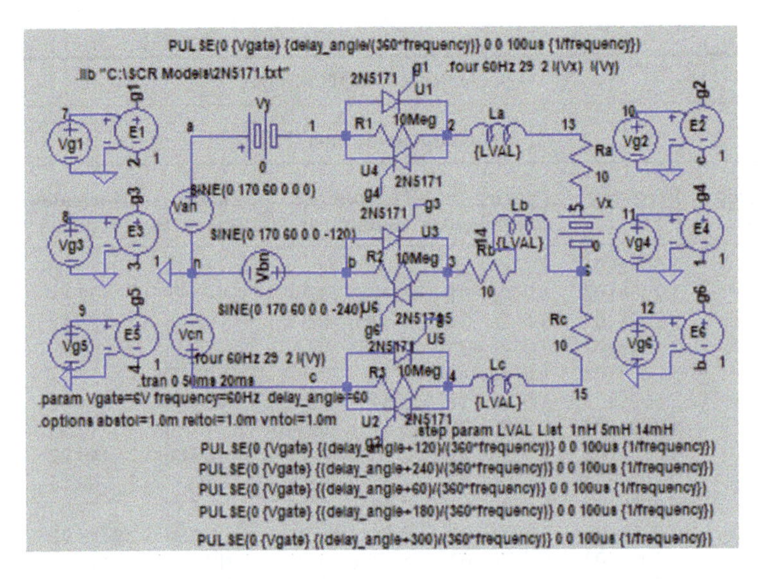

Figure 8.7 Schematic three-phase controller with an RL load

which, after substituting for impedance angle $\theta = \tan^{-1}(\omega L/R)$, gives

$$\sin(\beta - \tan^{-1}(\omega L/R)) = \sin(\alpha - \tan^{-1}(\omega L/R))e^{(R/L)(\alpha-\beta)/\omega} \tag{8.29}$$

For a three-phase controller, the maximum conduction angle of a thyristor is $\alpha_c = 180° - 30° = 150°$ allowing the time for the thyristor to turn-off, let us choose $\alpha_c = 148°$. Therefore, (8.18) gives the extinction angle as

$$\begin{aligned}\beta &= \alpha_c + \alpha \\ &= 148° + 60° = 208° \equiv 3.63 \text{ rad}\end{aligned} \tag{8.30}$$

Equation (8.29) can be written as

$$\sin(\beta - \tan^{-1}(\omega L/R)) - \sin(\alpha - \tan^{-1}(\omega L/R))e^{(R/L)(\alpha-\beta)/\omega} = 0 \tag{8.31}$$

which, for $\beta = 3.63$, $R = 10\ \Omega$, $\omega = 376.991$, and $\beta = 3.63$, gives by the Mathcad iterative method of solution as $L = 14.251$ mH. Let us select $L = 14$ mH.

The plots of the input phase voltages and output phase current are shown in Figure 8.8(a) for L = 14 mH. The effects of load inductances are shown in Figure 8.8(b) for L = 1 nH, 5 mH, 14 mH. As expected, the distortion of the load current is reduced with the increasing values of inductances.

8.6.1.1 LTspice Fourier analysis

The LTspice command (.four 60 Hz 29 1 *I*(*Ra*)) for Fourier analysis gives the Fourier components of the load current *I*(*Ra*) at the fundamental frequency of 60 Hz as follows:

Fourier components of *I(vx)*					
DC component: 0.0877144					
Harmonic number	**Frequency (HZ)**	**Fourier component**	**Normalized component**	**Phase (degrees)**	**Normalized phase (degrees)**
1	6.000e+01	1.244e+01	1.000e+00	30.16	0.00
2	1.200e+02	6.747e−01	5.422e−02	−143.23	−173.39
3	1.800e+02	5.562e−01	4.470e−02	−96.72	−126.88
4	2.400e+02	4.054e−01	3.258e−02	−35.13	−65.28
5	3.000e+02	1.878e+00	1.509e−01	−25.27	−55.43
6	3.600e+02	3.038e−01	2.442e−02	104.74	74.58
7	4.200e+02	6.859e−01	5.512e−02	14.10	−16.05
8	4.800e+02	2.937e−01	2.360e−02	−119.06	−149.22
9	5.400e+02	2.985e−01	2.399e−02	−59.37	−89.53
10	6.000e+02	2.583e−01	2.076e−02	−7.88	−38.04
11	6.600e+02	3.308e−01	2.658e−02	68.94	38.78
12	7.200e+02	6.517e−02	5.237e−03	146.16	116.00
13	7.800e+02	6.118e−02	4.917e−03	−90.06	−120.22
14	8.400e+02	1.852e−01	1.489e−02	−50.90	−81.06
15	9.000e+02	1.730e−01	1.390e−02	−6.30	−36.46
16	9.600e+02	1.022e−01	8.215e−03	44.90	14.74
17	1.020e+03	1.637e−01	1.316e−02	108.00	77.85
18	1.080e+03	1.030e−01	8.280e−03	−98.18	−128.33
19	1.140e+03	2.979e−02	2.394e−03	−92.80	−122.96
20	1.200e+03	1.474e−01	1.185e−02	−3.47	−33.63
21	1.260e+03	1.047e−01	8.412e−03	46.52	16.36
22	1.320e+03	4.684e−02	3.764e−03	119.13	88.97
23	1.380e+03	1.061e−01	8.525e−03	−122.29	−152.45
24	1.440e+03	9.453e−02	7.597e−03	−47.33	−77.49
25	1.500e+03	9.321e−02	7.490e−03	−7.22	−37.38
26	1.560e+03	8.325e−02	6.690e−03	52.84	22.69
27	1.620e+03	4.017e−02	3.228e−03	120.84	90.68
28	1.680e+03	4.261e−02	3.424e−03	−108.03	−138.19
29	1.740e+03	9.345e−02	7.510e−03	−71.77	−101.93
THD: 18.974166% (0.000000%)					

Note: THD is reduced from 33.817% to 18.972% due to the addition of an inductor *L*.

8.7 Summary

An AC voltage controller converts a fixed AC voltage to a variable AC voltage and uses thyristors commonly as switching devices. It uses a pair of thyristors connected in anti-parallel – one thyristor is turned on to allow current flow in one direction and other thyristor is turned on to allow current flow in the reverse direction. The output voltage is varied by varying the delay angle when the thyristor is tuned on. Thus, the voltage and the current on the input side are of AC types, and the voltage and the current on the output side are also of AC types. The

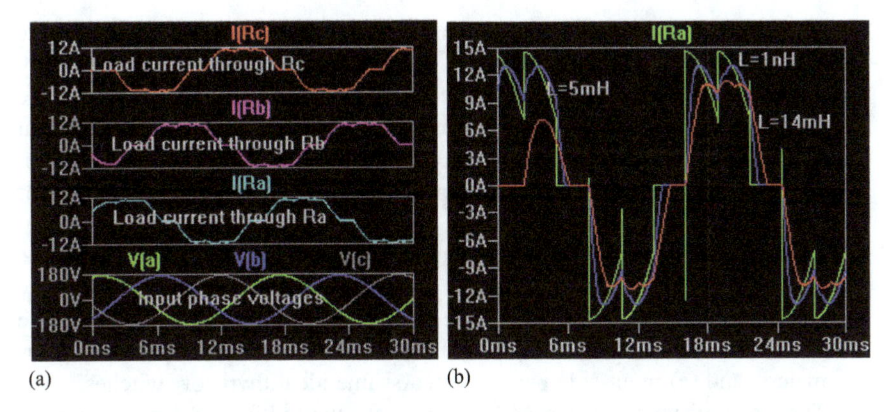

(a) (b)

Figure 8.8 Single-phase controller with an RL load: (a) input voltages and load currents for L = 14 mH and (b) effects of load inductances

rms output voltage can be varied from zero to the maximum value of the AC supply voltage by varying the delay angle. However, the input power factor depends on the delay angle and the output voltage. An inductor L is often connected in series with a resistive load to limit the ripple contents of the output voltage and the current. For a resistive load, the load current can be discontinuous depending on a higher value of the delay angle. For an inductive load, the load current becomes continuous. The output voltage of three-phase AC voltage controllers is much higher than that of a single-phase AC voltage controller. A thyristor can be turned on by applying a pulse for a short duration. Once the thyristor is on, the gate pulse has no effect, and it remains on until its current is reduced to zero. It is a latching device. The thyristors of the AC voltage controllers are turned off due to the natural behavior of the AC sinusoidal input voltage. The input power factor of AC voltage controllers depends on the delay angle and is low for higher delay angle.

Problems

1. The single-phase full-wave AC voltage controller in Figure 8.1 has a resistive load of $R = 5\ \Omega$. The input voltage is $Vs = 120$ V (rms), 60 Hz. The delay angles of thyristors T_1 and T_2 are equal, $\alpha_1 = \alpha_2 = \alpha = \pi/6$. Calculate the performance parameters: (a) the rms output voltage V_o, (b) the rms output current I_o, (c) the average current of thyristors, I_A, (d) the rms current of thyristors, I_R, and (e) the input power factor, PF. Assume ideal thyristor switches.
2. The single-phase full-wave AC voltage controller in Figure 8.1 has a resistive load of $R = 5\ \Omega$. The input voltage is $Vs = 120$ V (rms), 60 Hz. The delay angles of thyristors T_1 and T_2 are equal, $\alpha_1 = \alpha_2 = \alpha = \pi/6$. Calculate the performance parameters: (a) output-side parameters, (b) controller parameters, and (c) input-side parameters. Assume ideal thyristor switches.

3. The single-phase full-wave AC voltage controller in Figure 8.1 has a resistive load of $R = 5\ \Omega$. The input voltage is $Vs = 120$ V (rms), 60 Hz. The delay angles of thyristors T_1 and T_2 are equal, $\alpha_1 = \alpha_2 = \alpha = \pi/3$. Calculate the performance parameters: (a) the rms output voltage V_o, (b) the rms output current $I_{o.}$, (c) the average current of thyristors, I_A, (d) the rms current of thyristors, I_R, and (e) the input power factor, PF. Assume ideal thyristor switches.
4. The single-phase full-wave AC voltage controller in Figure 8.1 has a resistive load of $R = 5\ \Omega$. The input voltage is $Vs = 120$ V (rms), 60 Hz. The delay angles of thyristors T_1 and T_2 are equal, $\alpha_1 = \alpha_2 = \alpha = \pi/3$. Calculate the performance parameters: (a) output-side parameters, (b) controller parameters, and (c) input-side parameters. Assume ideal thyristor switches.
5. The single-phase full-wave AC voltage controller in Figure 8.1 has a resistive load of $R = 5\ \Omega$. The input voltage is $Vs = 120$ V (rms), 60 Hz. The delay angles of thyristors T_1 and T_2 are equal, $\alpha_1 = \alpha_2 = \alpha = \pi/2$. Calculate the performance parameters: (a) the rms output voltage V_o, (b) the rms output current $I_{o.}$, (c) the average current of thyristors, I_A, (d) the rms current of thyristors, I_R, and (e) the input power factor, PF. Assume ideal thyristor switches.
6. The single-phase full-wave AC voltage controller in Figure 8.1 has a resistive load of $R = 5\ \Omega$. The input voltage is $Vs = 120$ V (rms), 60 Hz. The delay angles of thyristors T_1 and T_2 are equal, $\alpha_1 = \alpha_2 = \alpha = \pi/2$. Calculate the performance parameters: (a) output-side parameters, (b) controller parameters, and (c) input-side parameters. Assume ideal thyristor switches.
7. The single-phase full-wave AC voltage controller in Figure 8.3 has an RL load of $R = 5\ \Omega$. The input voltage is $Vs = 120$ V (rms), 60 Hz. The delay angles of thyristors T_1 and T_2 are equal, $\alpha_1 = \alpha_2 = \alpha = \pi/6$. Calculate the performance parameters: (a) the value of inductance L for the continuity of the load current, (b) the rms output voltage V_o, (c) the rms output current I_o, (d) the average current of thyristors, I_A, (e) the rms current of thyristors, I_R, and (f) the input power factor, PF. Assume ideal thyristor switches.
8. The single-phase full-wave AC voltage controller in Figure 8.3 has a resistive load of $R = 5\ \Omega$. The input voltage is $Vs = 120$ V (rms), 60 Hz. The delay angles of thyristors T_1 and T_2 are equal, $\alpha_1 = \alpha_2 = \alpha = \pi/6$. Calculate the performance parameters: (a) output-side parameters, (b) controller parameters, and (c) input-side parameters. Assume ideal thyristor switches.
9. The single-phase full-wave AC voltage controller in Figure 8.3 has an RL load of $R = 5\ \Omega$. The input voltage is $Vs = 120$ V (rms), 60 Hz. The delay angles of thyristors T_1 and T_2 are equal, $\alpha_1 = \alpha_2 = \alpha = \pi/3$. Calculate the performance parameters: (a) the value of inductance L for the continuity of the load current, (b) the rms output voltage V_o, (c) the rms output current I_o, (d) the average current of thyristors, I_A, (e) the rms current of thyristors, I_R, and (f) the input power factor, PF. Assume ideal thyristor switches.
10. The single-phase full-wave AC voltage controller in Figure 8.3 has a resistive load of $R = 5\ \Omega$. The input voltage is $Vs = 120$ V (rms), 60 Hz. The delay

angles of thyristors T_1 and T_2 are equal, $\alpha_1 = \alpha_2 = \alpha = \pi/3$. Calculate the performance parameters: (a) output-side parameters, (b) controller parameters, and (c) input-side parameters. Assume ideal thyristor switches.

11. The single-phase full-wave AC voltage controller in Figure 8.3 has an RL load of $R = 5\ \Omega$. The input voltage is $Vs = 120$ V (rms), 60 Hz. The delay angles of thyristors T_1 and T_2 are equal, $\alpha_1 = \alpha_2 = \alpha = \pi/2$. Calculate the performance parameters: (a) the value of inductance L for the continuity of the load current, (b) the rms output voltage V_o, (c) the rms output current I_o, (d) the average current of thyristors, I_A, (e) the rms current of thyristors, I_R, and (f) the input power factor, PF. Assume ideal thyristor switches.
12. The single-phase full-wave AC voltage controller in Figure 8.3 has a resistive load of $R = 5\ \Omega$. The input voltage is $Vs = 120$ V (rms), 60 Hz. The delay angles of thyristors T_1 and T_2 are equal, $\alpha_1 = \alpha_2 = \alpha = \pi/2$. Calculate the performance parameters: (a) output-side parameters, (b) controller parameters, and (c) input-side parameters. Assume ideal thyristor switches.
13. The three-phase AC voltage controller in Figure 8.5 has a wye-connected resistive load of $R = 5\ \Omega$/phase. The input voltage is $Vs = 120$ V (rms), 60 Hz. The delay angle of thyristor T_1 is $\alpha_1 = \alpha = \pi/3$. Calculate the performance parameters: (a) the rms output voltage V_o, (b) the rms output current $I_{o.}$, (c) the average current of thyristors, I_A, (d) the rms current of thyristors, I_R, and (e) the input power factor, PF. Assume ideal thyristor switches.
14. The three-phase full-wave AC voltage controller in Figure 8.5 has a resistive load of $R = 5\ \Omega$. The input voltage is $Vs = 120$ V (rms), 60 Hz. The delay angles of thyristors T_1 and T_2 are equal, $\alpha_1 = \alpha_2 = \alpha = \pi/3$. Calculate the performance parameters: (a) output-side parameters, (b) controller parameters, and (c) input-side parameters. Assume ideal thyristor switches.
15. The three-phase AC voltage controller in Figure 8.5 has a wye-connected resistive load of $R = 5\ \Omega$/phase. The input phase voltage is $Vs = 120$ V (rms), 60 Hz. The delay angle of thyristor T_1 is $\alpha_1 = \alpha = \pi/2$. Calculate the performance parameters. Assume ideal thyristor switches. (a) the rms output voltage V_o, (b) the rms output current $I_{o.}$, (c) the average current of thyristors, I_A, (d) the rms current of thyristors, I_R € the input power factor, PF
16. The three-phase full-wave AC voltage controller in Figure 8.5 has a resistive load of $R = 5\ \Omega$. The input voltage is $Vs = 120$ V (rms), 60 Hz. The delay angles of thyristors T_1 and T_2 are equal, $\alpha_1 = \alpha_2 = \alpha = \pi/2$. Calculate the performance parameters: (a) output-side parameters, (b) controller parameters, and (c) input-side parameters. Assume ideal thyristor switches.
17. The three-phase AC voltage controller in Figure 8.5 has a wye-connected resistive load of $R = 5\ \Omega$/phase. The input voltage is $Vs = 120$ V (rms), 60 Hz. The delay angle of thyristor T_1 is $\alpha_1 = \alpha = 2\pi/3$. Calculate the performance parameters: (a) the rms output voltage V_o, (b) the rms output current $I_{o.}$, (c) the average current of thyristors, I_A, (d) the rms current of thyristors, I_R, and (e) the input power factor, PF. Assume ideal thyristor switches
18. The three-phase full-wave AC voltage controller in Figure 8.5 has a resistive load of $R = 5\ \Omega$. The input voltage is $Vs = 120$ V (rms), 60 Hz. The delay

angle of thyristors T_1 is $\alpha_1 = \alpha = 2\pi/3$. Calculate the performance parameters: (a) output-side parameters, (b) controller parameters, and (c) input-side parameters. Assume ideal thyristor switches.

19. The three-phase full-wave AC voltage controller in Figure 8.7 has a wye-connected RL load of $R = 5\ \Omega$/phase. The input phase voltage is $Vs = 120$ V (rms), 60 Hz. The delay angle of thyristor T_1 is $\alpha_1 = \alpha = \pi/3$. Calculate the performance parameters: (a) the value of inductance L for the continuity of the load current, (b) the rms output voltage V_o, (c) the rms output current $I_{o.}$, (d) the average current of thyristors, I_A, (e) the rms current of thyristors, I_R, and (f) the input power factor, PF. Assume ideal thyristor switches.
20. The three-phase full-wave AC voltage controller in Figure 8.7 has a wye-connected RL load of $R = 5\ \Omega$/phase. The input phase voltage is $Vs = 120$ V (rms), 60 Hz. The delay angle of thyristor T_1 is $\alpha_1 = \alpha = \pi/3$. Calculate the performance parameters: (a) the value of inductance L for the continuity of the load current, (b) Output-side parameters, (c) controller parameters, and (d) input-side parameters. Assume ideal thyristor switches.
21. The three-phase full-wave AC voltage controller in Figure 8.7 has a wye-connected RL load of $R = 5\ \Omega$/phase. The input phase voltage is $Vs = 120$ V (rms), 60 Hz. The delay angle of thyristor T_1 is $\alpha_1 = \alpha = \pi/2$. Calculate the performance parameters: (a) the value of inductance L for the continuity of the load current, (b) the rms output voltage V_o, (c) the rms output current I_o, (d) the average current of thyristors, I_A, (e) the rms current of thyristors, I_R, and (f) the input power factor, PF. Assume ideal thyristor switches.
22. The three-phase full-wave AC voltage controller in Figure 8.7 has a wye-connected RL load of $R = 5\ \Omega$/phase. The input phase voltage is $Vs = 120$ V (rms), 60 Hz. The delay angle of thyristor T_1 is $\alpha_1 = \alpha = \pi/2$. Calculate the performance parameters: (a) the value of inductance L for the continuity of the load current, (b) Output-side parameters, (c) controller parameters, and (d) input-side parameters. Assume ideal thyristor switches.
23. The three-phase full-wave AC voltage controller in Figure 8.7 has a wye-connected RL load of $R = 5\ \Omega$/phase. The input phase voltage is $Vs = 120$ V (rms), 60 Hz. The delay angle of thyristor T_1 is $\alpha_1 = \alpha = 2\pi/3$. Calculate the performance parameters: (a) the value of inductance L for the continuity of the load current, (b) the rms output voltage V_o, (c) the rms output current I_o, (d) the average current of thyristors, I_A, (e) the rms current of thyristors, I_R, and (f) the input power factor, PF. Assume ideal thyristor switches.
24. The three-phase full-wave AC voltage controller in Figure 8.7 has a wye-connected RL load of $R = 5\ \Omega$/phase. The input phase voltage is $Vs = 120$ V (rms), 60 Hz. The delay angle of thyristor T_1 is $\alpha_1 = \alpha = 2\pi/3$. Calculate the performance parameters: (a) the value of inductance L for the continuity of the load current, (b) Output-side parameters, (c) controller parameters, and (d) input-side parameters. Assume ideal thyristor switches.

References

[1] A.A. Zekry, G.T. Sayah, SPICE model of thyristors with amplifying gate and emitter-shorts, *IET Power Electronics*, 7(3), 724–735, 2014. https://doi.org/10.1049/iet-pel.2013.0158.

[2] L.J. Giacoletto, Simple SCR and TRAIC PSPICE computer models, *IEEE Transactions on Industrial Electronics*, 36(3), 1989, 451–455.

[3] M.H. Rashid, *Power Electronics: Circuits, Devices, and Applications*, 4th ed., Englewood Cliffs, NJ: Prentice-Hall, 2014, Chapter 11.

[4] M.H. Rashid, *SPICE and LTspice for Power Electronics and Electric Power*, Boca Raton, FL: CRC Press, 2024.

[5] M.H. Rashid, *Power Electronics Handbook*, Burlington, MA: Butterworth-Heinemann, 2010, Chapter 18.

Chapter 9
Protecting semiconductor devices

9.1 Introduction

The power semiconductor devices are turned "on" and "off" to converter and transfer energy from the source to the load. These devices are normally off. When the supply voltage is switched on, the device is in the off-state and withstands the supply voltage, ideally no current flows through the device. The rapid rate of the rise of the voltage across the device from zero voltage to the full supply voltage causes an internal current flow through the internal junction capacitance of the device and this may cause device failure. Due to the switching action of the devices, there are instances when one device is turning off while another device is turning on to maintain the continuity of the current flow. Ideally, when one device is turned off, another device should turn on immediately. However, practical devices have limitations and require finite time to fully turn on and turn off. As a result, a situation may occur when the voltage source is short-circuited during the turn-on and -off time intervals of the two devices. The rate of the rise of the current must be limited, not to exceed the device *di/dt* specifications. Also, a finite amount of power loss occurs across the semiconductor device due to the finite on-state resistance of the semiconductor device. The heat generated due to the power loss must dissipated to keep the junction temperature within the specified limits. These devices must also be protected due to the rising current under faulty conditions with the power electronic systems.

9.2 Types of protections

Figure 9.1(a) shows a DC–DC step-down converter with a voltage-controlled switch S_1 to connect the DC supply voltage to the converter. The switch is controlled by a pulse voltage of 5 V, delayed by 100 μs. When the switch control voltage rises to 0.5 V, the transistor switch offers a resistance of 0.1 mΩ and the transistor switch is on. When the switch control voltage falls to 0 V, the switch is turned off, The LTspice statement of the model parameters of the control switch is [3]

.models mod SW (Ron = 0.1m Roff = 1Meg Vt = .0.5V Vh = −0V)

Thus, the power semiconductor devices are subjected to changing voltages across the devices and changing current flow through the devices. The rate of the

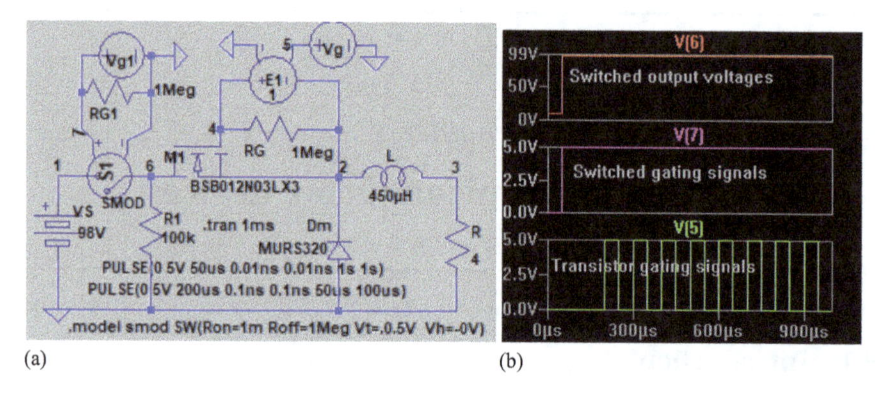

Figure 9.1 DC–DC step-down converter with supply-side voltage-controlled switch: (a) LTspice schematic and (b) gating and switch voltages

changes of the voltages and the current must be limited to certain values specified by the manufacturers. The semiconductor devices must be protected from the following types of operating conditions

- *dv/dt* protection,
- *di/dt* protection,
- Thermal protection, and
- Fault protection.

9.2.1 dv/dt *Protection*

When the supply voltage switch is turned-off, the voltage at node 6 is zero. The voltage at node 2 is also zero. As soon as the supply voltage switch is turned-on, the voltage at node 6 reaches to the supply voltage of $V_s = 98$ V, ideally instantaneously. Until the gating signals of the transistor are turned on, the voltage at node 2 is still zero, Thus, at the instant of turning-on the switch S_1, the voltage across the transistor switch M_1 changes from 0 to V_s (= 98 V) and the change of the voltage is $\Delta V = 98 - 0 = 98$ V. Thus, the current flowing the internal junction capacitor C_j of the transistor M_1 due to ΔV is

$$C_j \frac{dv}{dt} = I_j \tag{9.1}$$

which gives the circuit *dv/dt* during the turning on of the switch as

$$\frac{dv}{dt} = \frac{I_j}{C_j} \tag{9.2}$$

To limit the *dv/dt*, an external capacitor known as *snubber capacitor* C_s is normally connected across the switching device so that the circuit *dv/dt* does not exceed the rated *dv/dt* of the device, typically 1000 V/μs.

When the gating signal is fully turned on, the transistor switch voltage falls from V_S to its saturation voltage $V_{Sat} \approx 0$ V and the switch current rises to the load current I_L and continues to conduct. When the switch is turned-off at a time $t_1 = kT$, the voltage across the switch switches from the on-state of V_{Sat} to the off-state voltage of V_S. Thus, the rate of the rise of the voltage during the transistor switch turn-off is given by

$$\frac{dv}{dt} = \frac{V_S}{t_r} \tag{9.3}$$

where t_r is the rise time of the switch voltage.

The transistor is turned-on in the normal operation of on and off modes. During turn-off, the capacitor C_s charges by the load current $I_{L.}$ The capacitor voltage appears across the transistor switch and the *dv*/*dt* is

$$\frac{dv}{dt} = \frac{I_L}{C_s} \tag{9.4}$$

Equating (9.3) and (9.4) gives

$$\frac{V_S}{t_f} = \frac{I_L}{C_s} \tag{9.5}$$

which gives the required value of the C_s to limit the *dv/dt* of the transistor switch

$$C_s = \frac{t_r I_L}{V_S} \tag{9.6}$$

For example, $V_S = 100\ \text{V}, \quad I_L = 10\ \text{A}, \quad t_r = 1.6\ \text{V}/\mu\text{s}, \quad t_f = 3\ \text{V}/\mu\text{s}$

$$C_s = \frac{t_f I_L}{V_S}$$
$$= \frac{1.6 \times 10^{-6} \times 10}{100} = 0.16\ \mu\text{F}$$

Note: The total capacitance $C = C_j + C_s \simeq C_s$. C_j is much smaller than C_s and can often be neglected.

When the transistor is turned on in the next switching cycle and with the snubber capacitor connected across the transistor switch, the capacitor discharges through the transistor in addition to the load current I_L. To limit the discharge current, a resistance R_s known as the *snubber resistance* is connected in series the capacitor. Therefore, the peak current through the transistor switch is

$$I_{sw} = I_L + \frac{V_S}{R_s} \tag{9.7}$$

When choosing the value of R_s, the discharge time, $R_sC_s = \tau_s$ should also be considered. A discharge time of one-third the switching period T_s is usually adequate. That is,

$$3R_sC_s = T_s = \frac{1}{f_s} \tag{9.8}$$

which gives the value of R_s as

$$R_s = \frac{1}{3C_s f_s} \tag{9.9}$$

For $f_s = 10$ kHz and $C_s = 0.16$ μF, (9.9) gives

$$R_s = \frac{1}{3C_s f_s}$$
$$= \frac{1}{3 \times 1.6 \times 10^{-6} \times 10 \times 10^3} = 208\ \Omega$$

Equation (9.7) gives the peak transistor current as

$$I_{sw} = I_L + \frac{V_S}{R_s}$$
$$= 10 + \frac{100}{208} = 10.48\ \text{A}$$

In every switching cycle, the snubber capacitor charges and discharges, the energy stored in the snubber capacitor is dissipated in the resistance R_s. Thus, we can find the switching loss as given by

$$P_s = \frac{1}{2} C_s V_S^2 f_s \tag{9.10}$$

For $C_s = 1.6$ μF, $V_S = 100$ V, $f_s = 10$ kHz, (9.10) gives the power loss as $P_s = \frac{1}{2} \times 1.6 \times 10^{-6} \times 100^2 \times 10 \times 10^3 = 8$ W (for $R_s = 208\ \Omega$, 8 W)

As the voltage across the transistor M_1 switches from high *Vs* to low 0 and the voltage across the diode D_1 switches from low 0 to high *Vs*. Both the transistor and the diode must be protected from *dv/dt* with a snubber circuit as shown in Figure 9.2.

9.2.2 di/dt *Protection*

In a highly inductive load, the freewheeling diode D_m as shown in Figure 9.1(a) conducts when the transistor M_1 is turned-off at $t_1 = kT$. At the instant of the time

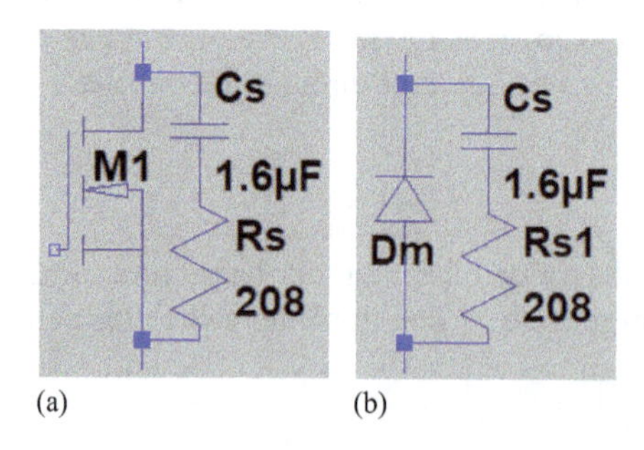

Figure 9.2 Snubber circuits: (a) for transistor and (b) for diode

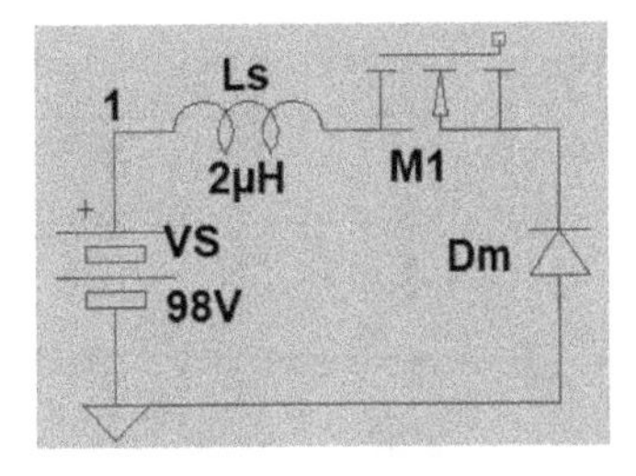

Figure 9.3 Series snubber for limiting di/dt

when the transistor M_1 is turned-on at the beginning of the next cycle, $t = 0_+$ M_1 starts conducting while D_m still conducting due to the reverse recovery time needed for the diode to stop conducting. There exists a condition such that there is almost a short circuit formed by the DC source V_S, transistor M_1, and the diode D_m causing a reverse recovery current flow through the diode as shown in Figure 9.3.

To limit the rate of the rise and the magnitude of the reverse current, an inductor known as the *series snubber inductor* L_S is normally connected in series with the transistor M_1 such that

$$L_s \frac{di}{dt} = V_S$$

which gives

$$\frac{di}{dt} = \frac{V_S}{L_s} \tag{9.11}$$

The inductor value should be such that the circuit *di/dt* in (9.11) does not exceed the rated *di/dt* of the device, typically 1000 A/μs.

For example, for a circuit $\frac{di}{dt} = 50$ A/μs and $V_S = 100$ V, (9.11) gives

$$L_s = \frac{V_S}{di/dt} = \frac{100}{50 \times 10^6} = 2\ \mu\text{H}$$

Note: In high power and current applications, the stray inductance of the wiring cables may be adequate to protect the device from *di/dt.*

For a diode reverse recovery time t_{rr}, the reverse current flowing through the diode is

$$I_{RR} = \frac{di}{dt} t_{rr} \tag{9.12}$$

For example, $\frac{di}{dt} = 50$ A/μs and $t_{rr} = 2$ μs, (9.12) gives the reverse recovery current I_{RR} reaches at $I_{RR} = \frac{di}{dt} t_{rr} = 50 \times 2 = 100$ A the end of the reverse recovery time, t_{rr}, the diode stops conducting. The reverse diode current I_{RR} forces

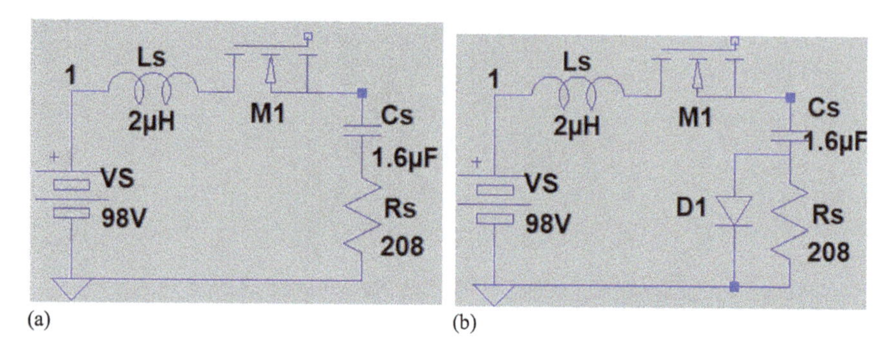

Figure 9.4 Effects of reverse recovery current: (a) series and RC snubbers and (b) series and RC snubbers with a diode

the current to flow through the RC-snubber circuit as shown in Figure 9.4(a) and the snubber capacitor over changes above V_S. The energy stored in the inductor $W_s = 0.5L_sI_{RR}^2$ is partly dissipated into R_s and partly overcharges the voltage of the snubber capacitor C_s. The current can be described by

$$L_s\frac{di}{dt} + R_si + \frac{1}{C_s}\int_0^t idt + v_c(t = 0)$$

For initial conditions, $i(t = 0) = I_{RR}$ and $v_c(t = 0) = V_S$. Another diode may be connected across R_s such that the diode bypasses the resistor R_s during changing of capacitor C_s as shown in Figure 9.4(b), but it limits the discharging current through the transistor when the transistor switch is turned on in the next cycle. However, it would add additional costs.

9.2.3 Thermal protection

Due to the on-state and the switching losses in the transistor, heat is generated within the power device. This heat must be transferred from the device to a cooling medium to maintain the operating junction temperature within the specified limit. Although this heat transfer can be accomplished by conduction, convection, radiation, or natural or forced-air, convection cooling is commonly used in industrial applications. The heat must flow from the device to the case and then to the heat sink in the cooling medium. For an on-state voltage of V_{on} across the switching device while the switch is carrying an average on-state current of I_{on}, the average power dissipation in the device is given by

$$P_{on} = V_{on}I_{on} \tag{9.13}$$

With P_{on} *as* the average power loss in the device, the electrical analog of a device, that is mounted on a heat sink, is shown in Figure 9.5. The junction temperature of a device T_J is given by

$$T_J = P_{on}(R_{JC} + R_{CS} + R_{SA}) + T_A \tag{9.14}$$

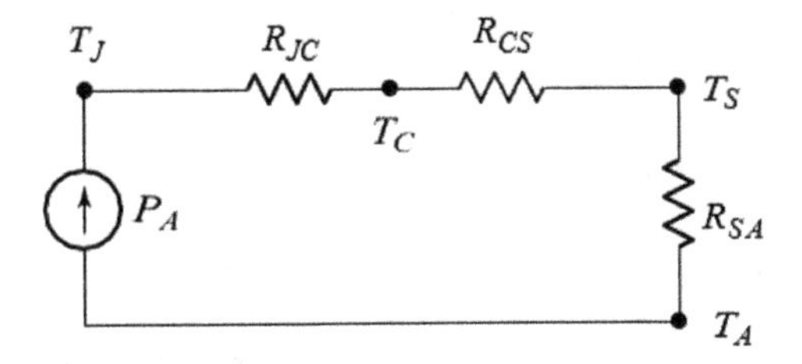

Figure 9.5 Electrical analog of heat transfer

where R_{JC} = thermal resistance from junction to case, °C/W; R_{CS} = thermal resistance from case to sink, °C/W; R_{SA} = thermal resistance from sink to ambient, °C/W; and T_A = ambient temperature, °C.

R_{JC} and R_{CS} are normally specified by the power device manufacturers. Once the device power loss P_{on} is known, the required thermal resistance of the heat sink can be calculated for a known ambient temperature T_A. We can choose a heat sink and its size that would meet the thermal resistance requirement.

For example, $V_{on} = 0.7$ V, $I_{on} = 100$ A, $T_A = 40$°C, $T_J = 200$°C, $R_{JC} = 0.35$ °C/W, and $R_{CS} = 0.25$ °C/W. Equation (9.13) gives the average power dissipation as

$$\begin{aligned} P_{on} &= V_{on} I_{on} \\ &= 0.7 \times 100 = 70 \text{ W} \end{aligned}$$

Equation (9.14) gives the required thermal resistance of the heat-sink as

$$\begin{aligned} R_{SA} &= \frac{T_J - T_A}{P_{on}} - R_{JC} - RCS \\ &= \frac{200 - 40}{70} - 0.35 - 0.25 = 1.686°\text{C/W} \end{aligned}$$

In addition to the average power dissipation in (9.13), the device is subjected to switching voltages and currents. As a result, there is a switching power loss during the turn-on and turn-off intervals of the devices.

Turn-on switching loss: During the turn-on time, the voltage across the transistor switch changes from V_S to on-state voltage V_{on} and the switch current changes from off-state of approximately zero to I_L.

Assuming a rise time of t_r as the transistor switch current and the fall time of the switch voltage, and assuming the transistor current rises linearly and the transistor voltage falls linearly, we get the transistor current, and the voltage as given by

$$i_{sw-on}(t) = \frac{I_L}{t_r} t \tag{9.15}$$

$$v_{sw-on}(t) = V_S - \frac{V_S}{t_r}t \tag{9.16}$$

Thus, the instantaneous power loss is given by

$$P_{sw-on}(t) = i_{sw-on}(t)\, v_{sw-on}(t) = \frac{I_L}{t_r}t\left(V_S - \frac{V_S}{t_r}t\right) \tag{9.17}$$

The peak power occurs when $\frac{d}{dt}(P_{sw-on}(t)) = 0$ and the time t_{rm} for the peak power is given by [1]

$$t_{rm} = \frac{t_r V_S}{2(V_S - V_{on})} \tag{9.18}$$

Substituting $t = t_{rm}$ in (9.17), the peak power during the rise time is given by

$$\begin{aligned} P_{rm} &= \frac{V_S^2 I_L}{4(V_S - V_{on})} \\ &= \frac{100^2 \times 100}{4 \times (100 - 0.7)} = 251.7 \text{ W} \end{aligned} \tag{9.19}$$

Turn-off switching loss: During the turn-off time, the voltage across the transistor switch changes from on-state voltage V_{on} to off-state voltage V_S and the switch current changes from on-state of I_L to the off-state of approximately zero.

Assuming a fall time of t_f as the switch current and the rise-time of the switch voltage,

$$i_{sw-off}(t) = I_L\left(1 - \frac{t}{t_f}\right) \tag{9.20}$$

$$v_{sw-off}(t) = \frac{V_S}{t_f}t \tag{9.21}$$

Thus, the instantaneous power loss is given by

$$P_{sw-off}(t) = i_{sw-off}(t) v_{sw-off}(t) = I_L\left(1 - \frac{t}{t_f}\right)\frac{V_S}{t_f}t \tag{9.22}$$

The peak power occurs when $P_{sw-on}(t)\frac{d}{dt}(P_{sw-on}(t)) = 0$ and the time t_{rm} for the peak power is given by

$$t_{fm} = \frac{1}{2} \tag{9.23}$$

And substituting $t = t_{fm}$ in (9.22), the peak power during the fall time is given by

$$\begin{aligned} P_{fm} &= \frac{V_S I_L}{4} \\ &= \frac{100 \times 100}{4} = 250 \text{ W} \end{aligned} \tag{9.24}$$

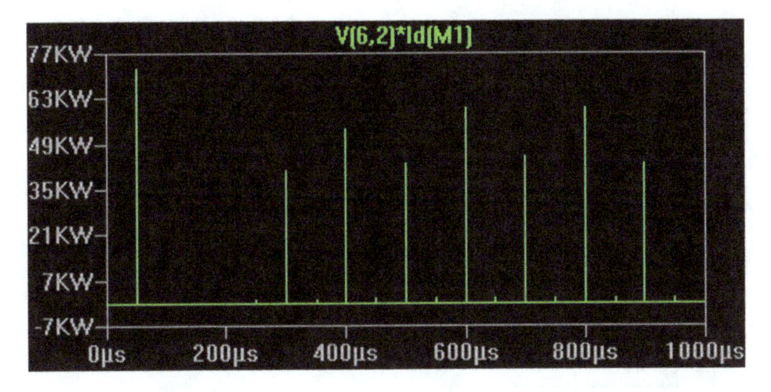

Figure 9.6 Instantaneous transistor power loss

Therefore, the total power dissipation is

$$\begin{aligned} P_T &= P_{rm} + P_{on} + P_{fm} \\ &= 251.7 + 70 + 250 = 571.2 \text{ W} \end{aligned} \tag{9.25}$$

We should note the losses during the turn-on and turn-off of the transistor switching device is significantly higher than the loss due to the transistor switch on-time. Thus, the switching device must be protected from the instantaneous junction temperature due to the instantaneous power dissipation. Figure 9.6 shows the LTspice instantaneous power across the switching transistor as shown in Figure 9.1 (a). Due to switching transients, the power loss is significantly higher during the transistor switching from on to off state and vice versa.

9.2.4 Fault protection

The power converters may develop short circuits or faults, and the resultant fault currents must be cleared quickly [1]. Fast-acting fuses are normally used to protect the semiconductor devices. As the fault current increases, the fuse must open and clear the fault current in a few milliseconds. Every semiconductor device must be protected for fault conditions. The semiconductor devices may be protected by carefully choosing the locations of the fuses. However, the fuse manufacturers recommend placing a fuse in series with each device. The individual protection that permits better coordination between a device and its fuse allows superior utilization of the device capabilities and protects thee devices from short through faults

When the fault current rises as shown in Figure 9.7, the fuse temperature also rises until $t = t_m$, at which time the fuse melts and arcs are developed across the fuse. Due to the arc, the impedance of the fuse is increased, thereby reducing the current. However, an arc voltage is formed across the fuse. The generated heat vaporizes the fuse element, resulting in an increased arc length and further reduction of the current. The cumulative effect is the extinction of the arc in a very short time. When the arcing is complete in time t_a, the fault is clear. The faster the fuse clears, the higher is the arc voltage [2]. The clearing time t_c is the sum of melting

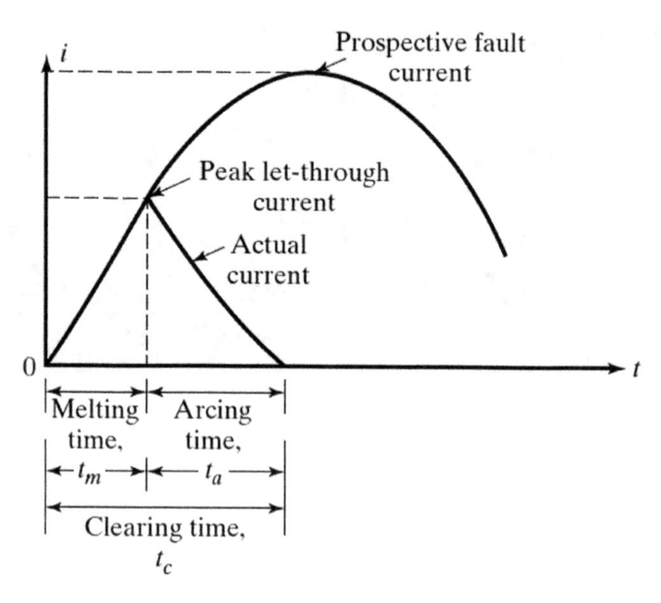

Figure 9.7 Fuse and fault currents [1]

time t_m and arc time t_a. t_m is dependent on the load current, whereas t_a is dependent on the power factor or parameters of the fault circuit. The fault is normally cleared before the fault current reaches its first peak, and the fault current, which might have flown if there was no fuse, is called the *prospective fault current*. This is shown in Figure 9.7. The current–time curves of devices and fuses may be used for the coordination of a fuse for a device.

If R is the resistance of the fault circuit and i is the instantaneous fault current between the instant of fault occurring and the instant of arc extinction, the energy fed to the circuit can be expressed as

$$W_e = \int Ri^2 dt \tag{9.26}$$

If the resistance R remains constant, the value of i^2t is proportional to the energy fed to the circuit. The i^2t value is termed as the *let through energy* and is responsible for melting the fuse. The fuse manufacturers specify the i^2t characteristic of the fuse. In selecting a fuse, it is necessary to estimate the fault current and then to satisfy the following requirements:

- The fuse must carry continuously the device rated current.
- The i^2t let-through value of the fuse before the fault current is cleared must be less than the rated i^2t of the device to be protected.
- The fuse must be able to withstand the voltage, after the arc extinction.
- The peak arc voltage must be less than the peak voltage rating of the device.

Note: As a general rule of thumb, a fast-acting fuse with a rms current rating equal to or less than the average current rating of the semiconductor normally can provide adequate protection under fault conditions.

Figure 9.8(a) shows the LTspice schematic of the DC–DC step-sown converter as shown in Figure 9.1(a) with a short-circuit at the load terminal of the load R at node 3. The short-circuit is simulated with a voltage-controlled switch. The i^2t of the fault current through the transistor switch and the fault current flowing through the load inductor $I(L)$ are shown in Figure 9.8(b). As expected, the i^2t of the switch current is increasing with time. The fuse must be selected to match the fault current and clear the fault while the fault inductor current decays through the free-wheeling diode. The load inductor L limits the i^2t value.

Figure 9.9(a) shows the LTspice schematic of the DC–DC step-sown converter as shown in Figure 9.1(a) with a short-circuit at the load-side of the transistor at node 2. The short-circuit is simulated with a voltage-controlled switch. The i^2t of the fault current through the transistor switch and the fault current flowing through the load inductor $I(L)$ are shown in Figure 9.9(b). As expected, the i^2t of the

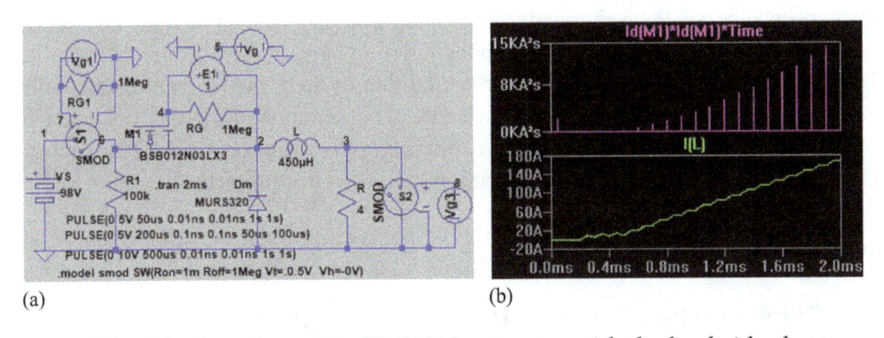

Figure 9.8 LTspice schematic of DC–DC converter with the load-side short-circuit: (a) LTspice schematic with an load-side fault and (b) fault currents

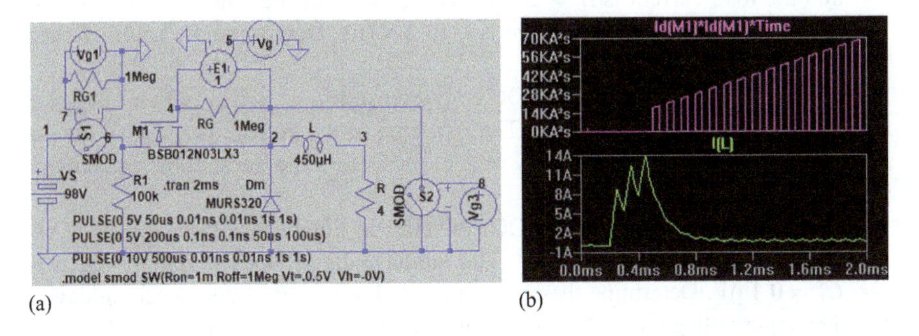

Figure 9.9 LTspice schematic of DC–DC converter with the output short-circuit: (a) LTspice schematic with an output-side fault and (b) fault currents

transistor switch current increasing with time and the fuse must be selected to match the fault current and clear the fault while the fault inductor current decays through the free-wheeling diode. The i^2t value is significantly higher because there is no inductance to limit the current. In practical current with a limiting *di/dt* inductor L_s, the i^2t value will be limited by the *di/dt* inductor.

9.3 Summary

The power semiconductor devices are turned on and off to converter and transfer energy from the source to the load. These devices are normally off. The rapid rate of rise of the voltage across the device from zero voltage to the full supply voltage causes internal current flow through the internal junction of the device and may cause device failure. Practical devices have limitations and require finite time to fully turn on and turn off. As a result, a situation may occur when the source is short-circuited during the device turn-on and -off time periods. The rate of rise of the current must also be limited, not to exceed the device specifications. A finite amount of power loss occurs across the semiconductor device due to the finite on-state resistance of the semiconductor device and the heat generated due to the power loss must the dissipated to keep the junction temperature within the specified limits. These devices must also be protected due to the rising current under faulty conditions of the power electronic systems. The semiconductor devices must be protected from the following types of operating conditions: (1) *dv/dt* protection, (2) *di/dt* protection, (3) thermal protection, and (4) fault protection.

Problems

1. The input voltage to a DC-to-DC converter in Figure 9.1(a) is $V_S = 48$ V and the load current is $I_L = 50$ A. The rise time of the voltage is $t_r = 4$ V/μs and the rise time of the voltage is $t_f = 2$ V/μs. Determine the snubber capacitance C_s in Figure 9.2(a).
2. The input voltage to a DC-to-DC converter in Figure 9.1(a) is $V_S = 120$ V and the load current is $I_L = 20$ A. The rise time of the voltage is $t_r = 4$ V/μs and the rise time of the voltage is $t_f = 2$ V/μs. Determine the snubber capacitance C_s in Figure 9.2(a).
3. The input voltage to a DC-to-DC converter in Figure 9.1(a) is $V_S = 24$ V and the load current is $I_L = 40$ A. The rise time of the voltage is $t_r = 4$ V/μs and the rise time of the voltage is $t_f = 2$ V/μs. Determine the snubber capacitance C_s in Figure 9.2(a).
4. The input voltage to a DC-to-DC converter in Figure 9.1(a) is $V_S = 48$ V and the load current is $I_L = 50$ A. The snubber capacitance in Figure 9.2(a) is $C_s = 0.1$ μF. Determine the operating switching *dv/dt* of the switching device.
5. The input voltage to a DC-to-DC converter in Figure 9.1(a) is $V_S = 100$ V and the load current is $I_L = 20$ A. The snubber capacitance in Figure 9.2(a) is $C_s = 0.1$ μF. Determine the operating switching *dv/dt* of the switching device.

6. The input voltage to a DC-to-DC converter in Figure 9.1(a) is $V_S = 100$ V and the load current is $I_L = 40$ A. The snubber capacitance in Figure 9.2(a) is $C_s = 0.1$ μF. Determine the operating switching *dv/dt* of the switching device.
7. The input voltage to a DC-to-DC converter in Figure 9.1(a) is $V_S = 100$ V and the load current is $I_L = 40$ A. The snubber capacitance in Figure 9.2(a) is $C_s = 0.1$ μF and snubber resistance $R_s = 100\ \Omega$. The switching frequency is $f_s = 10$ kHz Determine the (a) peak current through the switch and (b) the switching power loss P_{sw}.
8. The input voltage to a DC-to-DC converter in Figure 9.1(a) is $V_S = 48$ V and the load current is $I_L = 100$ A. The snubber capacitance in Figure 9.2(a) is $C_s = 0.1$ μF and snubber resistance $R_s = 100\ \Omega$. The switching frequency is $f_s = 10$ kHz Determine the (a) peak current through the switch and (b) the switching power loss P_{sw}.
9. The input voltage to a DC-to-DC converter in Figure 9.1(a) $V_S = 100$ V. Determine the value of series snubber L_s in Figure 9.3 to limit the *di/dt* to $\frac{di}{dt} = 100\ \text{A}/\mu\text{s}$.
10. The input voltage to a DC-to-DC converter in Figure 9.1(a) $V_S = 48$ V. Determine the value of series snubber L_s in Figure 9.3 to limit the *di/dt* to $\frac{di}{dt} = 500\ \text{A}/\mu\text{s}$.
11. The input voltage to a DC-to-DC converter in Figure 9.1(a) $V_S = 100$ V. Determine the value of series snubber L_s in Figure 9.3 to limit the *di/dt* to $\frac{di}{dt} = 500\ \text{A}/\mu\text{s}$.
12. The power dissipation of a switching devices that is a mounted on a heat sink is $P_{on} = 500$ W. The thermal resistances are $R_{JC} = 0.35$ °C/W, $R_{CS} = 0.25$ °C/W and the junction temperature $T_J = 200$ °C. Determine the thermal resistance R_{SA} for an ambient temperature of $T_A = 25$ °C.
13. The power dissipation of a switching devices that is a mounted on a heat sink is $P_{on} = 500$ W. The thermal resistances are $R_{JC} = 0.35$ °C/W, $R_{CS} = 0.25$ °C/W and the junction temperature $T_J = 200$ °C. Determine the thermal resistance R_{SA} for an ambient temperature of $T_A = 40$ °C.
14. The power dissipation of a switching devices that is a mounted on a heat sink is $P_{on} = 600$ W. The thermal resistances are $R_{JC} = 0.35$ °C/W, $R_{CS} = 0.25$ °C/W and the junction temperature $T_J = 200$ °C. Determine the thermal resistance R_{SA} for an ambient temperature of $T_A = 25$ °C.
15. The power dissipation of a switching devices that is a mounted on a heat sink is $P_{on} = 600$ W. The thermal resistances are $R_{JC} = 0.35$ °C/W, $R_{CS} = 0.25$ °C/W and the junction temperature $T_J = 200$ °C. Determine the thermal resistance R_{SA} for an ambient temperature of $T_A = 40$ °C.
16. When a transistor switch is fully turned on, it connects a DC voltage of $V_S = 100$ V to a load inductor of $L_s = 4$ μH. Determine (a) the *di/dt* of the current, (b) the inductor current at $t = 1$ ms, and (c) i^2t of the transistor current at $t = 1$ ms.
17. When a transistor switch is fully turned on, it connects a DC voltage of $V_S = 48$ V to a load inductor of $L_s = 4$ μH. Determine (a) the *di/dt* of the

current, (b) the inductor current at $t = 1$ ms, and (c) i^2t of the transistor current at $t = 1$ ms.

18. When a transistor switch is fully turned on, it connects a DC voltage of $V_S = 98$ V to a load inductor of $L_s = 2$ μH. Determine (a) the *di/dt* of the current, (b) the inductor current at $t = 1$ ms, and (c) i^2t of the transistor current at $t = 1$ ms.

References

[1] M. H. Rashid, *Power Electronics: Circuits, Devices, and Applications*, 4th ed., Englewood Cliffs, NJ: Prentice-Hall, 2014, Chapter 11.

[2] A. Wright and P. G. Newbery, *Electric Fuses*, London: Peter Peregrinus Ltd., 1994.

[3] M. H. Rashid, *SPICE and LTspice for Power Electronics and Electric Power*, Boca Raton, FL: CRC Press, 2024.

Index

www.ingramcontent.com/pod-product-compliance
Lightning Source LLC
LaVergne TN
LVHW022102240725
817018LV00003B/107